THE PILL

THE PILL

A Biography of the
Drug That Changed the World

BERNARD ASBELL

Random House
New York

Grateful acknowledgement is made to the following for permission to reprint previously published material:

COPYRIGHT MANAGEMENT, INC.: Excerpt from "The Pill" by Loretta Lynn, T. D. Bayless, and Don McHan. Copyright © 1973 by Coal Miners Music and Guaranty Music (administered by Copyright Management, Inc.). All rights reserved. International copyright secured. Used by permission.

The New York Times: Excerpts from "Why a Lag in Male-Oriented Birth Control?" by Jane E. Brody (October 16, 1983). Copyright © 1983 by The New York Times Company. Reprinted by permission of *The New York Times.*

ADDISON-WESLEY PUBLISHING COMPANY AND THOMAS A. BASS: Excerpts from *Reinventing the Future* (pages 92–93, 97–98) by Thomas A. Bass. Copyright © 1994 by Thomas A. Bass. Reprinted by permission of Addison-Wesley Publishing Company and the author.

THE ESTATE OF MARGARET SANGER: Excerpts from *Margaret Sanger: An Autobiography* by Margaret Sanger. Reprinted by permission.

IANTHE SWENSEN: "The Pill Versus the Springhill Mine Disaster" from *The Pill Versus the Springhill Mine Disaster* by Richard Brautigan. Copyright © 1968 by Richard Brautigan. Reprinted by permission.

Library of Congress Cataloging-in-Publication Data

Asbell, Bernard.
The Pill: a biography of the drug that changed the world/
Bernard Asbell.—1st ed.
p.   cm.
Includes bibliographical references and index.
ISBN 0-679-43555-7
1. Oral contraceptives—Research—History.   2. Oral contraceptives—Social aspects.   I. Title.
· RG137.5.A76   1995
613.9'432—dc20   94-23185

Manufactured in the United States of America
24689753
First Edition

BOOK DESIGN BY DEBORAH KERNER

To JODI
and
To JEAN

CONTENTS

x

Contents

THE VOICES

A letter to Margaret Sanger:

Englishtown, N.J.
January 5, 1925

Dear Mrs. Sanger

I received your pamplet on family limitation. . . . I am 30 years old have been married 14 years and have 11 children the oldest 13 and the youngest one year. I have kidney and heart disease, and every one of my children is defictived and we are very poor. Now Mrs. Sanger can you please help me. I have miss a few weeks and I dont know how to bring myself around. I am so worred and I have cryed my self sick and if I dont come around I know I will go like my poor sister she went insane and died. My Doctor said I will surely go insane if I keep this up but I cant help it and the doctor wont do anything for me. Oh Mrs. Sanger if I could tell you all the terrible things that I have been through with my babys and children you would know why I would rather die then have another one. Please help me just this once and I will be all right. Oh please I beg you. Please no one will ever know and I will be so happy and I will do anything in this world for you and your good work. Please please just this time. Doctors are men and have not had a baby so they have no pitty for a poor sick Mother. You are a Mother and you know so please pitty me and help me. Please Please.

Sincerly yours
[J.M.]

p.s. Please tell me how to get the pessary rubber womb cap. Not even the Doctors here know about them that is what thay tell me.

P r o l o g u e :   T h e   V o i c e s

Another letter to Sanger:

Bronx, N.Y.
Oct. 31, '32

Dear Madam,

   In answer to your letter of recent date, I know you will think of me
as being a very foolish women, when I give the reason why I did not return
to the clinic.

   The thing is this: That I have not as yet used the method. The first
few times I tried to insert it, I couldn't somehow get it right. And my
husband became so wildly impatient that I decided not to use it. Of course
it is foolish of me. . . . Not doing things in a natural way, I am very
nervous. Because my husband also is out of patience with me. Of course in
my heart and soul I know he is right; but I've borne 8 children within 12
years and the fear of going thru it again, just freezes the blood in my veins.
Because I am very sensitive about giving the children the right upbringing
and envirement; but when we have to squeeze in three bedrooms and crowded
into a small kitchen, all of us, it is enough to drive anyway crazy. And
even tho I would use that contraceptive, still I would have that stark fear in
me. Still I feel that I would say to him, "Be careful! Are you thru! etc."
Please don't be impatient with me and say I am foolish. I know I am acting
foolish, but I will try again to see if I can insert it if I can't I will come
again to the clinic. But it is hard for me, as the older children go to school,
and I have the 3 younger ones to leave at home. . . .

Very sincerely yours,
[M.P.M.]

From a reminiscence in a 1990 conversation with Rick Parsons,
a retired English cabinetmaker in his eighties:

*After I'd left my parents' home [in the 1920s] . . . I was
apprenticed to a relative who made furniture in a part of London where they
were all tenement houses. People were very very poor and the women used to
have old rundown shoes and stockings round the bottom of their feet and old*

clothing on, and they had a kid almost every year. Marie Stopes had opened a shop, a dismal sort of place, in one of the empty shops at the end of the road, and I remember these women queuing up outside for it to open, and to go in and get information, suppositories, caps, things like that. She was really the pioneer but she got no end of hostility in the press, in the church, and everywhere else. She was the subject of many trial cases. It's incredible that in my seventy-odd years I've seen such a change. . . .

I married my wife at twenty-one. When I got married was the first time that we had sex together. We'd come very very close to it. The teaching, from the church and the people all around, was so strong that it overrode all natural feeling, and this had a terrible effect, I think, on our development. We were subject to terrific temptation because my wife had left home and I obtained her a room and I had a room. It was in Somerstown. She hadn't any training whatsoever, so I had to help support her. I used to go around there for food, so we were often together in a room entirely on our own, and this went on for some months before eventually she pointed out that the economics were so bad that we would be financially better off if we got married and shared the room, which we did, you see. But during the whole of this time, although, as I say, we came very, very close to—we fondled and, um, fondled in depth—I forget the exact expression we used at that time—we never actually consummated the act.

It was partly the fear of pregnancy. It was partly fear of how you could possibly cope on the poor wages you had. It was also fear of what people would say. Those poor devils in the community who fell victim to their temptation, they were looked upon as sort of semi-criminals.

If there was a pregnancy when they didn't want one, well, people either got rid of it, and often in the process they destroyed their own health, or they had it, poor as they were. There were various means for trying to get rid of it. The first was taking Epsom salts with some gin. Now, that's probably why gin is known as "Mother's ruin." You drunk this, then you sat in a hip bath. Nine out of ten people didn't have an ordinary bath in those days, so you brought the hip bath in from the hook where it was hanging in the scullery, and you sent the kids out. The woman sat in that and kept on heating the water up, hoping for the best. And then, of course, there were all sorts of back street abortionists, and if they got caught they used to get very very heavy sentences, but in some ways I suppose they were attempting to do a social service. They charged, but got very very little money for what they did. The whole tragedy of the thing is that they were such uneducated women that they caused more damage than they ever did good, I'm quite sure of that.

There were a lot of deaths. I have an experience of one death. I was in the Army and on the very day I was to go home, I received a telegram and it wasn't very literate and I thought it said my wife died. When I got home I found that she was alive but had just come back from her sister's funeral. Her sister had become pregnant and had gone at her husband's instigation to an abortionist who lived in a street a little way away, and things had gone wrong and she got peritonitis and they couldn't save her. Her father said she turned completely black all over.

Pretty soon, my wife conceived. I had been unemployed for something like nine months. During that period I had literally pawned the quilt off our bed, the sewing machine I had proudly bought her, rugs, things like that. I had enough pawn tickets to make a pack of cards, over fifty. Well, it was obvious that we couldn't afford a child and had to do something about it—which, um, I personally did. Without describing the mechanics of the thing, I was intelligent enough to know that absolute cleanliness was necessary and I even wore a handkerchief around my mouth. Immediately afterwards when my wife started to hemorrhage, I ran round to the local doctor. He was a young fellow I had never seen before. He asked me particulars about what I had done, and he looked a bit grave and then said, "Well, you know, actually I should report this to the authorities. But I'm going to treat it as though it were just an ordinary accident and say no more about it." She was very very frightened. I had my arm round her and one arm on her breast while he was examining her below, and I remember him saying to her, "He's very good to you, isn't he?" And she said yes.

He came back after two or three hours, and the afterbirth had arrived, and he assured me everything was eliminated and that he would be sending around some medicine. I asked, "How much do I owe you?" My heart was in my mouth because I didn't have a penny. He said, "That will be seven and sixpence." I said, "I'm awfully sorry, I haven't got it now, but I'll bring it round this evening." As soon as I could, I went round to the pawn shop taking with me my wood plane which was one of my essential—one of my last remaining tools. I pawned that for the seven and six and took the money to him.

Fear is something that permeated the whole of the working classes to one degree or another. Throughout my childhood the earliest recollection I have is love from my parents, and then, as more children came, it was fear. My father made me feel that it was my responsibility to look after my brothers and sisters. I would stay awake with that dreadful element of responsibility mixed with fear.

And that same fear was what kept me from consummating before marriage. It had nothing to do with being good, but sheer fear.

Butuan City, Agusan, Philippines
Sept 7, 1951

Dear Mrs. Sanger,

. . . I perused [an article in Reader's Digest*] several times and after understanding everything it gave me the courage to write to you directly and personally. I know that you are a heaven sent to mothers and the answer to every parent's prayer.*

Mrs. Sanger, I am a mother of five children and by profession a nurse . . . for four hundred students and fifteen faculty members including their families. I am also giving Health Education classes to nine classes. . . . The present trend of education in my country is emphasize on Adult Education, thus after my duties in the school, I am required to give lectures on Health, Sanitation, Prenatal and Post natal care, Care of the Baby etc. Also I am to report problems encountered by the community and suggest any solution. We have many uninformed adults outside the school populace.

. . . Now and then I am also asked to tell the secret. Unwanted children constitute a main problem of our community. The average children of every family is seven. There are so many having fourteen, or fifteen. I myself belong to a family of fourteen and my husbands thirteen. At present we are only seven and my husband family three. Our brothers and sisters all died before they were a month old. My mother used to deliver yearly and she was worn out, old and broken before her time. My case is the same. Altho I am a nurse I do not know what to do. We have no such birth control clinics or leagues. I feel embarrass whenever someone asks me about the secret. I tried to experiment on orthoginol, rubber caps, rhythm and

diaphragm, but they were of no avail to me. I tried self induced abortion but I failed. I have to go to a fifty peso quack abortionist and if maybe I was not a nurse, my life was already ended. But you see Mrs. Sanger I am wondering why I fear pregnancy more than the loss of my life. Being on the family way again would mean a loss of a years work, and aging of a considerable years. At times I am afraid my husband would desert me as I can not give him full satisfaction. I can not help but avoid his coming near.

Through out my Adult Education work, I found out 70% of home problems is similar to mine. Therefore Madam I did not hessitate to write you. Please help us. If only I can go to you for instruction, or guidance, I could be happy again. Nevertheless I will be content only of what you will give me through the mails and I will always look forward for your answer.

Sincerely
[R.R.S.]

From a conversation in 1990 with Dilys Cossey, who in the late fifties in London saw an advertisement in the *New Statesman* for a secretary wanted by a birth control organization. "That's something that needs doing," she said to herself, and got the job for two pounds a week. Thirty years later, as chair of the United Kingdom's Family Planning Association, she evaluates the biggest change she witnessed over those years:

When the Pill came on the scene in the early 1960s, it was a godsend, an absolute godsend. That was the thing that made the big difference between when I was in my twenties and today's generation. There was no Pill. There was nothing you could do. If your boyfriend didn't have a condom, then you really had to rely on him being careful, and we all know what that means.

The Pill made contraception an item of talk—of dinner-table talk. You could say to somebody, "What kind of Pill are you taking?" Whereas you couldn't say to somebody, "What brand of sheath are you using?" That's because you put the Pill in your mouth. It had nothing to do with your genitals. It had nothing to do actually with having sex.

Prologue: The Voices

You knew you were protected the whole time you were taking the Pill. Women on the Pill could choose when to have sex, whether to have sex. They could choose when to have children or whether not to have children.

Now that was deeply frightening to many people, and it was a revolution. Suddenly we had women able to express their own sexuality as they wished. That was something that I couldn't—that a lot of people couldn't—grasp. So I think that that is really why the Pill continues to get a bad press today, when you get reports of the Pill being dangerous, or causing cancer, those enormous headlines. But look at the number of women who are on the Pill, who are happy to be on the Pill, whose lives have been transformed.

I think the Pill gets a bad press from some feminists. Everybody hailed the Pill at first as the great savior, the be-all and end-all, and then of course we realized that the dosage was wrong and that people had died from it, and then that it made women "available" to men so that men "could have their will with women whenever they wanted it." Which I think actually does not take into account that women could choose for themselves. They could always say no. I read a book recently by an American feminist of the sixties generation. At first she swallowed the Pill whole, but she is very clearly anti-Pill now. She said she realized she was only making herself available for men. Well, you know, that was up to her, I think.

Book I

1

THE CONCEPTION

Victory has a hundred fathers,
but defeat is an orphan.
—COUNT GALEAZZO CIANO, DIARY, SEPTEMBER 9, 1942

At a neighborhood supermarket in the university town where I live, now and then I see a man shuffling through the checkout line, smallish, bald, stooped with age but spry. He grips a paper bag that droops with a grocery item or two, scarcely enough for a single meal. His dark brown suit jacket doesn't match his darker pants, and a pair of solemn eyes dart this way and that, stirred by some restless hunger. I wonder, looking around the huge store, if a single shopper among the young collegians and profs and townies has any sense of how this almost invisible ninety-two-year-old has touched their lives intimately and has helped redirect moral beliefs and sexual practices in most of the nations of the world.

THE PILL

Since Russell E. Marker lives alone and avoids cultivation of friends, remarkably few townsfolk know who he is or what he has done. At the mention of his name, those who know something of him are likely to look distantly into their memories and say, "Marker. Isn't he the one they call the Father of the Pill?"

Indeed many do, but they are not quite right. Fifty summers ago, Marker, a chemist who never earned a Ph.D. and whose name is scarcely known beyond a small circle of scientists, took a creaky overnight bus to where a small bridge spanned a lethargic river in a tropical wilderness of Mexico. From that desiccated soil he dug out cartloads of stinking thick roots and hauled them away until he had piled fifty tons of the grimy brush behind a potter's shed he had rented to use as a crude lab. Then, during weeks of sweltering, lonely drudgery, Marker extracted from those roots a syrupy potion. That substance led, along a round-about path, to one of the life-changing inventions of the century. Without his discovery and his toil, the birth control pill as we know it could not have been created.

Russell Marker did not invent the Pill. He has repeatedly tried to make that clear, but his denial refuses to take root among people who talk of him. Again and again he has said that prevention of pregnancy as a goal never crossed his mind. His quest was simpler: a low-cost elixir in a new family of medicines called steroids. As in many of the great detective stories of science, Marker, with no thought of creating anything like a social revolution, discovered something on a path to something else, which provided someone else with a clue to something *else*, each goal along the way smaller, far less ambitious than the final breakthrough.

Nor did a pill to prevent pregnancy enter the thoughts of another ambitious scientist, young Carl Djerassi, who soon deciphered a puzzle that became the next essential step on the way to chemical birth control. But although some loosely label him too the "Father of the Pill," Djerassi, like Marker, has said again and again that he did not dream of contraception as a goal of his work.

Nor did it occur to one Frank B. Colton of a Chicago drug company who solved the same puzzle as Djerassi in an almost identical way at the almost identical time.

The Conception

• • •

The contraceptive pill was born unplanned—at least by the scientists most widely named as its parents—and it has lived a life full of surprise. Of many consequences that its creators never imagined, one is symbolically more remarkable than any other: To identify this extraordinary drug among all the thousands of potions, powders, capsules, caplets, tablets, and nostrums for sale at any pharmacy anywhere, all you need to do is spell the word *pill* with a capital *P*.

Over the past thirty years, the Pill has been swallowed as a daily routine by more humans than perhaps any other prescribed medication in the world. Its takers have counted in the scores of millions, and they have now downed the Pill by the hundreds of billions. Yet its eager consumers are, in general, quite well. Most take it neither to cure an illness nor to guard against one. The Pill has been called "the first medicine ever destined for a purely social, rather than a therapeutic, purpose."

The short, paradoxical, twist-and-turn history of the Pill is never more elusive and fickle than when we try to answer the most ironic question of its existence: Who brought the thing into the world?

And *why*?

During the 1960s, the first decade of the Pill's life, the surest answers were sounded by voices of religion: The Pill was conceived by the devil to sow wickedness.

By its second decade, the Pill, having become swiftly and hugely profitable, broke out with a marked rash of medical side effects, inviting a new explanation of its existence: The Pill was a plot authored by greedy drug companies with the compliance of submissive doctors. That reading, however, does not sit well with a retrospective look at the facts. When drug companies were first presented with the prospect of a contraceptive pill, scarcely without exception they couldn't run from it fast enough. Furthermore, the men who ran those drug houses (yes, virtually all men) could not believe that many women would choose to swallow chemicals to prevent babies. But one company, then another, did dare market the Pill, and they were soon happy to discover how to reduce its active chemicals so that earlier side effects largely disappeared.

And in its third decade, recently ended, the Pill aroused perhaps the most surprising resistance of all. Many daughters of the post-1960 women's revolution, which is almost exactly as old as the Pill itself, began demanding to know why it must be the *woman* who is called upon to swallow exotic concoctions and take medical risks. Those wonderful men of science who so ardently committed themselves to seeking a chemical "liberation" of women, why have they not troubled themselves to invent a pill *for men*?

Good questions, not yet satisfactorily answered, but they too harbor a factually incorrect assumption. The vision of a contraceptive pill was not men's in general or any individual man's, although at least five men of science have been publicly designated "the Father of the Pill."

Even the "father" who could stake the strongest claim of *deliberate* paternity, a man named Gregory Pincus, didn't think of inventing a birth control pill until the arresting scheme was proposed to him by two women who had a clear vision of exactly what they wanted—a "perfect contraceptive"—and who plunked down hard cash for producing it.

Those two women "stand by themselves as the indisputable mothers of the pill," in the words of Loretta McLaughlin, the biographer of a fifth frequently named "father," Harvard gynecologist John Rock. "From the moment they came on stage in the saga of its development, they took command of the scene. They did no less than *commission* the eminent male scientists who were to be the principals in its emergence, to make or find them an oral contraceptive. The women made it plain that what they wanted was a pill—like an aspirin—that would be cheap, plentiful, and easy to use. Moreover, the women virtually directed the men to be quick about it."

Of those two women, one had devoted a long life to transforming her personal cause into a public crusade; the other cherished and guarded her obscurity.

Margaret Sanger at the age of thirty-four coined a radical and inflammatory expression—*birth control*—and went on to found the International Planned Parenthood Federation. In 1950, after turning seventy-one, Sanger, still in full battle dress, joined forces with seventy-five-year-old Katharine McCormick, and together they set out to enlist science in their bold quest for a contraceptive that could be swallowed. Within a decade after commissioning that minute technology, it was physically produced, field-tested for effectiveness and safety, and approved by government. Thereafter it changed how human beings behave.

Millions today embrace the Pill as a salvation, while other millions shun it as sinful or a time bomb of dormant cancers. As a new century approaches, the tiny Pill persists in driving wedges between major religions, schools of medical opinion, factions of feminists. The Pill has brought the Roman Catholic Church into its most threatening confrontation with science since Galileo and its hierarchy into its most embarrassing defeat. Many attribute to the Pill more than to anything else the sexual revolution of the 1960s, although that judgment is far from secure. Perhaps more than any other influence, the Pill helped prepare the ground for later battles, both legal and moral, that have surrounded abortion, battles that clearly will continue to flare for many years.

As it was intended to do, the Pill has disconnected fear of pregnancy from the pursuit of sexual pleasure. But, intended or not, it has done far more. The Pill has led each of us, women and men alike and in a most personal sense, into a new era of potential mastery over our bodies and ourselves. As the first *systemic* contraceptive, it altered the routine functioning of the healthy human body. It opened the gateway to what I shall call the Era of BioIntervention, which is already taking us beyond reregulation of the human reproductive system. It is taking us beyond medicine into an eventual ability to modify—genetically—other body functions as well as our physical form itself. Perhaps the story of science's accomplishment in creating the Pill, and of some of its social repercussions, may help us better understand the ramifications of what we are poised on the rim of being able to do.

Already we know how to bring forth a baby from a mother who provides the womb while another furnishes the egg. We rapidly progress toward dispensing with fathers—seeming to need only their sperm deposited in plastic vials. Now that we can ascertain the sex of a baby before its birth, how far ahead lies the advance *choice* of its sex? Will humans want to—and be able to—resist that gift of science? And how soon will our new ability to splice and rearrange our genes—in a sense, the reinvention of ourselves as a species—offer us exotic choices, thus inviting social, religious, and ethical trials, conflicts, and crises that we cannot yet comprehend?

The moment of conception of the Pill is most often pinpointed as early 1951 at a dinner in New York arranged by Margaret Sanger. Her guests were an ally in the birth control movement since the 1920s, gynecologist Abraham Stone, and Dr. Gregory Pincus of Shrewsbury, Massachusetts, a research scientist of minor notoriety. Pincus, the world's foremost authority on the female component of fertility, the mammalian egg, had once managed the fertilization in a test tube of the eggs of a rabbit, and the eggs started dividing as they would in a uterus. His friends called the process "Pincogenesis." No one succeeded in reproducing the experiment independently, a standard requirement for proof of a claim in science. (Pincus acquired even more sinister fame—as "America's Count Frankenstein"—when a typographical error in a newspaper report of the rabbit experiment omitted a vital word: "Dr. Pincus said emphatically that he is [*not*] planning to carry it on to find out whether human babies can be made by test tube methods.") Pincus had since started and now directed a small, struggling, private research laboratory grandly titled the Worcester Foundation for Experimental Biology.

In the common way of history, nobody at the dinner was fully aware of the importance of what they were doing. Nobody connected with the event noted it in a journal or diary or transmitted word of it in any known surviving letter. For that reason, nobody knows exactly when the fateful meeting took place, although there is general agreement among those who have tried to track it down that it was in January or February. Nor does anyone know just where it was. Previous writings have placed the meeting in

Margaret Sanger's New York apartment. The problem with that is that in 1951 she had no New York apartment. Perhaps it was at the Waldorf-Astoria, where Sanger usually stayed when she visited New York from her home in Tucson, Arizona, or perhaps at the Carlyle Hotel, where she sometimes occupied an apartment kept by her son's wealthy mother-in-law.

The meeting, wherever it was, seems to have been spurred by a letter Sanger had received a few weeks earlier from Katharine McCormick, the daughter-in-law of Cyrus McCormick, inventor of the mechanical reaper. Mrs. McCormick was, in the words of the Harvard gynecologist John Rock, "as rich as Croesus. She had a *vast* fortune. Her lawyer told me she couldn't even spend the interest on her *interest*. She built dormitories, she built churches, she built hospitals. And she was *hepped* on birth control. So, inevitably, Mrs. Sanger got in touch with her."

Actually, it was McCormick who got in touch with Sanger. Their paths had crossed casually over the years, but in October 1950 McCormick wrote earnestly to Sanger to address "two questions that are much with me these days: A) Where *you* think the greatest need of financial support is today for the National Birth Control Movement; and B) What the present prospects are for further birth control research, and by research I mean contraceptive research."

No dawdler, Sanger fired back a reply that might disconcert a good many who revere her: "I consider that the world and almost our civilization for the next twenty-five years, is going to depend upon a simple, cheap, safe contraceptive to be used in poverty stricken slums, jungles and among the most ignorant people. I believe that now, immediately there should be national sterilization for certain dysgenic types of our population who are being encouraged to breed and would die out were the government not feeding them."

Sanger suggested a start-up research fund of twenty-five thousand dollars "definitely to be applied for contraceptive control." She said it should be divided among five or six universities "in this country, in England or in Germany." (Germany so soon after World War II? Because of that country's experience in sterilizing undesirables? Sanger would soon be persuaded to drop sterilization of "dysgenic types" from her pronouncements.)

In conversation with her dinner guests, Sanger chose her moment to look directly at Pincus and pose the question of the evening: What would it take to enlist science in finding the perfect answer to contraception?

His reply, while forever unrecorded, clearly encouraged her. Before long, Sanger and her friend Katharine McCormick were visiting Pincus in his laboratory in a redbrick box of a building on a winding, tree-shaded avenue of Shrewsbury, a suburb of Worcester.

They made a contrasting trio. Pincus's dense bush of graying hair and piercing black, ominously shadowed eyes almost a caricature of the menacing scientist's, played against his gentle and observant look of sympathy, a private brooding that bespoke his Russian-Jewish forebears. Except for those luminous dark eyes, he wore a remote, almost haughty air. Pincus was described by a young French scientist who met the famous endocrinologist during a Paris conference. Étienne-Émile Baulieu, who would one day develop an abortion pill called RU 486, recalls: "We all crowded into the hall for a look. . . . Pincus merely nodded his Einsteinian head. His thick, bushy eyebrows bunched in a frown. He was not tall, but he stood ramrod straight, hardly noticing anyone around him. After a few perfunctory handshakes, he was gone."

Had Pincus not already met Sanger and McCormick separately, he might easily have mistaken one for the other, as a colleague did, taking the one who stood almost six feet tall with the military shoulders, swooping-brim hat, and ankle-length matron's skirt to be Sanger, the embattled lifelong radical. But that one was Katharine McCormick, whose tastes in clothing had frozen almost a half-century earlier, in 1904, the year she had married. Pincus was aware that at about the time of that marriage she had become one of the first women to earn a degree in biology at the Massachusetts Institute of Technology, although she never used her training professionally. Standing beside her, the visitor with the elegance and restrained aura of wealth was the one who had molded herself from poverty, indeed from shame. Margaret Sanger was slight, scarcely five feet tall, with a striking crown of auburn hair that permitted confessions of gray, a cautious gaze through wide-apart gray eyes, and a subdued voice that drew her audience close. "But woe be to him," once warned Mabel Dodge, an early birth control ally and hostess of a famous New York salon, "who was misled by this calm exterior into ignoring the tremendous fighting spirit, the self-generating energy and the relentless drive that lay beneath it."

Could Pincus and his coworkers, the women asked again, produce a *physiological* contraceptive—not just a physical barrier between sperm and ovum—that would be safe and be available *soon?*

What, Pincus inquired, did they mean by soon?

One sees McCormick deferring to Sanger: The next few years.

Pincus hedged: Yes, probably, he could do it, but nobody could guarantee it.

Could he come close to guaranteeing it if he were assured of adequate money for laboratory staff, for supplies, for gathering and reinvestigating every scrap of relevant information from the scientific literature?

Known for his file-cabinet mind and index-card memory—and by his coworkers for reading mystery novels at the rate of almost one a night—Pincus silently riffled through possibilities, then speculated aloud that the perfect answer to Sanger's "perfect" contraceptive probably lay in clever use of a hormone. The trouble was, he knew some of the formidable obstacles blocking that path to the perfect. If it was intended to be used by many women, where was a large supply of the hormone to come from at a cost that even the rich among them could afford? And how was a woman to take it?

That, pressed Sanger, was what she was hoping Pincus might become interested in finding out.

Mrs. McCormick unabashedly pointed out that she was prepared to pay for finding it out—soon.

The suggestion had a tantalizing providential undertone. In 1951 Pincus's most absorbing mission was to keep his Worcester Foundation for Experimental Biology financially afloat.

That was a very severe word, *guarantee*, Pincus responded. But yes, he thought so.

Would he commit himself and his foundation to it if he got the money?

Pincus replied: Yes, he would.

Then a barrage of discomforting questions: What would it cost? When would he get started? When would he deliver?

Science, resisted Pincus, does not work that way.

Mrs. McCormick's life had taught her that most human undertakings *do* work that way. She pressed for an estimate.

According to a recollection years later by his widow, Pincus surrendered to a guess: "Right off the top of my head, I'd say one hundred twenty-five thousand dollars to start."

Mrs. McCormick drew a checkbook from her purse. Writing a draft for forty thousand dollars, she explained, "This is the end of the fiscal year. I'll talk to my financial man, and you'll get the rest."

He did, and before many years, Mrs. McCormick's investment expanded to two million dollars.

2

THE UNWORTHY SEARCH

*Formerly, when religion was strong and science weak,
men mistook magic for medicine; now, when
science is strong and religion weak,
men mistake medicine
for magic.*

—THOMAS SZASZ, *THE SECOND SIN*, 1973

W hen Margaret Sanger first challenged him to find the "perfect contraceptive," Pincus's mind surely called up a thin trail of experiments on a subject that for many years had seemed unworthy of important scientific attention: the female reproductive system. Virtually nothing was known of that system, at least nothing that would be acceptable today as science, until the 1840s, when two men of curiosity, Theodor Bischoff in Germany and Felix Pouchet in France, almost simultaneously found that the ovaries spontaneously and regularly release an egg. The great advance of that discovery was accompanied by a great error. Both men assumed that the egg was dropped at the time of menstruation and concluded that women were most fertile at that cyclical moment and least fertile between those events.

Until that discovery, a woman's part in conception was traditionally seen as passive. Women accommodated the male "seed" by providing it a warm and cozy place in which to grow until it could be hatched as a human being. The male *created* life. The woman simply provided a good bed to grow in. That viewpoint had begun with the discovery in 1677 by a Dutchman named Leeuwenhoek that "little animals" were swimming under his microscope in what he called *semine masculine.*

Members of the Royal Society of London suspected that these tiny creatures had something to do with contributing to pregnancy, but none could figure out how. Leeuwenhoek submitted the details of his finding to the secretary of the Royal Society with the sense of embarrassment and shame that, then and now, have often characterized the medical profession's timidity in discussing the undainty business of sex: "And if your Lordship should consider such matters either disgusting, or likely to seem offensive to the learned, I earnestly beg they be regarded as private, and either published or suppressed, as your Lordship's judgement dictates." (One Englishwoman of the era was, in a primitive sense, a step ahead of the scientists. In a paternity suit, she charged that she had conceived "by attracting the sperm or seminal effluxion" of a man who shared the use of her bath.)

The earliest modern investigation into female reproduction was by a gynecologist in Vienna, Emil Knauer, who in 1890 performed a curious procedure. Knauer cut open the abdomens of several mature female rodents and carefully plucked out their ovaries. Then he transplanted the tiny organs into other females that were not yet mature or had been castrated. The undeveloped females rapidly bloomed teats and other characteristics of sexual maturity. Knauer had suspected such a consequence, but he had to guess at its explanation. Those "feminizing" changes, he suggested gingerly and vaguely, seemed to result from some sort of "generative ferment" stirred by an unknown secretion of the ovary. Before long, a possible explanation was discovered by other researchers. They learned that animal glands put out a variety of mysterious chemicals, which are transported through the bloodstream to other specific organs. Like messengers, these chemicals seem to instruct their target organs to produce specific effects. By 1905, these messengers, having appeared in sufficient variety to warrant a generalizing label, were given the name *hormones*, from a Greek word that means "urging on," or "roused to activity."

The next discovery about the female reproductive system waited thirty-one years, until 1921, when another Austrian, Ludwig Haberlandt of Innsbruck, performed a transplant. He extracted the ovary from a pregnant rat and placed it into one that was not pregnant. The rat receiving it thereupon stopped ovulating. Like Knauer's, Haberlandt's result appealed for explanation.

Were Knauer's and Haberlandt's laboratory sorceries both the work of those mysterious hormones? How does a particular organ know which hormones to listen to and which to ignore? How does it decipher their messages?

By 1928, seven years after Haberlandt's published report, two Americans at the University of Rochester, George W. Corner, Jr., and Willard M. Allen, had delved deeper. For the first time they described intricate details of how the female reproductive system releases an egg, or *ovum*, then goes about preparing to release a new one. They found that when the egg is released, it leaves behind a small, empty sac, or follicle. Upon the egg's release, the follicle reshapes itself into a tiny vessel called a *corpus luteum*, or "yellow body." That yellow body immediately sets to work producing a minute quantity of a liquid—a hormone, clearly the one that had caused Knauer's "generative ferment." Corner and Allen identified their newly isolated hormone as central to the reproductive cycle: After release of the egg, the corpus luteum steps up its normal rate of production of the hormone, and as long as it keeps doing so in this expanded quantity, *no new eggs ripen.* Combining the Latin words *pro* (in favor of) and *gestare* (to bear), Corner and Allen called their newly discovered hormone *progesterone.*

A year later, in 1929, Edward Doisy of Washington University at St. Louis reported that the "feminizing" changes Knauer had discovered were not among the consequences that could be attributed to progesterone. Doisy had found a different liquid in the ovary follicles of pigs, extracted some, and treated young rats with it. The rats matured virtually before his eyes and rapidly went into pronounced heat. To name his new hormone, Doisy coupled the Greek word *oistros* (frenzy, or mad desire) with the word *gennein* (to beget), and called it *estrogen.*

Meanwhile, Philip E. Smith of Columbia had determined that male and female sex glands do not give orders directly to other organs through their self-produced hormones. Instead, the hormones carry their messages to the clearinghouse of all the body's mechanisms, the brain. That control center then sends out its own hormone messengers.

A complex system of feedback communication was gradually becoming clear. The menstrual cycle begins with low levels in the ovary of both estrogen and progesterone. The low level of estrogen "commands" the pituitary gland to start producing follicle-stimulating hormone (FSH). The follicle starts producing more estrogen. The rising level of estrogen then notifies the pituitary to start making luteinizing hormone (LH). Then, as all three of these hormones—estrogen, FSH, and LH—build up to a certain quantity, a single egg, which has ripened during all that activity, is released. The empty follicle is then reshaped by the LH into the corpus luteum. Thereupon the corpus luteum goes into serious production of progesterone while the ovary continues to put out estrogen. After fourteen days, the levels of these two hormones spill over and their production virtually halts. The whole female sex-hormone system seems to shut down. The soft lining of the womb crumples, shreds, and disintegrates, washed away by the bleeding of menstruation. The level of estrogen, now low, is the signal to the pituitary to start the cycle over again.

But what if a vigorous, ambitious, inquisitive sperm had thrust itself into the scene and successfully fertilized the ovum? That fulfilled egg, a *conceptus,* would arrive in the soft, welcoming bed of the womb, and a wholly different scenario would be played out. The corpus luteum, instead of disintegrating, would continue to pump out progesterone.

For many years the few scientists interested failed to understand how the uninterrupted supply of progesterone would block the busy "feedback" cycle of message-sending. But the discovery by Haberlandt in 1921 at last solved the puzzle: A female system that contains a fertilized egg stops ovulating because nature does not want one pregnancy to superimpose itself upon another. Of course!

In 1937 science added a subtle but critically important wrinkle. At the University of Pennsylvania, three researchers, A. W. Makepeace, G. L. Weinstein, and M. H. Friedman, began to believe that pregnancy might not be the only way for hormone signals to block a new pregnancy. Perhaps the body could be fooled into sending out a signal that it was pregnant even if it wasn't. They decided to test their idea on rabbits because a female rabbit, unlike other mammals, does not ovulate on a cyclical schedule. She does so in response to the stimulation of a male's mounting her and ejaculating. Approximately ten hours after the copulation, the female drops her egg to be fertilized by the warm, waiting semen. So the experimenters watched for their rabbits to mate, then early in that ten-hour "holding period" injected the females with progesterone. Sure enough, the rabbits failed to release their eggs. Just as the scientists had anticipated, the injected progesterone sent out a signal—a *false* signal—that the rabbit was already pregnant, that a new ovum was not wanted.

During the same year, Raphael Kurzrok, a Columbia University gynecologist, observed a different "false signal" of pregnancy. This one, however, had not been induced by an intruding scientist but was devised by nature herself. When a female breast-feeds her newborn, her system "pretends" it is still pregnant for several months, usually until the suckling baby no longer depends solely on nursing. Kurzrok soon published an arresting speculation: "If such cycles could be produced at will, we would have available a safe contraception method."

He was not the first to risk an extraordinary idea of that kind. In 1930, Haberlandt, as though playing at science fiction, had suggested in a scientific paper that someday the extract from ovaries of pregnant animals might be put to work as a human contraceptive; in fact, a woman might be able simply to *swallow* the extract in some form and thus avoid pregnancy. Haberlandt's speculation got no visible reaction from other scientists.

In a 1945 essay, Fuller Albright, an endocrinologist of Harvard and Massachusetts General Hospital, proposed a term "birth control by hormone therapy." His thought became known in hindsight as Albright's Prophecy:

> Since preventing ovulation prevents pregnancy, one could employ the same principles on birth control. Thus, for example, if an individual took 1 milligram of diethylstilbesterol [synthetic estrogen] by mouth daily from the first day of her period for the next 6 weeks, she would not ovulate during the interval; if she wanted to continue the birth control further, she could continue the stilbesterol and take a course of progesterone to cause menstruation. One could juggle the above therapeutics to make the menstrual period come on the least undesirable day.

Albright's Prophecy was brief and stated with extreme circumspection in an essay devoted mainly to the subject of serious menstrual disorders. In 1945, discussion of birth control by a Harvard professor, even in a scientific framework, was politically unsafe. Under Massachusetts law, indeed, it could be ruled a crime, punishable by jail. Just one year earlier, in 1944, Massachusetts voters had refused a proposal to relax the state's ban on disseminating birth control information.

Makepeace, Weinstein, and Friedman, in discovering the "false" messages of pregnancy, appeared to lock up the case that progesterone, the "pregnancy hormone," doubles as the pregnancy-*preventing* hormone. But by the year 1937 (and in 1945), a researcher entertaining the alien thought of seeking deliberately to *prevent* human pregnancies by inverting the normal use of the pregnancy hormone would meet an awesome obstacle. At the time of the Makepeace experiment, there was no known way to extract either from humans or from animals enough progesterone to inject a fraction of the women who needed it for alleviating a tendency toward miscarriages, easing menstrual pains, or other cyclical disorders. Because of its scarcity, its market price would be astronomical. Who would dare misappropriate a pinpoint of the precious fluid for the disreputable goal of avoiding pregnancy?

But now, in 1951, Gregory Pincus's conversation with Sanger and McCormick urgently lured him to dig into the details and possible meanings of a recent accomplishment by an obscure—and reputedly eccentric—chemist. Pincus knew immediately that he had to look into the work of Russell Marker.

3

Two Mothers

*The idea, in this country, never seems to enter any one's mind
that having or not having a family, or the number of which
it shall consist, is wholly amenable to their own control.
One would imagine that children were rained down
upon married people, direct from heaven, without
their being art or part in the matter.*
—Notes on the Population Question, by "Anti-Marcus"
(probably John Stuart Mill), London, 1841

To think of the Pill as arising solely from the intellects of its two mothers is to underestimate the force of its birth. The idea crashed its way out of the deepest, harshest secrets of each of their lives.

When Margaret Louise Sanger was born in 1879 in Corning, New York, a smoky center of railroading and glassmaking, her mother, Anne Purcell Higgins, had already brought forth five children and had come down with tuberculosis. After Margaret was born, Mother Higgins, frail, emaciated, and much of the time bedridden, produced five more babies and suffered seven miscarriages before her consumption killed her at age fifty on Good Friday, 1899. At the funeral in St. Mary's Catholic

Church at State and First streets, the heart and hub of Corning's Irish Hill, nineteen-year-old Margaret confronted her father. Michael Hennessy Higgins, more fondly known in the city's alehouses as "Granite" Higgins, was a philosophizing, rebellious, church-needling, binge-drinking, affection-loving refugee of the Irish potato famine who worked, when there was work, as a freelance carver of angels and saints on the gravestones of Corning's Catholics bound for Hope Cemetery.

Facing her father over her mother's coffin, so a family story goes, Margaret quietly hurled at him, "You *caused* this. Mother is dead from having so many children."

The Higgins brood, growing up in one stove-heated cottage after another in the extreme eastern end of Corning, "knew not where we belonged," Margaret would later write. "Everything we desired most was forbidden. Our childhood was one of longing for things that were always denied. We were made to feel inferior to teachers, to elders, to all."

In the young mind of Margaret, families overweighted with children meant "poverty, toil, unemployment, drunkenness, cruelty, fighting, jails." In contrast, the growing girl perceived that the families of executives of the Corning Glass works and of the railroads lived the good life on The Hill, the section of town overlooking—literally looking down upon—St. Mary's Church. They "owned their own homes; the young-looking mothers had time to play croquet with their husbands in the evenings on the smooth lawns. Their clothes had style and charm and the fragrance of perfume clung about them. They walked hand in hand on shopping expeditions with their children, who seemed positive in their right to live. To me the distinction between happiness and unhappiness in childhood was one of small families and of large families rather than wealth and poverty. I often watched them at play as I looked through the gates in passing."

Now free of nursing her mother through the final stages of illness, Margaret found herself the target of control by her father, who "suddenly metamorphosed from a loving, gentle, benevolent parent into a most aggravating, irritating tyrant." She especially resented his interference with the young men in her life. One memorable night when she returned home after his specified curfew, he slammed the front door and locked it, barring her entrance. An undercurrent of a deeper fear of him is described in her autobiography, with perplexing incompleteness, as "the only memory I have of any sex awakening or consciousness of sex." She wrote that one night as a child she was feverishly ill when her father, as he sometimes did at bedtime, visited to talk. He was drowsy, perhaps drunk, and apparently fell asleep. "I heard heavy breathing beside me. It was father. I was terrified. I wanted to scream out to Mother to beg her to come and take him away. . . . I lived through agonies of fear. . . . I was petrified. . . . I was cold; I began to shiver; blackness and lights flickered in my brain; then I felt I was falling, falling—and knew no more."

But in tracing the powerful influences of her father on the shaping of Margaret, those fearful memories, taken alone, mislead. Margaret adored Michael Higgins's high spirit and wit, his adventurousness and rebelliousness, which she later called "the spring from which I drank." Revering no masters, rejecting worship of any gods, Michael preached freely and endlessly to any audience on the virtues of socialism and feminism, as well as reading aloud to them from Henry George's *Progress and Poverty* and Irish poetry. He exhorted Margaret, the most intellectually inclined of his offspring, to subjugate her emotions to the higher power of reason. "I realized," she later wrote, "I was made up of two Me's, one the thinking Me, the other, willful and emotional, which sometimes exercised too great a power; there was danger in her leadership and I set myself the task of uniting the two by putting myself through ordeals of various sorts to strengthen the head Me."

Michael introduced his daughter to the excitement and secret glow of the virtue to be found in public scorn when he arranged with a Catholic priest for rental of a hall in Corning, then invited one of his idols, the religious freethinker Robert Ingersoll, to speak. Upon learning that the speaker was to be a notorious heretic, the priest literally barred the door. When Michael, holding Margaret's hand, arrived at the locked entrance with Ingersoll, "hoodlums" pelted them with "tomatoes, apples, and cabbage stumps," She was later to recall. For Margaret, there were shades of delight in a new importance. "Thereafter, we [the Higgins brood] were known as children of the devil. On the way to school, names were shouted, tongues stuck out, grimaces made; the juvenile stamp of disapproval had been set upon us. . . . No more marble angels were to be carved for local Catholic cemeteries, and, while father's income was diminishing, the family was increasing."

Before her mother died, Margaret enrolled at a Methodist boarding school run by Claverack College in the Catskills, as close to New York City as to home. Two of her older sisters, Mary and Nan, pooled portions of their meager salaries to cover most of her tuition, and Margaret worked in the school kitchen to complete the $225-a-year cost. She set for herself the goal of going to medical school at Cornell but soon settled for nurse's training. During those first two years at Claverack, Margaret too developed a tubercular infection that was to drain her energy for the next twenty years, at one point imperiling her life.

Soon after losing her mother, Margaret left Corning, never to live there again. (In later years as a constantly traveling agitator, Margaret pulled the shades down whenever she passed through Corning on the New York Central, to avoid "feeling physically ill," she told family members. The disdain was reciprocated, according to Dick Peer, the city's informal historian and former newspaper editor: "To this day, in polite conversation Corning Catholics avoid mentioning the names of two of their own. One is Ed Davis, the New York state executioner who invented the electric chair. The other is Margaret Sanger.")

After her first glimpse of a world larger than Corning, she was to write, "I wanted a world of action. I longed for romance, dancing, wooing, experience." She applied to an acting school in New York but refused to give the measurement of her legs on the application form and thus lost her hope of an acting career.

For her final training as a nurse, Margaret Higgins transferred to the Manhattan Eye and Ear Hospital. Her textbook, which she kept, is devoid of underlining except for the chapter on obstetrics, in which there is underlining in every paragraph.

Margaret was repelled by the chief goal of most of her school-mates: "The thought of marriage was akin to suicide." That aversion soon was modified. At a hospital dance she met a young architect and painter, William Sanger. The next morning at seven thirty, after her night shift, he was waiting for her on the hospital steps—and the next morning and the morning after that. Indeed, it was the "romance, dancing, wooing" she had longed for. Six months later, on a day in 1902 when Margaret was scheduled for a two-hour break, they scurried to a minister and married, and she got back to work with time to spare. The wedding, such as it was, involved three jabs of rebellion, sure to appeal to Michael Higgins. Two of them: a Catholic bride taking a Jewish husband, then giving her vows before a Protestant minister. The third: Nursing school students being forbidden to marry, the twenty-two-year-old bride kept her wedding a secret.

Within months Margaret found herself pregnant with her first child, to be named Stuart. During those months her tubercular condition worsened. A year later the young couple built a home, from William's own plans, in Hastings, on the Hudson River. Stuart was followed in 1908 by a brother, Grant, then, a year later, by a girl, Peggy, for whom Margaret had longed. Margaret Sanger at last got her full taste of suburban housewifery in her own hillside home and clean dresses that smelled of perfume. Although she didn't play croquet with her husband, she did go to card parties and women's meetings and learned to shop for clothes in Manhattan. But after a few years, she would say, "this quiet withdrawal into the tamed domesticity of the pretty riverside settlement seemed to be bordering on stagnation." Stagnation seemed to settle into the marriage too.

In 1911 the couple gave up proper Hastings to plunge into the arts, intellectualism, and radical politics of Greenwich Village. The Sangers rented an apartment "way uptown" on 135th Street but spent their liveliest and happiest hours downtown in the MacDougal Street loft of the Liberal Club and the restaurant of Polly Halliday downstairs. Through a band of bohemian artists and radicals that William knew, Margaret basked in the dazzling conversation of Eugene V. Debs, the Socialist hero; John Reed, the Leninist journalist who would one day be buried in the

Kremlin wall; and Emma Goldman, the Socialist agitator for women's rights who, since 1900, had been advancing the revolutionary concept of "voluntary motherhood." The Sangers became regulars in the elegant and heady salon of Mabel Dodge at 23 Fifth Avenue, where they heard Big Bill Haywood extolling the Industrial Workers of the World (IWW)—the "Wobblies"—and Walter Lippmann explaining the explosive new ideas of Sigmund Freud, and Will Durant reporting on sexual liberation as newly advocated by Havelock Ellis and Richard von Krafft-Ebbing.

Bill Sanger had long been a member of the Socialist party, and now, inspired particularly by Emma Goldman, Margaret committed herself to activism too. "Almost without knowing it, you become a 'comrade,'" she wrote exultantly in a letter.

Margaret worked on the women's committee of the Socialist party and on October 31, 1911, was unanimously elected its "organizer" at a salary of $15 per week. Her duties were to supervise recruiting of women, oversee propaganda for socialism and women's suffrage, and organize naturalization classes for immigrant women. Margaret soon became dissatisfied with the suffragist position of demanding only political freedom for women. What she wanted was economic, social, and intellectual freedom. What was needed, Margaret concluded, was "direct action" as practiced by anarchists and the IWW and not the "political action" of the Socialists.

In 1912 she took part in an IWW strike in the textile mills of Lawrence, Massachusetts, where she helped evacuate the strikers' children. Returning to New York, in one of her first experiments with the bracing jargon of class struggle, Margaret wrote in *The Call*, a Socialist daily New York newspaper, that in Lawrence "capitalism shows its fangs of despotism and murder." Despairing for the young soldiers assigned to guarding the struck mills, she declared: "The time has come to educate these boys, to remind them to what class they belong, and when they realize this they will refuse to murder their own working brothers, to serve as hirelings to prop up the profit system, which bases its existence upon the blood of the famished workers." When the Wobblies organized a walkout in the mills of Paterson, New Jersey, she picketed with the strikers.

What Margaret would call her "Great Awakening" was her shift from economic and class-struggle radicalism to a new direction of overhauling gender relationships and the meanings of sex. It had been spurred mainly by a series of lectures in 1909 by Sigmund Freud at Clark University in Worcester, Massachusetts. The new interest led her to reading Havelock Ellis, the British pioneer in the psychology of sex, and especially the Swedish feminist Ellen Key, who stimulated Margaret to reexamine the traditions of marriage and womanhood. Key had dared to write that morality was subjective, a matter of personal choice. What made marriage holy and durable, Key argued, was not law or tradition, but only sexual love. When a marriage denied sexual satisfaction to a woman (the very idea that a *woman* should expect satisfaction had been regarded as a heresy), such deprivation of pleasure took precedence over any mere matter of marital law and was justification for dissolving the marriage. Individual desire—satisfaction of the inner self—was sacred above all else, Key urged, and should be allowed full development of freedom of expression.

Sanger, deeply stirred, wrote in her journal in 1914: "I love being swayed by emotions, by romances, just like a tree is rocked to and fro by various breezes—but stands firmly by its roots." She rejected "the Christian, democratic, ascetic ideal," exalting instead "life in its fullness and all that is high, beautiful, and daring." Reversing her earlier journal entry that gave a higher value to the "thinking Me" than to her "willful and emotional" self, she now declared: "Those who restrain desire do so because the desire is weak enough to be restrained. Reason usurps its place and governs the weak and unwilling and being restrained, it by degrees becomes passive till it is only the shadow of desire." Reason is external, the leftover result of past mistakes. "Emotion however is that which urges from within, without consciousness of fear or consequences."

While Margaret's radicalism brought her into closer affinity with her father, her switch in focus from socialism to the role of sex in women's lives did not sit well with him. "Margaret," he asked her one day, "can't you find some other subject in the world to talk about besides the bedroom?"

Going to work as a visiting nurse for a New York social-welfare agency, Sanger most often took obstetrical patients. They rarely were emergencies and thus permitted her to control her time to care for her own children.

Assignments on the Lower East Side especially drew her: "Below Fourteenth Street, I seemed to be breathing a different air, to be in another world and country." The tenth ward was New Israel, where 76,000 Jews huddled in 1,179 tenements; the Bloody Sixth was Little Italy. More than 508,000 souls crowded in less than a square mile below Fourteenth Street and east of Broadway. "These submerged, untouched classes were beyond the scope of organized charity or religion," she later wrote. "No labor union, no church, not even the Salvation Army reached them."

The Lower East Side opened Sanger's eyes to immigrant mothers' ignorance about their own bodies and those of their families, and their consequent susceptibility to the scourges of syphilis and gonorrhea. In 1912 she wrote a series of articles for *The Call* on venereal disease and feminine hygiene, as though that Socialist-party organ was the direct channel to the unlettered masses. She titled it "What Every Girl Should Know." The explicitness of it shocked the Post Office, which banned *The Call* from the mails. The paper's editors canceled the series but got back at the Post Office by printing boldly:

WHAT EVERY GIRL SHOULD KNOW

N

O

T

H

I

N

G

!

BY ORDER OF THE POST OFFICE DEPARTMENT

As a nurse, Sanger saw many women fall victim to ignorance, but the tragedy of one slum dweller in particular burned into her memory. Sadie Sachs was a slight twenty-eight-year-old Russian-

Jewish mother of three small children. Her husband, Jake, a truck driver, found her one stifling day in mid-July of 1912 on the kitchen floor of their cramped Grand Street tenement, unconscious and bleeding. She had attempted a self-induced abortion. For many days, her doctor and Sanger, who had been assigned the call at random, fought to control the septicemia that threatened the woman's life. When the battle was won, Mrs. Sachs pleaded: "Doctor, what can I do to stop having babies?"

The doctor was not an unkindly man, Sanger later wrote, "but such incidents had become so familiar to him that he had long since lost whatever delicacy he might once have had." Gruffly, perhaps to conceal his own frustration, he picked up his hat and bag and said, "Better tell Jake to sleep on the roof."

Sanger fought back a deep revulsion. Yet she did not know what she would have told the woman. Other women like Sadie had implored her to tell them what to do to prevent another child. "Tell me the secret—tell me what rich women use," they would plead. The young nurse had to answer that she knew no "secret." One woman offered her money to tell. The nurse knew that only doctors were "qualified" to talk about diaphragms and whatever, that most poor working-class women couldn't afford doctors, and if they did go, they were too ashamed to ask, and when asked most doctors were too aloof and embarrassed to tell, and that husbands rejected using "rubbers." Trying to press a few coins into the nurse's hand, the woman pleaded, "But that's all I have."

Three months later, Sanger got another call to the same address. This time Sadie Sachs had run to a five-dollar abortionist. Within ten minutes of Sanger's arrival, Sadie Sachs was dead. It was not the first such death the nurse had known—she had heard an estimate of 100,000 illegal abortions a year in New York alone, leading to only God knew how many deaths misattributed to other causes—but this was the one that changed her forever:

> I walked and walked and walked through the hushed streets. . . . I looked out my window and down upon the dimly lighted city. Its pains and griefs crowded in upon me, a moving picture

rolled before my eyes with photographic clearness; women writhing in travail to bring forth little babies; the babies themselves naked and hungry, wrapped in newspapers to keep them from the cold; six-year-old children with pinched, pale, wrinkled faces, old in concentrated wretchedness, pushed into gray and fetid cellars, crouching on stone floors, their small scrawny hands scuttling through rags, making lamp shades, artificial flowers; white coffins, black coffins, coffins, coffins interminably passing in never-ending succession. The scenes piled one upon another. I could bear it no longer.

As I stood there the darkness faded. The sun came up and threw its reflection over the house tops. It was the dawn of a new day in my life also. . . .

I was resolved to seek out the root of the evil, to do something to change the destiny of mothers whose miseries were as vast as the sky.

Later she wrote of the Sadie Sachses of the world: "I can still see them, those poor, weak, wasted, frail women, pregnant year after year like so many automatic breeding machines. . . . I came to a certain realization that my work as a nurse and my activities in social service were entirely palliative and consequently futile and useless to relieve the misery I saw all about me."

Sanger quit nursing to spend what she describes in her autobiography as "almost a year" studying all she could find about contraception in the Library of Congress, the Library of the New York Academy of Medicine, and the Boston public library. Finally she traveled to France in 1913 to find "reliable" contraceptive information. "She could not have looked very carefully," one of her biographers, David Kennedy, suspected, thus adding another evidence of Sanger's lifelong penchant for tailoring "facts" to fit a mannequin of how she wanted her life story to appear. Kennedy pointed out that at least fourteen years before her search, in 1898, the *Index Catalogue of the Library of the Surgeon General's Office* listed nearly two full pages of citations on contraception. Those writings specifically described uses of the condom, vaginal douching, suppositories, tampons, and the cervical pessary, or "womb veil."

"By making birth control seem an innovation she personally had imported from France," Kennedy taunted,

she conveniently suppressed some historical facts: that contraception was widely practiced among certain social groups in the United States as early as the nineteenth century; that the medical profession did have some, admittedly undiffused, contraceptive knowledge available. . . . Mrs. Sanger usually used the myth of the year's fruitless search to advance her position in relation to the medical profession, which balked at her nonmedical leadership in a medical area. Claiming that before her researches in France doctors knew little or nothing about contraception was one way to establish her legitimacy with regard to organized medicine.

To thirty-three-year-old Margaret Sanger the leap from industrial radicalism to contraception was not a complete trading in of one world for another, especially in the political atmosphere of Europe. During that year, 1913, Rosa Luxemburg in Germany and Anatole France in France were calling upon workers to undertake a "birth strike." By refusing to bear children, went the radical-theorist argument, workers would stop the flow of exploited manpower into the industrial and military machines and thus help jam the wheels of capitalism. But the radicals were not arguing for contraception.

Sanger groped for her own mode of repudiating orthodoxy. " 'Virtue,' 'Marriage,' 'Respectability,' " she pondered in a journal entry after returning from Paris, "they are all alike. How glorious too and how impudent the present society—which dares to shut up young girls or women in their 'homes' because that girl defies conventions and fills the longings of her nature. For this she is an outcast. The whole sickly business of society today is a sham. . . . One feels like leaving entirely or going about shocking it terribly."

As Margaret Higgins of Corning had done, Katharine Dexter of Dexter, Michigan, and Chicago also fell in love with and married an artist. Even the time of the wedding was almost the same—

Margaret's in 1902, Katharine's in 1904. The resemblances ended
there.

Rather than squeezing her wedding into a two-hour lunch
break, Katharine gave her vows at her mother's magnificent
Château de Prangins in Geneva, which had been built originally
for Joseph Bonaparte. Wedding guests had filled out the passen-
ger rosters of trans-Atlantic luxury liners for weeks. The cost of
the great party was hardly a strain on her mother, Josephine
Moore Dexter, a former schoolteacher of West Springfield,
Massachusetts, who had recently inherited the fortune of her husband, Wirt Dexter, a man of major undertakings who had descended from men of major undertakings. As a Chicago attorney
Dexter had led the rebuilding of his city after its disastrous fire.
His father had helped found the University of Michigan. The
Dexter name had been planted in America in 1642, and Dexters
paraded for uninterrupted generations through Harvard and
Harvard Law School. Katharine's great-grandfather, a leader of
the Massachusetts Federalists, became a charter member of the
United States Senate. He left the Senate when the infant nation's
second president, John Adams, appointed him Secretary of War,
then Secretary of the Treasury.

Dexter women were encouraged to become "doers" as well as
ladies. Katharine's lively intelligence clearly marked her for college, with goals more demanding than mere husband-catching.
When biology caught her fancy, her father encouraged her to enroll at the Massachusetts Institute of Technology, where few
women had preceded her. In 1904 she earned her bachelor of science degree upon submitting a thesis entitled "Fatigue of the
Cardiac Muscles in Reptilia."

If her mother had "married well," Katharine did even better
only weeks after her graduation. Her groom, who had played var-
sity tennis at Princeton and graduated with honors in 1895, was
Stanley McCormick, youngest son of Cyrus McCormick who
had invented the reaper and created the industrial empire of the
International Harvester Company. The elder McCormick had
died in 1884, and of course young Stanley's success in the "family
business" was automatic. His executive talents were real, how-

ever, and he soon rose to the position of corporate comptroller. When simultaneously he began winning recognition as an artist, a rich life for the talented couple seemed assured.

The splendid marriage was not two years old when the gifted husband began coming apart with bewildering symptoms of schizophrenia. His illness, striking suddenly, appeared to his wife and doctors to be the work of some inborn defect waiting to seize him. That view was consistent with a revival at that time of the Mendelian theory that madness might be inherited. He soon was declared legally and hopelessly insane.

From the brilliant social whirl that had been her natural surround, Katharine now withdrew and took up the life of a semi-recluse. In summer she took her sick husband to her mother's château in Switzerland, and they wintered at her mother's townhouse on Boston's Commonwealth Avenue. Soon Katharine built a storybook castle in the coastal hills near Santa Barbara, California. She called it Riven Rock, and to ensure that her husband's days were ornamented with beauty and peace, she hired forty gardeners to tend to the plantings and six musicians to play for him.

His illness and spells of dementia forged in Katharine a resolve never to have a child of her own. The costs, practical and emotional, are suggested by her devotion over the remaining half century of her life to finding a "perfect" way to control conception.

In 1909, when Stanley was declared legally incompetent, lawyers for the Cyrus McCormick estate, alarmed that trust funds, like a huge river, were passing multimillions to a crazed heir, began a battle to restrain Katharine's power to spend the money by requiring court approval of her decisions.

At about the same time that Margaret Sanger was abandoning the bourgeois life of the suburbs for Greenwich Village, Katharine McCormick also chose a radical outlet for her energies. She threw herself into the women's suffrage movement.

McCormick spoke on the Bedford Common at the first open-air demonstration for suffrage in Massachusetts. She testified before a state legislative committee in 1911 and worked with Carrie Chapman Catt in the National American Woman Suffrage Association, serving as treasurer and vice president. When the

suffragette *Woman's Journal* ran out of money, McCormick provided six thousand dollars to keep it going. The campaigners
were resourceful. When forbidden to speak on the Nantasket
beach, they waded into the water with their VOTES FOR WOMEN
banner and spoke from the sea to the women on the shore.

For Margaret Sanger, the growing crusade for women's suffrage
had no attraction. The new set of friends she had made in
Greenwich Village, anarchists and Marxists, had given it the back
of their hands, the vote having little or nothing to do, in their
view, with liberation of the working classes from their industrial
oppressors.

Upon returning to the United States, the cause of women, beyond the suffrage issue, spurred her. Sanger founded a small
newspaper, the *Woman Rebel*, in March 1914. It lasted until
October. The masthead of its first edition proclaimed "No Gods
No Masters," an IWW slogan. In a signed article, she urged
women "to look the whole world in the face with a go-to-hell
look in the eyes; to have an ideal; to speak and act in defiance of
convention." The *Woman Rebel*, she promised, would "stimulate
working women to think for themselves and to build up a conscious fighting character." But until then, women were enchained
"by the machine, by wage slavery, by bourgeois morality, by customs, laws and superstitions. . . . It will also be the aim of the
Woman Rebel to advocate the prevention of conception and to im-
part such knowledge in the columns of this paper."

In her first issue, another article urged prevention of conception—what amounted to a "birth strike"—in order to frighten the
"capital class." Later she prescribed seven circumstances in which
birth control ought to be practiced: when either spouse has
a transmittable disease; when the wife suffers a temporary infection of lungs, heart, or kidneys, the cure of which might be re-
tarded in pregnancy; when a mother is physically unfit; when
parents have subnormal children; if the parents are adolescents; if
their income is inadequate; and during the first year of marriage.
Neither article, however, specified how a conception was to be
prevented.

The paper found its readers mainly through the small but militant locals of IWW. Emma Goldman took copies on her lecture tours, distributing fifty in Minneapolis and five hundred in Los Angeles. The April issue assailed John D. Rockefeller, Jr., as a "blackhearted plutocrat whose soft, flabby hands carry no standard but that of greed." Workers would never be free "while there is private property which prevents all men and women having free access to the means of life. . . . All must possess together—in common, that is." Marriage, "the most degenerating influence in the social order," was a form of property regulation enslaving women. The Socialist revolution must transform marriage into a "voluntary association."

Because she was a regular reader of another radical journal, *The Masses*, Sanger surely read Max Eastman's criticism: "The *Woman Rebel* has fallen into that most unfeminist of errors, the tendency to cry out when a quiet and contained utterance is indispensible." Because of its "style of overconscious extremism and blare of rebellion for its own sake those who incline to the life of reason will be the last to read it."

Although the *Woman Rebel* never gave specific instructions for preventing conception, the postmaster for New York City declared early in 1914 that there was sufficient cause to ban it from the mail under Section 211 of the Criminal Code of the United States. Section 211 was part of the Comstock law, which had been passed hurriedly in 1873 following a remarkably noisy campaign by Anthony Comstock and his Society for the Suppression of Vice. It prohibited mailing, transporting, or importing "obscene, lewd, or lascivious" matter, while providing no criteria for judging those offenses. The law also specifically banned all devices and information for "preventing conception." That particular provision was not one that Sanger and her newspaper were charged with violating, but the postmaster refused to stipulate what his censors did find objectionable.

Sanger published her April issue without interference, but the censors declared unmailable, one after another, the issues of May, July, August, September, and October. (The June issue, not banned, planted a historical marker. In it Margaret Sanger used for the first time the term "birth control.")

In August 1914, two federal agents appeared at Sanger's home to serve her with an indictment for nine violations of the Comstock law. Rather than specifying particular instances of advocating contraception, the Justice Department classed them all under the blurry offense of "indecency." If convicted, she could receive up to forty-five years in prison. That posed for Sanger a tormenting political question. Should she risk going on trial for a charge that allowed her no real victory? Even if she won her case on the "obscenity" charge, which did not seem likely in the prevailing climate, the victory would still leave untested the right to speak out specifically for contraception.

To her great disappointment, Socialists failed to come to her support in great numbers. Most feminists remained more interested in obtaining the vote for women. Those political strains contributed to a marked deterioration of her marriage. Sanger made a fateful choice: She decided to flee the country, leaving the children with her husband, and with no clear view of when she might be able to return.

First she took a midnight train to Montreal. Under the assumed identity of "Bertha Watson," she boarded the R.M.S. *Virginian* for Liverpool, where a sister lived. It was a dangerous trip. War had broken out in Europe three months earlier, and the ship was loaded with munitions. Seeking emotional shelter in her journal, she wrote about her new feelings on the place of women in marriage, and particularly about her place in her own: "The man who shouts loud about his liberal ideas and thinks himself advanced finds the servile submission of his wife charming and womanly. . . . How closely he keeps her within the boundary of his own, like a priest who watches and weeds the young ideas to keep them forever within the enclosure of the church."

After a few days in Liverpool, through acquaintances of her New York Socialist and anarchist friends, Sanger was invited to tea by Mr. and Mrs. Charles V. Drysdale, who as leaders of the English Neo-Malthusian League were active propagandists for the prevention of world overpopulation. "On the appointed afternoon," the Drysdales' secretary noted, "we awaited with curiosity, and also a little apprehension, the visit of the 'Woman Rebel,' but we were hardly prepared for the surprise given us by the soft-voiced, gentle-mannered, altogether charming 'rebel' who tapped at the door at four o'clock." After telling of her harrassment under the Comstock law and absorbing their applause for her courage, Sanger was later to note, "That afternoon was one of the most encouraging and delightful of my life."

The Drysdales urged her to visit the British Museum in London and read up on Malthusianism, its conflicts with Socialist theory, and its connection with the rationale of birth control. Soon that path led to tea at the Brixton apartment of Havelock Ellis, the world-famous author of the seven-volume *Studies in the Psychology of Sex*. Tall and lean with long, tailored white hair and whiskers, Ellis was "fascinated by the young woman's story, by her courage, her devotion to an ideal, her fire, her vitality and beauty," wrote a biographer of Ellis. "He had never been so quickly or completely drawn to a woman in the whole of his life." Sanger left his apartment feeling for her new friend, she later wrote, "a reverence, an affection, and a love which strengthened with the years." She had been escorted "into a hitherto undreamed-of world."

Ellis and the Drysdales argued that Margaret's New York radicalism had been too broadly aimed, too unstructured. They pressed her to focus on the single issue of birth control "and leave the denunciation of capitalism, churches, and matrimony" to others willing to settle for more anger and arm-flailing harangues and less effectiveness. "I have been carefully reading the *Woman Rebel,*" Ellis soon wrote her. "You know I think you are splendid. I do not always agree with you when you attack, but I always agree when you define. . . . It is no use, however, being too reckless and smashing your head against a blank wall, for not one rebel, or even many rebels, can crush law by force." Changing the law "needs *skill* even more than it needs strength."

Ellis and Sanger began meeting daily at the British Museum reading room. He drew up reading lists for her, recited the work of others in birth control, answered her questions like a tutor. She studied the early population theories of Malthus, John Stuart Mill, and Robert Owen. He introduced her to theories of improvement of the human species through selective breeding and eugenics, and the supposedly "scientific breeding" of the American Oneida Community. Also, he led Sanger to a book that influenced her life profoundly, *Elements of Social Science* by George Drysdale, the brother of Charles, who had received her on her arrival in England. In it he argued that only artificial contraception could expand the amount of love in the world. Thus George Drysdale was probably the first to attempt to link science with the enhancement of romantic love.

That mentorship by Ellis blossomed into a passionate love affair, a romantic peak never to be matched in Sanger's lifelong series of affairs that probably numbered as many as a hundred, in the estimation of Margaret Lampe, Sanger's granddaughter and, later, her frequent traveling companion. The affair was all the more noteworthy because Ellis was sexually dysfunctional. Although he was married devotedly, his wife was a lesbian, and they usually occupied separate homes.

In January 1915, Charles Drysdale dispatched Sanger with a letter of introduction to Dr. Johannes Rutgers, director of the government-supported birth control clinic at The Hague, Holland. Dr. Rutgers persuaded her to give up a view she had expressed in a pamphlet, *Family Limitation*, that women could "teach each other" contraceptive methods, and even that they could base their main learning of the subject on mere pamphlets. Birth control was more than simply a matter of instruction, he urged. A woman must be medically examined and individually fitted for a diaphragm, or pessary, the most modern of the devices at that time. Contraception was a medical matter. Sanger became converted to that principle and so altered the future course of the birth control movement.

Meanwhile, at home the legal complications got worse. A man rang William Sanger's doorbell in New York, identified himself as a Mr. Heller, a birth control advocate, and asked for a copy of *Family Limitation*. He said he wanted to have the pamphlet "to distribute among the poor people he worked with." Sanger promptly obliged and accepted payment for the pamphlet. A month later, Anthony Comstock himself showed up at Sanger's door and ar-

rested him for distributing "obscene, lewd, lascivious, filthy, indecent and disgusting" literature. Comstock's true purpose, many
thought, was to force his real target, Margaret Sanger, back to the
United States to face trial.

Conducting his own defense, William Sanger declared, "The
law is on trial here, not I." The presiding judge, not seeing it that
way, pronounced Sanger a "menace to society." He gave Sanger a
choice of a $150 fine or thirty days in jail. With the same bravado
that led him to conduct his own defense, Sanger responded: "I
would rather be in jail with my conviction than be free at a loss of
my manhood and my self-respect."

His choice, threatening, in effect, to orphan the Sangers' three
children, now forced Margaret to return to America and face her
own incarceration. Margaret wrote to her sister Nan resenting
that "Bill had to get mixed up in my work after all, and of course
make it harder for me and all of us!" In September, hazarding a
submarine attack, she sailed from Bordeaux.

Less than four weeks after Margaret's return to America, her
young daughter, Peggy, died of pneumonia. A mix of grief and
guilt disabled the shocked mother. Already weakened by her tuberculosis, she ran a daily temperature of over 100° F., which spun
Margaret into what her doctor, Morris Kahn, noted as a mild
"nervous breakdown from excessive mental and physical strain."

But she recovered soon enough to face her trial and to conduct her own defense as her husband had done, expecting that
her martyrdom might inspire others to join her cause. Some of
her friends worried that the state of her emotions was leading her
into poor decisions. Several free-speech advocates and lawyers
among her friends urged her to plead guilty and accept what
surely would be a light sentence, thus enabling her to go on with
her work. But she refused, writing to "Friends and Comrades":
"The whole issue is not one of a mistake, whereby getting into jail
or keeping out of jail is of importance, but the issue is to raise . . .
birth control out of the gutter of obscenity and into the light of
human understanding."

From England, Sanger's new friends, novelist H. G. Wells and
Marie Stopes, the author of *Ideal Marriage*, a sex manual, among

scores of others, wrote letters directly to President Woodrow
Wilson urging his intervention on Sanger's behalf. Perhaps that
had something to do with the courtroom surprise of February 18,
1916. The government prosecutor entered a request of nolle
prosequi—canceling the prosecution without exonerating her of
the accusations.

In March 1915, while Margaret was in exile, a group had met
in New York to form the National Birth Control League (an an-
tecedent of the International Planned Parenthood Federation).
The founders were mostly educated and well-to-do reformers,
not revolutionaries. Although birth control was becoming a lively
topic in conversation and even in the press, Caroline Nelson, a
radical among the League's founders, wrote to Margaret Sanger in
June that "it is almost impossible to interest the workers." Neither
"the workers" nor "our very learned radical men" could talk about
the subject without "giggling and blushing." As a result "our
League here consists mostly of professional people . . . but no real
working people. . . . This outfit will simply keep it as a semi-fash-
ionable league. They are the people who don't need the information and never did, and how we are going to get it to the workers
is the problem that I constantly harp on."

As Margaret's national activities and prominence grew, the remains of her marriage withered away. Having separated from
William Sanger in 1914, she would divorce him in 1920.

She was soon to split from her radical comrades as well, but
that parting was less orderly. In 1916 she launched a new publication, the *Birth Control Review*, throwing in with a Socialist from
Cleveland, Frederick A. Blossom, who later became a Wobbly. As
its managing editor, he personally distributed the first issue at a
rally for Sanger at Carnegie Hall on January 29, 1917, prior to
her trial. By that time she had broken with the National Birth
Control League, which had set as its first priority the passage of
laws permitting birth control. Sanger's commitment was to involve the "masses" in direct action—breaking the laws as a way of
forcing change. Later she was to soften her views, looking more
favorably on legislative change.

In the spring of 1917, Blossom suddenly quit the *Birth Control*

Review, absconding with the magazine's meager funds, its account books, some office furniture, and worst of all, its list of subscribers. Perhaps the split had been hastened by Sanger's change of heart on legislative activity. More likely, each had a passion for controlling the birth control movement and saw the other as a rival and obstacle. When no appeals to Blossom's good sense worked, Sanger took her complaint to the district attorney.

No matter what sympathy her new plight had drawn from her leftist friends, that move finished her with them. They had pleaded with her to accept an investigation by a Socialist-party committee, but she refused. One leading party member declared: "The undisputed fact . . . remains that Mrs. Sanger had gone to an outside agency—the capitalist district attorney—to involve and perhaps imprison a comrade. In this respect she was guilty. We prepared for and actually succeeded in ousting her."

That break seemed to free her to forsake broad-brush radicalism and to focus on the goal of birth control, as Havelock Ellis and her other English friends had urged. She made peace with the National Birth Control League and began to court the fashionable, the rich, and the influential. Her audiences and her own influence rapidly expanded.

Although denied the martyrdom of an adverse court decision, Margaret Sanger launched a national speaking tour in the spring of 1916, systematically creating her own style of winning support. In St. Louis, a threatened boycott by Catholics closed the hall where she was to speak. Restirring the excitement of her father's confrontation with a Corning priest, Sanger relished such barriers: "As a propagandist, I see immense advantages in being gagged. It silences me, but it makes millions of others talk about me, and the cause in which I live." In Portland, Oregon, she was arrested but immediately released. In a later tour, an overflow crowd waiting outside New York's Town Hall for a birth control rally was locked out upon the demand of Archbishop Patrick Hayes. "Children troop down from Heaven because God wills it," the archbishop announced, and any attempt to prevent the increasing tide of humanity was "satanic." When Sanger tried to speak she was arrested. Boston authorities threatened to close any meeting at which she spoke. When they backed down and let her appear at Ford Hall Forum as scheduled, Sanger walked to mid-stage gagged. Arthur M. Schlesinger, Sr., the Harvard historian, read her prepared text to the delighted audience. In every speech of her 1916 tour—in Chicago, St. Louis, Detroit, Los Angeles, San Francisco, Portland—she retold the story of Sadie Sachs, decried the dangers of abortion, pleaded for new moral standards, and demanded a new freedom for women.

Despite these triumphs, Sanger dreaded speaking to large audiences. Disliking the bombastic style of the radical orators who had been her models, she engaged a speech coach to teach her to moderate her voice. Sanger memorized a set talk that she delivered again and again in a more soft-spoken way.

"She remembered the suffragettes, strident with horrid voices and bulky frames and making lots of noise. She never wanted to do that," Margaret Lampe recalls. "In Chicago, she was about to give a speech, and whoever was making the introduction was *booming* his voice out, and the people in front practically had their hands up, and the people in back couldn't hear anything for the echoes. Everyone was uncomfortable, and no one could tell what this man was saying. She didn't want to be like that. She always dressed in a very plain way so that her clothes and her movements would not detract from her message. She said she always wanted to appear feminine, ladylike and soft-spoken, never so beautiful that women didn't like her but attractive enough so that men wanted to give her money. Those are the stories she told me. She also said to me, 'There may be a time in your life when you will be in a group where you are the only woman and there are many, many men. I want to tell you a lesson I have learned. Do not speak for the first twenty minutes of the meeting. Then ask the hardest question of the day. When you think the debate on an issue is about halfway through, make your point in three minutes or less. When you feel the debate is almost over, restate your point in one minute or less, say no more.' It is the most effective way of boardsmanship, and I have used it my whole life.

"When she courted the medical profession without having the medical background, she would go to them saying, 'I need your help. This is my problem. Can you help me? Could you explain this?' Before it was over, they were eating out of her hand. If she had come and said, 'This is what you have to do,' nothing would have happened. So, those are the types of stories she told."

Sanger returned to New York from her first speaking tour determined to steer her work into a new, more ambitious phase. After visiting Dr. Rutgers in Holland, she had written to American friends from England: "Clinics is the watchword."

She would now defy the conventions and the law in the most direct and practical way. She would open a birth control clinic.

4

A STOREFRONT IN BROOKLYN

No cause ever fought has been fought against more stupid, blind social prejudice,
not even the cause of the people against the divine right
of Kings, nor the cause of equal suffrage,
nor any of the battles of freedom.
—PEARL BUCK, ON THE TWENTY-FIRST ANNIVERSARY OF THE
BIRTH CONTROL MOVEMENT

One fall day in 1916 in a driving rainstorm, Margaret Sanger, her sister Ethel Byrne, like Margaret a nurse, and a friend, Fania Mindell, tramped the immigrant-teeming streets of the Brownsville section of Brooklyn in search of a low-cost location for the first birth control clinic in America.

They found two first-floor rooms at 46 Amboy Street near Pitkin Avenue, which a sympathetic landlord, a Mr. Rabinowitz, gave them for fifty dollars a month. Mr. Rabinowitz personally painted the rooms snow-white and said proudly, "It's more hospital-looking." Then the three women knocked on doors of the neighborhood's tenements and rooming houses to spread the word of their new service. Instead of welcoming them, some strangers commanded sternly, "Don't come over here," "We don't want trouble," "Keep out." Undaunted, the nurses passed out five thousand handbills printed in English, Yiddish, and Italian:

Mothers! Can you afford to have a large family?
Do you want any more children?
If not, why do you have them?

DO NOT KILL, DO NOT TAKE LIFE, BUT PREVENT.

Safe, Harmless Information can be obtained of trained Nurses. . . .
Tell Your Friends and Neighbors. All Mothers Welcome.
A registration fee of 10 cents entitles any mother to this information.

Would people come?

Early on the morning of October 16, 1916, their scheduled opening day, Ethel Byrne peered out the window and called to Margaret in astonishment. Sanger recorded in her autobiography:

Halfway to the corner they stood in line, shawled, hatless, their red hands clasping the chapped smaller ones of their children. All day long and far into the evening, in ever-increasing numbers they came, over a hundred the opening day. Jews and Christians, Protestants and Roman Catholics alike made their confessions to us.

Every day the little waiting room was crowded. Women came from the far end of Long Island (the press having spread the word), from Connecticut, Massachusetts, Pennsylvania, New Jersey. They came to learn the "secret" which they thought was possessed by the rich and denied to the poor. . . .

Tragic were the stories of the women. One told of her 15 children. Six were living. "I'm 37 years old. Look at me! I might be 50!" One reluctantly pregnant woman who had borne eight children had had two abortions and many miscarriages. Worn out from housework and from making hats in a sweatshop, nervous beyond words, she cried hysterically, "If you don't help me, I'm going to chop up a glass and swallow it tonight!"

I comforted her the best I could, but there was nothing I would do to interrupt her pregnancy. We believe in birth control, not abortion.

On the tenth day, a severe-faced woman pushed her way through the waiting clients into the inside room and announced: "I am a police officer. You, Margaret Sanger, are under arrest."

"You're no woman!" Sanger is reputed to have hurled back. "You're a traitress to your sex."

Instantly, three plainclothesmen from the vice squad materialized and herded the roomful of waiting clients into line as they would the inmates of a brothel. Women wailed in terror; their infants began to cry. The raiders ransacked drawers, confiscating 464 case histories and the clinic's collection of condoms and diaphragms as well as instructional pamphlets.

"It was half an hour before I could persuade the men to release the poor mothers," Sanger wrote.

Meanwhile the event had brought masses of people spilling into the streets. Newspapermen and photographers joined the throng. White-hot with indignation, I refused to ride in the Black Maria. I insisted on walking the mile to the court, marching ahead of the raiders, the crowds following.

I spent the night in the Raymond Street jail, in a cell so filthy that I shall never forget it. The mattresses were spotty and smelly. I wrapped myself in my coat, fighting roaches, crying out at a rat which scuttled across the floor. It was not until afternoon that my bail was arranged. As I emerged from the jail, I saw the woman who had threatened to swallow glass; she had been waiting there all the time.

I went back at once and reopened the clinic, but now the police made Rabinowitz sign eviction papers on the ground that I was "maintaining a public nuisance." Again, I was arrested.

The district attorney now charged the sisters with violation of Section 1142 of the New York State Penal Code—Sanger and Byrne with illegal distribution of contraceptive information,

Mindell with distributing "obscene" literature promulgating contraception.

Ethel Byrne's case, if it was not already certain to end in conviction, was surely jeopardized by the startling defense advanced by her counsel. Section 1142 was unconstitutional, he argued, because it denied a woman "her absolute right of enjoyment of intercourse unless the act be so conducted that pregnancy be the result of the exercise. This clearly is an infringement upon her free exercise of conscience and pursuit of happiness." Shocking the court even more deeply, counsel contended that that right belonged to "married persons, or even single persons."

On the day of Ethel's trial, January 4, 1917, fifty women, many of them "prominent socially" according to *The New York Times*, took Margaret to breakfast at the Vanderbilt Hotel and accompanied her to the public seats of the courtroom.

The turnout did not prevent Ethel's sentence of thirty days on Blackwell's Island.

Ethel Byrne attracted four days of front-page attention by declaring that, like the suffragettes of England, she would not eat, drink, bathe, or work while in jail. The stories reported how jailers force-fed her with brandy, warm milk, and eggs through a tube inserted in her throat. Only the resumption of German submarine warfare drove Ethel's hunger strike from the front page. The *New York Tribune* editorialized presciently: "It will be hard to make the youth of 1967 believe that in 1917 a woman was imprisoned for doing what Mrs. Byrne did."

Sanger's trial opened a few days after Byrne's.

"One by one, the Brownsville mothers took the stand," according to Sanger's written description, which she adorned with accents and all. The district attorney asked one mother:

"Have you ever seen Mrs. Sanger before?"

"Yess. At the cleenic."

"Why did you go there?"

"To ask her to stop the babies."

"Did you get this information?"

"Yes, dank you, I got it. It wass gut, too."

Eager to prevent a public furor, the judge offered Sanger freedom for the mere promise that she would not repeat her offense.

Sanger responded that she had broken the law deliberately. She wanted hers to be a test case and could "not respect the law as it exists today." Sanger then turned down an alternative of paying a fine and was sentenced to thirty days in the Queens County Penitentiary in Long Island City. She, too, went on a hunger strike.

"The prisoners—prostitutes, pickpockets, thieves—had somehow heard about me and birth control," Sanger wrote. "One asked me to explain to them about 'sex hygiene.'" Sanger sought permission to gather the women into a small class, but the matron said, 'Ah, gwan wid ye. They know bad enough already.'" But Sanger persisted, got her way, and also began teaching some of the women to read and write. Having served her thirty days, she walked out of the prison gates to find a band of her comrades singing the *Marseillaise*.

Eventually, her case broke new legal ground. She appealed her conviction through several levels of courts, always failing to get the conviction overturned. But Judge Frederick Crane, writing for a unanimous New York Court of Appeals, cited an allowable exception to the anticontraception law and significantly broadened its meaning. Most doctors had read the statute's phrase that granted an exception "for the cure or prevention of disease" to mean only venereal disease. Perhaps unaware that he was foreseeing by more than half a century the advent of the Pill and chemical birth control, Judge Crane wrote: "This exception in behalf of physicians does not permit . . . promiscuous advice to patients irrespective of their condition, but it is broad enough to protect the physician who in good faith gives such help or advice to a married person to cure or prevent disease. 'Disease,' by Webster's International Dictionary, is defined to be 'an alteration in the state of the body or of some of its organs interrupting or disturbing the performance of the vital functions and causing or threatening pain and sickness; illness; sickness; disorder.'"

So be it. If the judge was thus labeling pregnancy a *disease* for the sake of justifying the prevention of it, Sanger was not about to argue.

Accompanying those courtroom maneuverings, a kettleful of family tensions had boiled up between Margaret and Ethel.

THE PILL

Margaret once confessed to Ethel's daughter Olive, who was to record it in a detailed unpublished memoir of the Higgins family, that she had been jealous of Ethel's soft brown eyes and auburn hair plus their mother's obvious preference for her youngest daughter.

Olive wrote:

Another bitterness between the sisters arose from an episode that occurred shortly after Margaret rushed home from England to be with her little daughter Peggy, who was very ill. Peggy crushed her mother by saying, "I want Aunt Ethel to hold me, not you." This absolutely devastated Margaret despite the fact that a two year absence is a big void in the life of a child. She told me years later that she "never, never got over" Peggy's rejection, which "was almost worse than her death." Both sisters were dramatic and each was selfish. There had been too little opportunity to stand out as individuals in their early lives and certainly an overabundance of sharing. Neither of them was a good wife or mother. They felt that marriage was an infringement on their private lives and their children were both a burden and an occasional joy. They used boarding schools to relieve the former and summer vacations to experience the latter.

That the two most rebellious Higgins sisters, Ethel and Margaret, virtually abandoned their children is clear enough, but more puzzling is that they were sometimes willing to consign them to the very kinds of environments that they despised and escaped. Olive, in early teens, was put by Ethel into a Rochester convent, apparently of the most conventional kind, while her mother was living in Greenwich Village among her exciting political friends. Olive's unpublished memoir reveals the paradoxes of the mothering style of the Higgins women:

It was in Rochester that my Aunt Margaret came to speak on birth control to a woman's group. Knowing that I was in school there, she telephoned to say she was coming to see me. Pandemonium broke loose. At first there was a firm refusal to allow her in the secret precincts of the convent. But Margaret was not a pioneer of women's rights for nothing. She threatened to call the police and charge abduction. At last the matter was referred to the Bishop, who reluctantly agreed that she could see me in the presence of the mother superior. I had seen Margaret only twice and hadn't much enthusiasm for her visit and I was ashamed and afraid that so much fuss would cause me further trouble.

But when I went into the superior's office, the warmest, the most delightful person greeted me with tender hugs and compliments, mentioning nothing about the fuss and with no complaints, but just gave out the charm and love I sorely needed. She was gay, pretty, and I thought she looked like a movie actress. She told me about my aunts and uncles and that my mother loved me and wanted me to be happy and to write her if I needed anything. At the end, she gave me ten dollars and a box of candy. But she also gave me a sense of "being someone," not just a kid nobody wanted.

Their jealousies as sisters, which contributed to the tensions between Margaret and Ethel, later took political as well as familial form. In describing "Ethel's feud with Margaret," Olive provides an account of Ethel's courtroom and jail experience that differs from the "authorized" story:

In the early days of the birth control movement they were close and worked hand in hand putting their efforts into establishing the first public clinic. When Ethel went on the famous hunger strike that brought the movement into the national public eye, Margaret promised the judge who presided over the case that Ethel would no longer be associated with birth control if he would release her from jail. Ethel was furious with that provision and wanted to go on as before. Margaret refused. The former assumed that Margaret was using that excuse to get rid of her mainly, she often asserted, because Margaret had gone "uptown" with the Movement and had no use for the Village people who started things in the first place. All this may have been true. The Village people were long on talk and short on cash, and Margaret knew where success came from—money, and the people who have it.

Margaret Lampe, Sanger's granddaughter, recalled for me further details of the sister rivalry: "They were mad at each other for years. It set in after Ethel's hunger strike in jail—and Mimi [Lampe's intimate family name for Sanger] took the credit for fasting in jail as something *she* [Margaret Sanger] originally did—as part of the glamorous image of *her*. I have a letter that no one has known about until this moment in which Margaret writes to Ethel asking her to ignore that she, Ethel, went on a jailhouse fast, because Hollywood might be doing a picture about Margaret Sanger, and she wanted it to be *her* story alone, not her sister's. Ethel resented Mimi's fabrication and wanted her to be more truthful about it. Ethel was much more honest and forthright about those things. You know, Margaret changed her birthdate. She was born September 14, 1879, but it's listed practically everwhere as 1883. That was Ethel's birthdate. She took her younger sister's birthdate. She changed it in the family Bible. That Bible is in Smith College. The ink she changed it with is now faded, and the original ink has come back. [An inspection of the Bible at Smith confirms that Margaret wrote the later birthdate over her mother's original inscription, but with time, the alteration faded, clearly revealing the original. The self-serving inaccuracy is also confirmed by a handwritten entry in the 1880 census for Steuben County, New York.]

"She was very vain. As she got older, she forgot what she had made up, and some of the real stories came out. That's just the way she was. She told lies about her background because she was embarrassed about it. She was embarrassed by the meagerness of it and wanted to paint a more pleasing image of herself. She was embarrassed by the occupations her brothers and sisters had. Her older sister Mary had been a maid for the wealthy Abbott family near Corning and was such a meek, quiet woman that when she had an appendicitis attack, she didn't tell anyone, didn't want to bother anybody, and she died. Another sister, Nan, became a Christian Scientist. One of her brothers was a prizefighter; one went into the navy. [Margaret's father, too, was one of her embarrassments, and she "improved" his background in her autobiography. She wrote that in the Civil War Michael Higgins marched with General Sherman's troops across Georgia and was cited for bravery. Actually, as a member of New York's 12th Regiment, he served uneventfully at Bachelor's Creek, North Carolina, then inexplicably was listed until the war's end as missing in action.] She was quite proud of only one of them, [her brother] Bob Higgins, who became a football coach at Penn State for many years. Ethel, the youngest of the girls, became an absolute rabble-rouser. She was a Communist. Not a Socialist, like Margaret had been, but a Communist. She subscribed to the *Daily Worker* in the late twenties.

"Margaret was the most complex person I've ever known. She was warm, loving, immensely kind to her grandchildren. Yet she was unreliable with them because she would promise them something, then she would be gone and it wouldn't come to pass." Yet Margaret Lampe liked her? "Oh, I loved her. Absolutely *loved* her. She adored me and I adored her. The most fascinating woman I ever met in my life. She taught me more about politics, about people, about looking at issues. She has given me so much strength and courage. I have done lots of interesting things in my life because of her."

Margaret's tense, rivalrous relationship with Ethel, for all its duplicities and manipulations, continued to entangle the two sisters for many years. They spent summers near each other in cottages at Truro on Cape Cod, a favorite colony of radicals and freethinkers. In New York Margaret occupied an apartment on the second floor of an old brownstone house at 246 West Fourteenth Street, on the northern edge of Greenwich Village. The rooms were large, with high ceilings. Its kitchen, a former clothes closet, housed a small gas stove and sink. A bathtub hid behind a large screen just outside the kitchen. Running a bath was a tedious chore of heating water on the kitchen stove and pailing it into the tub. A toilet was shared with the occupants of the front apartment.

Eventually Ethel replaced Margaret in the apartment, sharing

it with her lover, Robert Parker, a writer and editor with a withered arm and deformed leg, consequences of childhood polio.
Rob became important to Margaret Sanger also, by becoming the
ghostwriter of her autobiography, and thus an accomplice to the
improvements and refinements of the self-portrait she decided to
leave to history. "Many years later," wrote Ethel's daughter,
Olive, in her unpublished memoir, "Aunt Margaret Sanger told
me that Rob and Ethel had been married in Nantucket. My
mother pooh-poohed the report, saying that Margaret was
known to invent 'nice covers' for circumstances that she thought
might become a matter of scandal, embarrassing to herself."

After completing their collaboration on the book-length autobiography, Margaret rewarded Parker with a large basket of fruit
for Christmas. He had "expected something more negotiable
than the fruity token," Olive recalled. "He held the offending offering out for us to see and sadly said, 'Just what I wanted.' "

Sitting out her jail sentence, which followed Ethel's, Margaret
had a rare opportunity to think out her future. From her cell, she
wrote to Ethel: "I'll have to go west and find a widower with
money and settle down for life."

By 1922, she found J. Noah H. Slee, president of the Three-In-
One Oil Company, an enthusiast for birth control and twenty
years her senior, who left his wife to pursue Sanger. He followed
her everywhere, seeking her companionship, and each finally
won the other. Her victory was an extraordinary marital arrangement, inspired by the lives of Havelock and Edith Ellis. Like the
Ellises, Sanger and Slee each would have a separate home, although later they combined them into separate portions of the
same house. Each spouse, they stipulated in a written contract,
was to conduct business without influence by the other, except as
requested. When both were busy, each was to be satisfied with
communication through their secretaries. The bride was to keep
her first married name, which had become the famous brand
name of her activities. A final touch: The marriage was kept secret for almost two years, for reasons never made clear.

Margaret Lampe describes her step-grandfather as a "warm,

loving Dutchman who absolutely adored Margaret Sanger. J. Noah Slee told his family that Margaret was the most wonderful sexual delight he had ever known in his life." Between 1921 and 1926, he became the Birth Control League's largest contributor, giving more than $56,000.

By 1926 the league had acquired thirty-seven thousand members, drawn from Sanger's newer, higher-tone political allies. A survey showed that the typical member was a second-generation American white Protestant, with a higher average income—three thousand dollars a year—and better education than most Americans. (The league's most commonly recommended contraceptive device, the diaphragm, not yet mentionable in polite conversation, was sometimes called the "uppity-cuppity.") Not surprisingly, the league's highest proportion of membership was in New York, with other members distributed mostly in the urbanized parts of Illinois, Ohio, California, Michigan, and Texas.

In that year of 1926, Sanger went to Europe to organize a World Population Conference, consisting mainly of world-renowned scientists. Through the importance of their names, she hoped to impress the League of Nations with the urgency of birth control as a solution to the threat of overpopulation.

That conference theme—overpopulation—marked a major shift in Sanger's battle cry for birth control. No longer would her appeal center on rescuing overburdened mothers from overlarge families of children; even less on separating the fear of pregnancy from the pleasures of sex. She had accepted the political advantages inherent in the teachings of her neo-Malthusian friends in England. Her new constituency—the educated, rich, and influential—were more ready to accept the high-minded, impersonal ground of a worldwide social problem than the embarrassing subject of sex. The new focus on overpopulation at last freed the movement of taboo.

Still, Sanger's new respectability continued to show scars. When the World Population Conference opened in 1927 in Switzerland, a dispute broke out between Catholics and other delegates over whether to allow Sanger's name on the program, even though her husband had paid for much of the conference.

Her critics succeeded in preventing even a mention on the agenda of contraception.

A year later, in an outburst of name-calling and accusations of power-grabbing between Sanger and new leaders of the National Birth Control League, she resigned as its president. That, of course, ended Noah Slee's financial support, and by the 1930s the organization had shriveled in both size and influence.

While Sanger remained an icon of controversy, the atmosphere of repression surrounding birth control was lifting. One sign of the end of the Comstock era in the United States came in 1936 after Margaret Sanger ordered by mail a new model of a Japanese diaphragm, requesting that it be addressed to her friend Dr. Hannah Stone. The package was confiscated by U.S. Customs as indecent, as defined by the Comstock law. The case of what became the world's most famous diaphragm moved up through the courts as *U.S.* v. *One Package.* Finally, U.S. district court judge Grover Moscowitz threw out the government's suit and ordered the package delivered. The following year, in 1937, Franklin D. Roosevelt's attorney general, Homer Cummings, announced that the government would not appeal to the Supreme Court. Thus, the mails were thrown open irrevocably to the shipment of contraceptives. In the same year, the American Medical Association officially recognized birth control as part of legitimate medical practice. With their profession's stamp of approval, doctors, often accused of running together like blind herds, fell into line behind birth control, at least verbally. In the first poll of medical opinion on the subject, conducted by Dr. Alan F. Guttmacher of the Johns Hopkins School of Medicine, in 1944 fully 97.8 percent of doctors said they approved of birth control to "insure good health for mother and child," 79.4 percent for "economic reasons," and 86.5 percent favored it for child spacing. Considering the widely perceived timidity of doctors to discuss sex with their patients, let alone a topic as unpleasant as birth control, it was surprising that 72 percent of practitioners said they believed that planning families would help marital adjustment.

Local birth control leagues, almost all inspired by Sanger's work or directly organized by her, eventually unified as the Planned Parenthood Federation of America, which by 1951, the year of the fateful Sanger-Pincus private dinner, sponsored two hundred clinics across the United States. The PPFA, in turn, became the core of the International Planned Parenthood Federation, of which Sanger was elected first president.

Thus Sanger officially became a world figure, committing herself to eradicating the global famine of birth control knowledge. The immensity of the commitment defied comprehension. In solid array against her ill-equipped little army stood the great organized religions, cultural traditions, and, perhaps most unyielding of all, simple, entrenched mass ignorance. On her side, however, stood one powerful force that transcended national and religious boundaries: the ages-old fiery passion of women of the world to know "the secret."

Even educated, upper-class women sought this knowledge: In England, Dilys Cossey, the Family Planning Association chair, has told her version of youthful sexual innocence: "When I was a young woman at the university, I remember we were all sort of herded into a room one evening and were talked at by this—this frightful woman doctor. I am not quite sure what she was trying to tell us, but I think it was something to do with fetal development, and we were all very scornful, and I think what she was trying to tell us was *Don't get pregnant*. But she didn't tell us how *not* to. It was all quite extraordinary. I do remember when I was at school, we had a very brave biology mistress who told us the facts of life. We were in the lower fifth, and we had a double biology period, and we had these drawings on the board. We were told about the penis and the vagina, and we were all agog. But at least we had some straight information."

Organized instruction about birth control had begun in New York with Margaret Sanger's first clinic in 1916. Soon thereafter, a clinic in London was opened by Marie Stopes, who became famous on both sides of the Atlantic for her groundbreaking book *Married Love*. The book caused a public outcry in 1918 and became a runaway sensation, selling out six editions in a year and ulti-

mately more than a million copies in a dozen languages. Marie
Stopes was a paleontologist by training. After winning an annul-
ment of her marriage to an impotent husband, she devoted herself to insuring, through writing, that no woman would again
suffer from sexual ignorance and inhibition as she had. Her florid
prose is illustrated best in her description of orgasms: "The half
swooning sense of flux which overtakes the spirit in that eternal
moment at the apex of rapture sweeps into its flaming tides the
whole essence of the man and woman." But her more clinical de-
scriptions of coital positions and technique, although they
shocked at first, became a decade later the standard fare of marriage manuals.

"While she was a great pioneer," said Nancy Raphael, an early
organizer of the Family Planning Association in London, "she was
not a great founder of a movement. She was really more like an
artist, not a reasonable being, but rather like a witch with blazing
blue eyes. She prepared the ground. She got things moving. Our
aim before World War II was to stimulate the local health authorities to provide a family-planning service. But we fairly soon came
to the conclusion that it was hopeless. There was no public acceptance of this thing. We had to start founding the clinics ourselves.

"We soon found out that to call an organization the National
Birth Control Association made it doubly unpopular. It wasn't all
right to call it 'birth control.' They were dirty words. In the late
thirties there had been a population scare in this country.
Population began to drop. It could have well been the fear of war.
In a brilliant stroke, Margaret Pyke and one or two others
thought up the name Family Planning Association. We think of
'family planning' now as just a phrase in the English language. But
it wasn't then. They invented it. And it certainly did change the
way in which we were seen, I think, in a positive way. Helping
people to *plan* their families was good."

"Some people wouldn't have dreamt of going up to our first
clinic," adds Dr. Faith Spicer, one of the association's pioneer
physicians. "Their husbands would beat them up if they went. I
remember one awfully nice lady saying to me, 'He'd kill me if he

knew I was coming here, but anyway he's ever so good because he wants it only every other Friday.' So really, all she was was a sort of willing vessel for the bloke. I don't think many women then really accepted that there was a chance they could enjoy themselves. But there were always crowds of people in the clinic, rows and rows of women sitting and waiting. A nurse would come streaking along, saying, 'Knickers off, girls, stockings down.' We worked behind screens, one woman in that room and one woman in the next. It's so different now, you can hardly believe it was like that."

Lady Helen Brook further recalls: "Many patients came from Kings Cross and Euston and all around there, and I went to see what sort of backgrounds they had. I wanted to see for myself what it was all about. Around Euston and Kings Cross there was a desperate smell of poverty. Terrible back lavatories and those awful backyards. There we were in the clinics saying to them: The way you put your cap in is this and that and the other, and about six o'clock in the evening before your husband comes home, you know, put your leg up on the side of the bath and pop in the cap and then you're all relaxed for the evening. And then you found, as we did when we worked very hard to make a survey in Islington, that they didn't *have* baths. They were lucky if they went to the local bath once a week. And the husbands behaved so badly. Husbands would say they weren't getting their proper satisfaction if a woman wore a cap, and he would snatch it away and throw it in the fire, and the woman would come back and be very weepy about it, and then we would have to fit her up again.

"In our interviews, we found we had to take out all the business of asking about their religion. You had a Jewish woman who'd say, 'Do you think I'm being wicked?' You'd have a Roman Catholic woman saying, 'Do you think I'm being wicked?' I must say, they were very busy about their sins."

Dr. Sylvia Dawkins added: "We'd ask general medical questions, the length of marriage, the number of children, and what methods, if any, they had used—that was a nice one. This one woman said to me, 'I make my old man sleep with the boys. But, Doctor, he comes creeping back. He's ever such a creeper.' With some of them, I'd ask, 'Do you get any pleasure? Do you enjoy it?' I didn't use the word 'orgasm' because most of them wouldn't have understood it. Sometimes they used the term 'to come,' but many of them feared that that made pregnancy more likely and they used to hold back and resist.

"The cap made the woman able to accept her body, her own vagina, to learn where the sensitive areas were, about which they knew nothing. They'd been forbidden, remember, to touch. That's *wicked*. Most of all we were teaching that they had to examine themselves. This was terrifying to them. Because Mother had always said, 'Don't touch.' Here we were saying, 'You *must* touch,' so it was an eye-opener."

Many years before anyone seriously imagined an organized worldwide birth control movement, Sanger's partner-to-be, Katharine McCormick, had begun responding to her husband's sudden fall into hopeless insanity by investing heavily in neuropsychiatric research. In 1927 she established her own Neuroendocrine Research Foundation at the Harvard Medical School. She had to choose her benefactors carefully because as long as her husband remained alive her expenditures required approval of the probate court in Chicago. If McCormick stayed close to neuropsychiatric research, related to her husband's illness, her spending won easy approval. Upon his death she would gain full control of his money and do what she pleased with it.

McCormick also contributed to the Planned Parenthood Federation but kept the amounts small and inconspicuous, since they were for the "scandalous" purpose of contraceptive research.

Through her interest in birth control McCormick had first met Sanger in 1917, and in 1927 McCormick consented to lend the Château de Prangins in Switzerland, the scene of her wedding, to be used for a lavish reception for the three hundred delegates to Sanger's World Population Conference. During the 1920s McCormick became one of several sympathizers who, amidst their travels to Europe, helped smuggle diaphragms into the United States to supply Sanger's clinic. McCormick's devotion to women's suffrage kept the two women on somewhat separate paths. But in 1928 they began exchanging occasional letters in which they groped vaguely toward sharing their common interest in some kind of major breakthrough to contraception.

In 1947, Stanley McCormick died. Katharine, at seventy-one, had to wrestle in the courts against the lawyers for other McCormick claimants over control of the full inheritance from her husband; it would take her five years to win. Upon winning control, she sold five of Stanley's estates, including Riven Rock near Santa Barbara, to pay her inheritance taxes and reduce her enormous expenses. Finally, she could see ahead to a freedom that would allow her to spend her money, her time, and her devotion on what she felt was most important. In October 1950 she wrote the letter to Margaret Sanger that raised "two questions that are much with me these days": the present prospects of contraceptive research and the greatest need for financial support.

Thus came about Sanger's 1951 dinner with Gregory Pincus and, later, Katharine McCormick's visit to the Worcester Foundation for Experimental Biology.

While reproductive biology had been a central subject of attention at the struggling Worcester Foundation, contraceptive research did not play a large role there. Pincus had been receiving small sums from the Planned Parenthood Federation to study early development of mammalian eggs, but it was only a fraction of the amount required for any concerted effort to develop a hormonal contraceptive.

Sanger knew, of course, where to find money to redirect him. In March 1952 she wrote McCormick about Pincus's work for the federation. By May, McCormick, who had previously known Pincus, moved from Santa Barbara into her house in Boston, which put her closer to Worcester. After her first on-the-spot bank draft to Pincus (apparently her "end of the fiscal year" excuse had meant the pause she required until gaining full control of her husband's money), McCormick's commitment grew rapidly. She contributed sums ranging from $150,000 to $180,000 a year to the Worcester Foundation for the rest of her life and was to leave the foundation a million dollars in her will. Her money paid the way for the Pill. Not a single government dollar went into developing what might be regarded as the most revolutionary pharmaceutical invention of the century.

The enthusiasm of the two aging feminists for a contraceptive that would be "swallowed like an aspirin" stirred no visible excitement at Planned Parenthood in New York. Sanger especially resented the federation's apparent greater interest in its 15 percent slice out of all donations to pay for its operating "overhead." She was further irritated by PPFA director William Vogt's coolness toward Pincus's work. He showed no interest in visiting Worcester. When she called at the New York office, she got only "desultory conversation" rather than a visionary plan. McCormick wrote Sanger of her feelings of slight and offense when Paul Henshaw, the federation's part-time director of research, "offered to show me the office layout which Mr. Vogt said they were rapidly outgrowing, so I looked into two or three smaller rooms, was introduced to two office women and then took my leave. . . . It appears to me that no one there . . . is really concerned over achieving an oral contraceptive and that I was mistaken originally in thinking they were."

And so in the early 1950s two essentially political women hired their own scientist to pull out of a hat their envisioned perfect contraceptive. Neither Margaret Sanger nor Katharine McCormick knew a great deal about where or how Pincus might begin such a search. But he knew one place he must start. There was a man last known to be in or near Mexico City who seemed to have vanished from the world of science, and now Pincus wanted to find him.

5

WAITING

*Forgive, O Lord, my little jokes on Thee
And I'll forgive Thy great big one on me.*
—ROBERT FROST

Long before Margaret Sanger glared her father down at her mother's funeral, women had begun their unremitting search for some scheme, almost any scheme, that would enable choosing to not become pregnant. The search had begun not a century earlier, nor still another century before that. Since the dawn of known time women have pursued an escape from the "great big trick" played on them: being thrust into a life of inescapable sexuality, with the ever-hanging consequence their natural endowment to birth babies, whether a woman wants one at a given moment of her life or not. The awesome blessing is inseparable from the fearsome curse. Women's search for a way to disconnect the making of love from the making of a baby has persisted urgently, inventively, artfully, and passionately.

If such scientist-designers of the Pill like Russell Marker and Carl Djerassi were innocent of the revolution they were to foment in the lives of ordinary women and men everywhere, then what about those ordinary sexual women and men themselves? How did *they* dream of the coming revolution, those gonad-driven creatures so often ashamed and terrified of their own inner flames? How did they rehearse for the freedom they craved while waiting for some answer they did not know they were waiting for?

Before there was science, there was magic, often offering itself as religion, which we can trace only as far back as we can trace words on preserved paper.

Pliny (A.D. 23–79), the Roman compiler of *Natural History*, answered the timeless yearning with the first known advocacy of the world's most reliable means of birth control (also the most persistently resisted): abstention altogether from sex. And he warranted, at least for males, a method for doing so: "If a man makes water upon a dog's urine, he will become disinclined to copulation." Thirteen frustrating centuries would pass before ibn-al Baytar of Islam recommended a comparable remedy for Islamic females: "If a woman urinates in the urine of a wolf, she will never be with child." The Greek physician Dioscorides recommended: "The menstrual blood of a woman appears to prevent conception when they spread themselves with it."

The prescriptions were as panoplied as they were relentless. Sixth-century medical writings of the learned Aetios of Amida advised women to wear cat liver in a tube on the left foot or wear the testicles of a cat in a tube around the umbilicus. Or the woman should carry as an amulet around the anus the tooth of a child. Or she should wrap in stag skin the seed of henbane diluted in the milk of a mare nourishing a mule, carry it on her left arm, and take special care not to drop it.

Albert the Great (Albertus Magnus), the thirteenth-century Italian bishop who compiled two encyclopedias, was the teacher of Saint Thomas Aquinas, the principal author of Catholic doctrine on the sinfulness of contraception. But before he was swept by that current, Albert gave helpful hints for both temporary and permanent avoidance of pregnancy: "The ancients say that if a woman hangs about her neck the finger and the anus of a dead fetus, she will not conceive while they are there. It is also said that if one cuts off the foot of the female weasel, leaving her still alive, and if one puts this foot about the neck of a woman, she will not conceive while she wears it; and that if she takes it off she will become pregnant. If one takes the two testicles of a weasel and wraps them up, binding them to the thigh of a woman who wears also a weasel bone on her person, she will no longer conceive."

A Middle Ages legend advised European women that to avoid pregnancy one should spit three times into the mouth of a frog. Another way was simple and blunt: A woman was to go to the grave of her sister and call out three times, *"I don't want any more children."*

Moroccan folk wisdom, according to Westermarck in *Ritual and Belief in Morocco,* had a more elaborate variation: After a burial, a woman who lingered after other mourners left would thwart pregnancy "by stepping three times over the grave; but all the steps must be made in the same direction, since otherwise the return step would counteract the effect of the earlier step." They also believed that water used for the washing of a dead person, when drunk by a woman, would make her infertile.

Among medieval Serbs, if a bride riding to her wedding had family planning in mind, she needed to decide how many pregnancy-free years she wanted, then sit on that number of fingers for the whole of her carriage ride. If by some misfortune that usually dependable method failed, she was to dip the same number of fingers into her first child's bath water.

Throughout recorded history, wives and husbands yearning to choose their time of pregnancy searched for something more reliable than magic. Some of the earliest prescriptions were written by Chinese emperor Shen Nung, who was believed to have lived from 2737–2696 B.C. A Chinese man's chief anxiety arose from the demand that he produce sons. (Hence the most common "contraceptive" in China and some other Eastern cultures: killing an unwanted girl after its birth.) The best route to more sons, of course, was more pregnancies. A man's supply of seed was severely limited, he believed, and thus had to be hoarded; also that a woman's *yin* essence (her vaginal secretions) would strengthen his *yang* essence (his semen). So, paradoxically, the man tried to increase the likelihood of impregnation by forestalling pregnancy. If he bedded down with as many women as possible *without ejaculating*, he would strengthen his semen.

A book illustration from the Ming dynasty portrays the "stream of life" that results from the practice of withholding ejaculation. It depicts the path of a man's semen returning up his spinal cord to his brain. The man is to accomplish such rerouting by gripping his testicles tightly just before he loses control of the ejaculation. Emperor Shen Nung also suggested "gnashing the teeth a thousand times . . . [and] pausing nine times after every series of nine strokes," and pretending the woman was frightfully ugly. (As an alternate method he could nip his *p'ing-i* point, situated just above the right nipple.) Various branches of Buddhism and Hinduism also believed that refusal to squander semen helped males become one with the deity by enabling their sperm to return to the brain.

The technique of withholding ejaculation has come to be called **coitus reservatus**, not to be confused with **coitus interruptus**, or withdrawal before ejaculation.

The oldest known and longest-honored of all contraceptive techniques is chronicled in Genesis and remains popular today. Coitus interruptus, or withdrawal, calls upon the male to withdraw his penis from the vagina in a disciplined maneuver at the moment preceding ejaculation. The first recorded practitioner was Onan, who was branded a sinner for his trouble: "Then Judah told Onan to sleep with his [dead] brother's wife, to do his duty as the husband's brother and raise up issue for his brother. But Onan knew that the issue would not be his; so whenever he slept with his brother's wife, he spilled his seed on the ground so as not to raise up issue for his brother. What he did was wicked in the Lord's sight, and the Lord took his life."

Precisely *what* was wicked in the Lord's sight? The interpretation hardened by the ages, first proposed by Thomas Aquinas, is that the spilling of the seed was masturbation. Aquinas wrote in "Summa Contra Gentiles": "It is contrary to man's good that the seed be emitted in such a way that generation cannot follow: and if this be done deliberately, it must needs be a sin." For centuries, rabbis condemned withdrawal as "plowing in the garden and emptying upon the dunghill."

In later centuries, when it might be rationalized that a woman's health would suffer from a pregnancy, rabbis and most clergy, relaxing their previous condemnations, came to condone withdrawal—or, in the roguish phrase of Rabbi Eliezer (c. A.D. 100), allowing a man to "thresh inside and winnow outside."

In 1831, Robert Dale Owen, founder in New Harmony, Indiana, of an "ideal community," published his *Moral Physiology* in which he frankly and unsensationally proposed family limitation through use of withdrawal, as well as of the condom and vaginal sponge. He predicted that he would be answered with "abuse from the self-righteous . . . misrepresentation from the hypocritical . . . reproach even from the honestly prejudiced." He was correct. One Boston editor denounced him for "a mean, disgusting, and obscene book, filled with arguments that would disgrace the tenants of a brothel."

The practice of withdrawal has remained unsatisfying, and not only to those who have practiced it. One critic of the technique, Dr. George Drysdale, the late-nineteenth-century English birth control advocate whose writings influenced Margaret Sanger, blamed withdrawal for nervous disorders, sexual "enfeeblement," and "congestions in men." Eliza Duffey, an American contemporary of Drysdale, wrote that in women coitus interruptus caused tumors, inflammations, and ulcers of the uterine system. A physician, David Booth, asked a patient in 1906 to describe her feelings following coitus interruptus. She told him "she felt she wanted to sneeze and couldn't." According to Shirley Green, the

English author of *The Curious History of Contraception,* withdrawal fails eighteen out of every one hundred couples over a full year's practice, yet today, despite that rate, "in the United Kingdom it even rivals the Pill in popularity."

Withholding and withdrawal are self-disciplines that put the man in charge. But the centuries of searching have mostly dwelt on ways that keep the woman in control.

Hippocrates, the most famous of all Greek physicians (460–377 B.C.), in *On the Nature of Women,* prescribed: "After coitus, if a woman ought not to conceive, she makes it a custom for the semen to fall outside when she wishes this." Soranus, a Greek who practiced medicine in Rome between 98 and 138 A.D. and wrote forty known treatises, was more clinically specific in his *Gynaecology:* "The woman ought, in the moment during coitus when the man ejaculates his sperm, to hold her breath, draw her body back a little so that the semen cannot penetrate into the *os uteri:* then immediately get up and sit down with bent knees, and in this position, provoke sneezes."

A thousand years later the Islamic physician Rhazes echoed Soranus and prescribed in even greater detail: "Immediately after ejaculation, let the two come apart and let the woman rise roughly, sneeze and blow her nose several times, and call out in a loud voice. She should jump violently backwards seven to nine paces." He also advised expelling semen from the vagina by squeezing the navel with the thumb.

For centuries women were misled by the most learned of doctors and midwives who preached a method of timing intercourse with their menstrual cycles—pre-Vatican "rhythm"!—but who were teaching it exactly wrong. From the second century onward, with the writings of the eminent Soranus, women were sent into almost certain pregnancy: "In cases where it is more advantageous to prevent conception, people should abstain from coitus at . . . the time directly before and after menstruation."

In ancient China, Tung-Hsuan wrote (with males only in mind) in *Records of the Bedchamber:* "Every man who desires child should wait until after a woman has had her menstruation. If he copulates with her on the first or third day thereafter, he will obtain a son. If on the fourth or fifth days, a girl will be conceived. All emissions of semen during copulation after the fifth day are merely spilling one's semen without serving a purpose."

Beyond magic, baby murder, and tricks of timing, people discovered physical barriers to prevent contraception. The **pessary**—inserted in the vagina to prevent the sperm from reaching the ovum, and possibly with accompanying chemicals to kill the sperm—has tribal and ancient roots. Achenese women of Sumatra kneaded primitive barriers from local plants as though informed by chemical knowledge. The plants they used were found to be rich in tannic acid, a sperm neutralizer—forerunners of the modern diaphragm and jelly. Perhaps similar acids explain why the Egyptians described on papyrus, circa 1850 B.C., the scientific breakthrough of making pessaries from dried crocodile dung (also recorded by the Greeks as recently as 300 B.C.). Another papyrus dated around 1550 B.C. prescribed tips of acacia (gum arabic), rolled with honey and moistened lint, to be placed in the vulva. The tips of acacia shrub, when fermented, produce lactic acid, used today in contraceptive jellies.

A Sumatra tribe, the Karo-Bataks, devised a small ball of opium as a barrier. Natives of West Africa used crushed roots. Central Africans chopped grass or tore rags. On the Easter Islands women inserted clumps of seaweed. When the sticky mixture was rolled with lint, an effective physical barrier was produced as well.

Casanova (1725–1798) wrote in his *Histoire de ma Vie* of introducing to some of his partners the vaginal insertion of small exquisite gold balls. They weighed exactly sixty grams, measured eighteen millimeters in diameter, and were ordered from a particular goldsmith in Geneva at the high price of six quadrupels apiece. As a precursor of the modern diaphragm, Casanova also described cutting a lemon in half and extracting most of its juice, not for employing the spermicidal talent of the acids, but for using the rind as a cervical cap.

In the 1830s a German, Dr. Friedrich Wilde, introduced a small cap that fit snugly onto the cervix, to stay there until its wearer's period was due. Almost a half century later, the Wilde cap, which never became popular, made way for a larger cap that we know today as the diaphragm. It was invented by another German, Dr. Wilhelm Mensinga, and became popular when it was used by Aletta Jacobs in the world's first birth control clinic in Holland in 1882, thus earning its nickname among the English, the Dutch cap.

The first American patent for a pessary was headed "J. B. Beers, Preventing Conception, No. 4729, patented Aug. 28, 1846." It was for a "curved hoop attached to a handle by a springjoint." Active searching in the nineteenth century for an effective barrier led one American medical observer, W. D. Buck, to complain in 1866 against the "raid being made upon the uterus. . . . The *Transactions of the National Medical Association for 1864* has figured 123 different kinds of pessary, embracing every variety, from a simple plug to a patent threshing machine, which can only work with the largest hoops. They look like the drawings of turbine water wheels. . . . Pessaries, I suppose, are sometimes useful, but there are more than there is any necessity for. I do think that this filling of the vagina with such traps, making a Chinese toy-shop of it, is outrageous."

Aristotle was possibly the first to propose contraception by using chemicals found in nature. In *Historia Animalium* he advised women wishing to avoid pregnancy to anoint "that part of the womb on which the seed falls with oil of cedar, or with ointment of lead or with frankincense, commingled with olive oil." Later, Pliny, in his *Natural History*, recommends the use of natural substances for birth control, not as spermicides or fertility blockers but as tools for creating a disdain for sex—an approach that was to blossom with the growth of Christianity. "According to Osthanes," wrote Pliny, "if a woman's loins are rubbed with blood taken from the ticks upon a black wild bull, she will be inspired with an aversion to sexual intercourse."

A fourth-century Greek physician is the first known to have advised chemicals for use by the man. In his *Medical Collection*,

Oribasius wrote: "When one wants to prevent conception before copulation one anoints the virile part of the man with hedysome juice." But if it was too late for contraception, the task was the woman's again: "*After* coitus, her application of ground-up cabbage blossoms prevents the semen from congealing."

He was followed by a sixth-century Greek, Aetios of Amida, who wrote that "the man ought to smear his penis with astringents, as for example, with alum or pomegranate or gallnut triturated with vinegar; or wash the genital organs with brine, and he will not impregnate."

Aetios' prescription of vinegar introduced one of the most effective vaginal spermicides of all time, although it wasn't rediscovered for another thirteen centuries, and it was put into frequent use mainly by the French.

Islamic physicians were prolific in compiling lists of workable natural substances. Rhazes, in the ninth century, suggested at least seventeen, among them devices that contained cabbage, pitch, ox gall, animals' ear wax, and whitewash. Avicenna, of the eleventh century, added another with extra cautions: "The woman must also be careful to smear tar in the vagina before and after coitus and to anoint the penis with it."

Centuries before Margaret Sanger thought of a contraceptive "that could be swallowed like an aspirin," birth control drugs found in nature—and *taken orally*—were in lively use. Some were as simple as brewed teas. Pliny, typically seeking potions to encourage abstention, suggested that women "check libidinous tendencies" by brewing a potion of crushed willow leaves. He further advised men: "A most powerful medicament is obtained by reducing to ashes the nails of the lynx, together with the hide. . . . These ashes, taken in drink, have the effect of checking abominable desires in men." If the "abominable desires" of a woman persisted, she was to drink potions of parsley and mint, which would "curdle" her received semen into sterility.

Preceding Russell Marker by sixty years, a member of the British government medical mission in Fiji in the 1880s, one D. Blyth, described a folk prescription for contraception drawn from roots and barks: "Fijian women have a decided aversion to large families, and have a feeling of shame if they become pregnant too often, believing that those women that bear a large number of children are laughing-stocks to the community." So to avoid pregnancy they extracted medicines from the leaves and roots of the roqa tree and the samalo, denuding the root of its bark and scraping it down. The scrapings and leaves were soaked in cold water, then strained before the women drank the potion. More recently, in 1927, Pitt Rivers, an anthropologist, wrote of reports by missionaries returning from Australasia "that some mysterious contraceptive drug was used by the unmarried girls. Native herbs and roots, mixed together with all manner of magical substances, such as spider's eggs, skins of snakes, etc., are as a matter of fact made into concoctions and drunk by girls with this idea. I have myself collected such recipes from Melanesian and Papuan sorcerers and old women."

There has been a belief that a source of oral contraception is to be found in the lowly mule, the hybrid of a female horse and a male donkey, which is barren. Dioscorides suggested making a meal of the kidney of a mule. Soranus, always cautious, again suggested that "some people believe" that eating the uterus of a she-mule would bring sterility to the diner. Aetios wrote with greater certainty, "The burned testicles of castrated mules drunk with a decoction of willow constitute contraceptives." That advice, of course, was for men. For women he prescribed "a decoction of willow bark with honey to temper its bitterness."

A seventh-century Chinese manual called *Thousand of Gold Prescriptions* offered a recipe for what amounted to a **sterility pill**: "Take some oil and quicksilver and fry a whole day without stopping. Take one pill as large as a jujube seed on an empty stomach and it will forever prevent one from becoming pregnant."

More than a thousand years later—in 1728—the Chinese were prescribing another way: "Take 1 *sheng* (Chinese pint) of a leaven called *mien ch'u* [consisting of wheat flour, kidney beans, the juice of *shin liao*, and apricot kernels] and 5 *shengs* of liquor without dregs. Knead this into a paste and boil until there are but 2.5 *shengs* left. Use a silken cloth to strain and throw away the dregs. Divide the liquid into doses. Wait until menstruation is about to come and in the evening take one dose; on the following morning take another dose. The menstruation will then flow and for the rest of her life she will be without children."

Indian women of the eleventh to thirteenth centuries found more elaborate prescriptions in *Pancasayaka*:

The woman who drinks on a lucky day apalasa and . . . fruits as well as flowers of the salmali tree, together with melted butter, will certainly become unfruitful. If she drinks regularly of the decoction of the root of the pavaka tree and sour rice water, and keeps it up for three days after the end of the menstrual period, she will remain unfruitful until death . . . If a woman eats or drinks continuously for half a month a large pala of three-year-old molasses . . . she will surely be unfruitful to the end of her life. Two large karsa of the seeds of the rakasa tree, drunk with white rice water for seven days after the end of the menstrual period, causes certain unfruitfulness for those with gazelle eyes.

Among ancient Hebrews the permanent birth control potion was the "cup of roots," responsible for the traditional story of Rabbi Hiyya whose heart was broken around the year A.D. 200. After his wife, Judith, fed up with childbearing, drank the potion, the rabbi was said to have wept, "I wish you had given me at least one birth more."

From the beginning of time, when all else failed, an unwanted child has been "prevented" by **abortion.**

Physicians of ancient China and India appeared to make no important distinction between contraception, sterilization, and abortion. Certainly no moral distinctions were made between them. (Nor, apparently, did these ancient writers make a distinction between the vaginal cavity, the cervix, and the uterus.) Their bizarre and clearly painful methods for purging women of a conceptus were not notably different from the perilous practices of twentieth-century backstreet abortionists. The Muslim physician Rhazes wrote in his *Quintessence of Experience*:

If these methods do not succeed and the semen has become lodged, there is no help for it but that she insert into her womb a probe or a stick cut into the shape of a probe, especially good being the root of the mallow. One end of the probe should be made fast to the thigh with a thread that it may go in no further. Leave it there all night, often all day as well. Use no force: do not hurry: and do not repeat the operation or you will cause pain. Wait thus for one or two weeks until gradually the menses appear and the whole thing will slowly become open and clean. Some people screw paper up tight in the shape of a probe and after binding it securely with silk, smear over it ginger dissolved in water. . . . There is no better operation than this.

In the earliest days of belief in oral contraception, a faith in solutions containing metal traces was to prove prescient. Soranus, the second-century author of *Gynaecology*, noted that many people believed "the water from the firebucket of the [black]smith, when drunk continuously after every menstrual period, causes sterility." By the sixth century, Aetios of Amida wrote with greater certainty, "Copper water in which one extinguishes iron, drunk continually, and above all immediately after the end of menstruation, is anticonceptional." The conviction persisted into the thirteenth century when Arnald of Villanova, who practiced in Paris, Rome, and Barcelona, reported, "If a woman drinks in the morning for three days two *minas* of water in which smiths quench their forceps, she will be sterile permanently." Six hundred years later, in 1886, a Dr. Fossil noted that women of East Austria seeking to prevent pregnancy were drinking blacksmith's water after every period. As late as 1914 the National Birth Rate Commission of England heard a witness report with distress that "laboring women" in Selly Oak near the iron and steel center of Birmingham were "drinking the water in which copper coins had been boiled."

They had discovered a track that has lately found corroboration. Copper is the active agent used in the modern IUD because, according to John M. Riddle of North Carolina State University, author of a 1992 book, *Contraception and Abortion from the Ancient*

World to the Renaissance, trace amounts of it appear to "cause chemical changes harmful to sperm and the conception event."

The modern era of the timeless search to separate sex from pregnancy might be dated from the invention of the **condom.**

According to a 1709 issue of the gossipy, ribald British magazine *The Tatler,* the male-worn barrier was first devised by a patron of Wills Coffee-House: "A Gentleman of this House . . . observ'd by the Surgeons with much Envy; for he has invented an Engine for the Prevention of Harms by Love-Adventures, and has, by great Care and Application, made it an Immodesty to name his Name." The magazine tattles, however, that "this worthy Member of the Commonwealth . . . by giving his Engine his own Name, made it obscene to speak of him more."

A popular story cast the gentleman as a Dr. Condom, the physician of Charles II, and his mission to reduce the number of his king's illegitimate issue. (Dr. C. was a notable failure. Charles is said to have acknowledged fourteen bastards and probably overlooked more.)

But preceding that entertaining story by a year, in 1708 an anonymous English poem, "Almonds for Parrots," praised "matchless Condon" whose fame will "last as long as Condon is a Name." The poem referred to people selling condoms in St. James's Park, Spring Garden, the Play-House and the Mall, but only as protection against disease and with no mention of contraceptive use.

In the 1720s the son of the bishop of Peterborough, White Kennett, later a rector himself, extolled the condom for freeing women from "big Belly, and the squawling Brat."

In the early nineteenth century, a seventeen-year-old philosopher-to-be named John Stuart Mill scampered around London distributing the "diabolical handbills" published by Francis Place, a maker of leather breeches (and father of fifteen children) who battled for the rights of the poor and for liberating the working classes through birth control. Titled "To the Married of Both Sexes," the handbills gave instructions on the use of a contraceptive **sponge.**

Place probably learned about the sponge from the French. He

urged it upon his English countrymen and women because it was cheap, easily available, and was "most likely to succeed in this country, as it depends upon the female." The piece of sponge he suggested was "about an inch square, being placed in the vagina previous to coition, and afterwards withdrawn by means of a double twisted thread, or bobbin attached to it." It was harmless and did not "diminish the enjoyment of either party. The sponge should, as a matter of preference, be used rather damp, and when convenient a little warm."

His handbills to the "married of both sexes" had two companion pieces tailored to class cultures. One of them, "To the Married of Both Sexes in Genteel Life," assured the upper-class reader that the sponge was "an easy, simple, cleanly and not indelicate method." The other, "To the Married of Both Sexes of the Working People," put forth in paternal tones the long-range economic benefits of an agreeable birth control method: "When the number of working people in any trade or manufacturer has for some years been too great, wages are reduced very low, and the working people become little better than slaves. . . . You cannot fail to see that this address is intended solely for your good."

The sponge never quite caught on among the poorer classes. The *Trades' Newspaper and Mechanics' Weekly Journal*, while appreciating Place's sentiments, denounced his teachings: "Certain practices for regulating the population of the country, which, though represented as the fruits of *the soundest political economy*, are in fact . . . detestably wicked."

A gentleman signing his letter I.C.H. wrote to Place protesting that the sponge was a dangerous method, "for the orgasm is often so violent that any substance will be carried into the womb as greedily as a fish swallows the bait." But Richard Carlile acclaimed the sponge in his 1826 *Every Woman's Book* as the choice of "the females of the more refined parts of the continent of Europe, and with those of the Aristocracy of England. An English Duchess was lately instanced to the writer, who never goes out to dinner without being prepared with the sponge. French and Italian women wear them fastened to their waists, and always have them at hand."

• • •

In 1832 Charles Knowlton, a young Massachusetts physician who became convinced that most of the health problems of his impoverished patients resulted from the burdens of too many children, described in writing his invention of a **syringe**, which came into fairly widespread use. The device, to be used immediately after ejaculation (although "five minutes' delay would not prove mischievous"), had a soft metal barrel and a piston head tightened with a wrapping of flax or hemp. The woman was to assume a position (unspecified) that "common-sense cannot fail to dictate," and inject one of the following solutions: "alum" (aluminum potassium sulfate, a crystalline solid), mixed with a pint of water to create "a lump as large as a large chestnut"; or sulfate of zinc added to a pint of water, using "a large thimble full"; or "eratus" salt added to a pint of water, using "two common-sized *even* teaspoons full"; or four or five "greatspoons" full of "good vinegar" mixed with a pint of water; finally, four or five greatspoons of liquid chloride of soda dissolved in a pint of water. To these prescriptions, Knowlton thoughtfully added that in winter an added dash of spirits would keep the liquid from freezing, and he further solicited the woman to take care that the room wasn't too cold.

In 1856, almost a quarter century later, Dr. William Alcott reported that Knowlton's elaborate contraption "is in vogue, even now, in many parts of our country, and is highly prized." It rose to new levels of popularity in the 1870s when a rubber-goods dealer named Kendell went to prison for sending one of Knowlton's syringes through the mail. It thereupon became famous as the "Comstock syringe," named after the crusader for "decency" who had brought about Kendell's arrest.

No matter how women since the beginning of time quested for ways to defeat pregnancy, and no matter how many ancient doctors experimented and advised in the search, the medical profession would have no part of it. Even in the late nineteenth century, physicians were competing with the clergy to denounce contraception. In 1887 England's leading medical journal, *Lancet*, zestfully published an exposé by a physician on the traffic in contraceptive devices and information. It sent shudders through the medical community:

This abomination (i.e., contraception) has lately forced itself into notice in a manner which can no longer be ignored by clean people. In common with most medical men I have had some hazy notion that . . . there has been for many years an illicit traffic in various preventatives of pregnancy. Now and again such information is cunningly worked up into an advertisement, and meets the eye amongst such innocent company as the last fashion in sanitary undergarments and the latest fad in tinned beef. But I had yet to learn that the druggist's shop was the centre from which such drugs and instruments were now distributed, accompanied with the fullest directions in plain matter-of-fact language. . . . Catalogues of the various articles are issued, numbered in regular order. . . . The drugs are put in little boxes and large. . . . Travellers go about the country showing their samples, and . . . catalogues are distributed by post to probably every address in the trade directory; and . . . there is also a pamphlet for home reading, written as a dialogue between two men—the one prosperous and happy, and the other poor and needy in everything except a large family.

You, Sirs, may easily plead that this subject is not one a decent man would care to handle with a pair of tongs; but I trust you will agree with me in the hope that . . . the medical profession . . . must never identify itself in this matter, however indirectly; and that . . . if this evil is to continue, at all events it shall never exist as a sidewing of the healing art.

As late as 1914, after a book titled *The Misuse of Marriage* grudgingly approved marital lovemaking during the "safe" period, the bishop of Southwark rebutted: "I hold that if you relax the idea that intercourse has any other purpose ultimately behind it except the production of children . . . you open a door to the lowering of the whole idea of the union between the man and the woman." But that interpretation, someone pointed out, meant that a couple might be restricted to making love only seven or eight times during their entire marriage, whereupon the bishop, unflustered, replied, "Well, what is the harm of that?"

The era of the commercially manufactured birth control **suppository** opened in England in 1886 when *The Wife's Handbook* reported: "Mr. W. J. Rendell, Chemist, 26 Great Bath Street, Farringdon Road, London E.C., has invented some quinine pessaries which dissolve. They are sold at 2 shillings per dozen. . . . There is nothing but quinine and cacao-nut butter in these pessaries, consequently nothing to irritate either the woman's vagina or the male organ. It is but right to say that these pessaries are at present only on trial. Time will show whether they can be relied upon to prevent conception."

Time demonstrated, in fact, that Rendells, as they came to be called throughout England, proved quite popular at least until World War II, and were reasonably reliable, although always suspect. An Englishwoman with a long memory, Madge Wicke, who was in her twenties as World War II approached, has recalled: "These Rendells, you bought them in a tin, like a cough-drop tin, like tiny egg things. There was a rumor that a certain percentage from the factory *had* to be imperfect because the government had insisted on it to build up the population in case there was a war. Now, I don't know how true that was, but this was a very, very rife rumor. So people began using two at a time, just in case one of them was a dud."

Some leading American drug companies, eager to make a market out of the pervasive desire of women to avoid unexpected pregnancy, searched for ways to circumvent moral opposition. They invented one of the brilliantly evasive phrases of the early twentieth century: "feminine hygiene." The phrase entered the language without resistance until July 1950, when the crusading magazine *The American Mercury* published an exposé by Grace Naismith called "The Racket in Contraceptives."

Women, wrote Naismith, "are woefully susceptible to 'feminine hygiene' advertising, one of the most insidious euphemisms that has ever been perpetrated by big business. Many of the

douches, powders and suppositories that come under the classifi-
cation of feminine hygiene are highly unreliable as contraceptives. . . . The manufacturers all claim that 'these products are not
sold as contraceptives'! Of course not. There are laws prohibiting
the advertising of contraceptives, except in medical or drug publications. Then what are they advertised for—to the tune of more
than $1 million a year?"

For example, the July 1933 issue of *McCall's* had celebrated the
contraceptive promises of Lysol in an article by a "famous Parisian
gynecologist," under the title "A Husband . . . a Wife and Her
Fears":

> Without [feminine hygiene], some major physical irregularity
> [avoiding reference to the unspeakable: an overdue period] plants
> in a woman's mind the fear of a major crisis. Let so devastating a
> fear recur again and again, and the most charming and gracious
> wife turns into a nerve-ridden, irritable travesty of herself. . . . It
> all sounds very dreadful, doesn't it? But it needn't happen. The
> proper technique of marriage hygiene, faithfully followed, re-
> places fear with peace of mind. . . . What is the proper technique?
> To my practice I recommend the "Lysol" method. . . .

The largest seller among douches, Lysol in its advertising fairly
crooned its message about "the secret": "On the teeth of a clean,
dry, fine-tooth comb, put a drop of water. It remains a drop. Now
do the same with Lysol. Even on this dry surface, Lysol reaches
out and spreads down in between the teeth of the comb. . . .
Lysol's unusual spreading power enables it to spread and spread,
reaching down into tiny vaginal crevices."

In their 1951 book *Planned Parenthood, A Practical Guide to Birth
Control Methods,* Drs. Norman E. Himes and Abraham Stone tell of
a survey at Newark Maternal Health Center in New Jersey of
women using Lysol. Of 507 users, 250 became pregnant. The
doctors condemned Lysol, which is a strong poison in concentrated solutions, as less effective a spermicide than "many simple
household supplies such as lemon juice or vinegar."

In 1944 a line of products labeled Zonite was banned by the

Federal Trade Commission because of ads implying they were effective as contraceptives. The products, said the FTC, were "not effective, reliable, or dependable contraceptives in that they cannot and do not always contact all spermatozoa in the genito-urinary tract." That finding did not deter the manufacturer for long. It soon changed its advertising to read: "You know it is not always possible to contact all the germs in the tract. But you can BE SURE ZONITE DOES kill every *reachable* living germ."

"Actually," Naismith pointed out indignantly, "the germs normally present in the vaginal tract do not have to be killed. Not only are they quite harmless, but they are even beneficial. The word 'germ' is subtly intended to imply 'sperm.' The assumption is that the solution will kill sperm. This is a fallacy. . . . As a matter of fact, water itself is a spermicide."

Manufacturers of douches distributed "educational" booklets that confused, probably deliberately, the vague hints of contraception with additional appeals that played on women's alienation and fear of their own bodies. Against delicate pink and flowery decorations, these booklets spun friendly, empathic phrases: "It's wise not to take chances with such a fleeting thing as daintiness. Syringe regularly." "A marriage may lose all the glorious magic of romance . . . because of a wife's carelessness about feminine hygiene. You surely don't want to risk such a deeply personal tragedy." One pamphlet pictured a young wife, tears streaming down her face as her husband turns away. The text weeps, "Why are his kisses just 'pecks' now?"

The coin had another side. A manufacturer's ad in *The American Druggist* trumpeted: "Hit the Jackpot in this $212 million market. Push Zonite, one of the fastest money leaders in this rich feminine hygiene field."

The feminine-hygiene bonanza produced more than five hundred liquids, tablets, suppositories, and powders. One powder had to be blown into the vagina with an instrument resembling a bazooka. Until the FDA banned it, the same company also marketed a douche bag with a ballooning technique that forced fluid into the womb and fallopian tubes, often causing infection. Another instrument for injecting contraceptive jelly into the

vagina was shown by Dr. Clarence Gamble, a birth control pioneer and member of an AMA committee on contraception, to be
powerful enough to propel the jelly six feet into the air at a speed
of thirteen miles an hour. That power, said Dr. Gamble, was dangerous enough to force some of the jelly into the uterus.

Physicians themselves were often duped into recommending
dangerous contraceptives, probably the consequence of their
own skittishness about the unpleasant business. Many doctors
prescribed wishbone-shaped pessaries, stems, and rings. In the
American Journal of Obstetrics and Gynecology, Dr. S. Leon Israel in
1947 reported several cases in South Carolina alone of pelvic in-
flammation, one of critical peritonitis, and nationally more than a
dozen deaths attributable to such devices. In some instances
flanges of the wishbone pessaries were known to have perforated
the walls of the uterus. In others the contraption became embed-
ded in the cervix and had to be removed surgically. Several makes
of stem pessaries were seized by the FDA. "All contraceptive devices which are inserted into the cervical canal or into the uterus
are dangerous," warned Dr. Pendleton Tompkins. "They ought to
be rendered illegal." (This was said long before the appearance of
what became known as the IUD, the intrauterine device, which
was to become very popular but would produce its own set of
problems.)

The most reliable of all methods known in the pre-Pill era, the
diaphragm-and-jelly method, never caught on among working-
class women. In 1948 Dr. Earl Lemmon Koos of the University of
Rochester surveyed 156 working-class families on their contraceptive choices. Only 2 percent of working-class women, he
found, said they used the diaphragm. The two main reasons for
that low usage, Dr. Koos concluded, were cost and simple lack of
knowledge. Less than 6 percent of the working-class women got
their contraceptive information from doctors. Their most frequent advisor was the corner druggist. Another study showed
that 50 percent of diaphragm-and-jelly users abandoned that safe
method because they "did not like it," preferring to take chances
on less secure methods rather than bother with the interruption
and manipulation required.

In sharp contrast, Dr. Alan Guttmacher, the Johns Hopkins gynecologist and birth control pioneer, upon surveying two thousand fertile wives of professional and white-collar men at about the same time as the Koos working-class survey of the late forties, found that the diaphragm and jelly were used by 41.7 percent and the condom by 43.3 percent. At about the time of Guttmacher's survey, the Margaret Sanger Research Bureau in New York had distributed over 110,000 diaphragms and found them to be 95 percent effective.

Of an estimated $200 million spent each year in the 1950s for contraceptives, about half was spent on condoms. They were second in use to douches, although far more effective. Ten years earlier, around 1940, condoms were often produced so poorly that at least half were found to be defective. By 1947, however, five leading manufacturers had produced 720 million condoms, and tests showed them far improved. A condom cost about eight tenths of a cent to manufacture, but a packet of three generally sold for fifty cents or more. Manufacturers were able to extract this extraordinary markup, often 2000 percent, because the condom was a "whisper item"—the customer had to overcome shame simply to buy it.

6

AN ADVENTURE IN MEXICO

*I do not much wish well to discoveries, for I am always afraid
they will end in conquest and robbery.*
—SAMUEL JOHNSON, 1773

In Mexico, Gregory Pincus did indeed locate the man he was
looking for: the disappeared chemist Russell Marker. In 1992,
forty years after that encounter, Marker, at age eighty-nine,
in faltering voice and halting sentences but with young, keen,
ever-blinking black eyes, recalled the long-ago visit for
me: "His principal interest was whether, if he needed a large
amount of progesterone, it could be available. He didn't tell
me why he wanted large quantities. And I wasn't interested
in why."

To Pincus, everything about his visit to Marker was a peculiar
riddle. Why, in trying to locate this new hero of chemistry, had
Pincus needed to become a virtual sleuth? Why had Marker cut

his ties with all his former colleagues, not only in Mexico but in the United States?

This baffling man Marker, the sorcerer who had commanded from jungle roots the exact chemical imitation of human progesterone, had won notice as recently as August 1949, on the first page of *The New York Times* for another stunning discovery: a cheap source of cortisone for treating rheumatoid arthritis. Previously that extraordinary new drug, like progesterone, had to be taken from humans or animals, making it scarce and prohibitively expensive. And what was his new source for cortisone? The same seemingly worthless vegetation that yielded the progesterone, but with a delicately different set of chemical manipulations.

Just a short time before Pincus's 1952 visit to Marker, medical writer Leonard Engel marveled in *Harper's* at the "comparatively simple process for making cortisone from an inedible wild Central American yam. . . . The story of these developments . . . is replete with the stuff of which movies are made . . . an unrestrained, dramatic race involving a dozen of the largest American drug houses, several leading foreign pharmaceutical manufacturers, three governments, and more research personnel than have worked on any medical problem since penicillin." Three biographers would soon name Marker "the father of the Mexican steroid industry." Carl Djerassi, whose own major accomplishments in steroid research were soon to grow straightaway from the work of Marker, would one day declare, "I think he deserves the Nobel Prize."

So all the more puzzling to Gregory Pincus that Marker, at the age of forty-nine, would say he had "lost all interest" in science and scientists and had destroyed all his laboratory notes, letters, and reprints of his astonishing output of 213 contributions to scientific journals.

Mystified, Pincus pressed questions.

"He wanted to know what I was doing in Mexico [after quitting chemistry]," Marker said—indeed the same question Marker had refused to answer for anyone until he did for me four decades later—"and he tried to convince me to go back into production."

Then, in a caustic overstatement that slipped from an old, still-unhealed wound, Marker asserted that he would not return to chemistry "because I considered all chemists to be crooks."

Marker's mind-changing and world-changing Mexican adventure, indeed "the stuff of which movies are made," takes us deep into the grubby toil that often hides behind the glamorized curtain of science research; into unseemly scenes of accusations of greed and theft, of beatings of women, of shots in the dark of night with possible intent to murder; into some of the highest and lowest politics of Big Science, and to huge profits in corporate counting rooms. And at the spine of its narrative lies one chemist's unyielding tenacity.

Born in 1902 in a one-room log cabin on a farm seven miles east of Hagerstown, Maryland, young Russell Marker decided early he would not become a farmer like his sharecropper father. Over his father's opposition but with his mother's support, he enrolled in high school at a daily price of walking two miles at dawn to get a train, riding it for four miles, then walking another mile to the schoolhouse. With total absence of uncertainty, upon graduation he enrolled at the University of Maryland. What had instilled in him this admirable drive for learning? Marker's grim-faced, point-blank answer: "To get out of farm work."

"I decided first," Marker reminisced, "that I was going to take chemical engineering. I have no idea why I decided on that." He had had no high school chemistry, no physics. "After the first year, I changed to regular chemistry. I didn't know what a beaker was. I didn't know what a test tube was. I had to ask the man working next to me, because he had had it in high school. In my junior year organic chemistry was given, and I got very interested in it. I was told by someone who had taken the course that it was one of the hardest I would be subjected to. So I bought the book used in the course and during the summer went completely through it, working out all the problems. When I went back, organic chemistry was like rolling off a log." In his senior year he earned pocket money cleaning the lab for twenty-five cents an hour. He won his bachelor's degree in 1923.

His father then expected Russell to return home and help

"make the farm produce as much as it could," but Marker confounded him by signing up for graduate school. In two years he earned his master's degree and, after two more, completed his doctoral thesis, which was soon published in the *Journal of the American Chemical Society*. His advisor, Morris S. Kharasch, informed Marker that he lacked a course in physical chemistry required for his doctoral degree. All his life young Marker had trained into himself an unbending resistance to his father's attempts at controlling him. Now that training rose in him. Marker told Kharasch he would not take the course. He had already mastered the subject on his own in the laboratory, he said, and considered any time spent outside the lab as wasted. Alarmed, Kharasch offered to extend Marker's fellowship stipend for another year, warning his star student that without the doctorate he "would have no future and would end up as a urine analyst." Marker accepted the challenge and left the university in June 1925 without the degree, which he was never to earn.

Marker married a schoolteacher and went to work for the Ethyl Gasoline Corporation in a garage that served as a laboratory in Yonkers, New York. The job paid $2,600 a year. Assigned to try to improve the company's "no-knock" additive to gasoline, he soon developed the concept of the octane rating, still important today in the marketing of motor fuels.

His talent for piercing directly to the solution of problems soon won him an invitation, despite his lack of a Ph.D. "union card," to join the revered research faculty of the Rockefeller Institute (since renamed Rockefeller University) in Manhattan. Starting there in 1928, Marker worked straight through his two-month summer vacations, sometimes staying in the lab for forty-eight hours without break "mainly because I was interested in what I was doing." He and his senior colleague, Phoebus A. Levene, published more than thirty papers in the next few years.

At Rockefeller, chemistry researchers generally lunched together, then marched as a group to the library for an hour or two of looking at newly arrived journals that reported the latest experimental findings. Marker developed a special interest in reports from Germany where some young laboratory pioneers were breaking important ground in a new research field popular on both sides of the Atlantic: sterols—later known as steroids—solid alcohols widely occurring in animals and plants.

Research in steroids, and particularly an important subgroup, hormones, was severely inhibited by their scarcity and prohibitive cost. One such hormone, progesterone, was so expensive that even research scientists couldn't afford it. The minute amounts extracted from animals was bought up to improve fertility in world-class racehorses. Only a few European pharmaceutical companies ventured the laborious steps of extracting hormones from humans and animals and converting them to usable form. A few scientists there experimented with creating synthetic hormones by juggling the position of atoms in the molecules of other steroids, but none succeeded. Marker saw an opening.

"I wanted to see if I could work out a process for making hormones available in quantity," he has recalled as though that youthful ambition were routine. "I thought there would be a big demand for them eventually. I went to Levene about it. He asked what I thought they could be made from. I said, 'From vegetable material.' "

Levene replied that Dr. Walter A. Jacobs in the pharmacology department had worked on trying to draw hormones from sarsaparilla roots and had shown "definitely" that it couldn't be done. Besides, Levene added with the drive toward organizational order that often overcomes academic administrators, research in plant material was off limits because it was the domain of the pharmacology department. When Marker protested, the two men took their difference to Simon Flexner, president of Rockefeller, who backed Levene. Marker's fierce resistance to control taking hold again, he thereupon announced that if he couldn't work on steroids at Rockefeller, he would find a place where he could.

"Flexner got mad and pounded on the table. He said, 'It is a great honor to be a member of the Rockefeller Institute. No one leaves here unless his contract runs out or he is fired. You're going to work with Dr. Levene on whatever he says.' "

When Marker had worked at the Ethyl laboratory, he met

Frank Whitmore, the dean of physics and chemistry at The Pennsylvania State College, isolated deep in the Alleghenies. Impressed by Marker's work, Whitmore offered him a job "anytime you want one." So now Marker wrote to him and learned that all Whitmore could offer at that moment was a fellowship paying $1,800 a year. At Rockefeller he was earning $4,400. Already the abrupt defector from his graduate studies at Maryland, Marker now deepened a pattern that would shape his life by irately walking out of his prestigious appointment at Rockefeller. After eight years at Penn State, Marker would rise in 1942, without a Ph.D., to the rank of full professor, at a salary of $3,480. (A few years later, The Pennsylvania State College too was "promoted," to the "rank" of university.)

Marker's eagerness to work on steroids at Penn State was first met by letdown. Dean Whitmore provided him with neither assistants nor funds for special equipment. In the storeroom the only starting material Marker could find was a kilo of "very dirty cholesterol," which he used to "get the feel of steroid chemistry by repeating some of the experiments of the Swiss and German chemists" for degrading steroids to hormones.

In "getting the feel," Marker soon conjured from the urine of pregnant women thirty-five grams of pure progesterone. At that time it was the largest amount ever produced in one lot. "I handed my thirty-five grams in a small bottle to the president of Parke-Davis [a drug company that had begun contributing partial financing for Marker's work], and he said, 'We'll put a price of a thousand dollars a gram on this.' One of the things known then was that women who habitually had miscarriages were cured by taking small doses, one or two milligrams, of progesterone. So a gram would give five hundred doses. Later I found you could get progesterone on a much better scale from bulls' urine."

Such conversion, as Leonard Engel explained, required

the most intricate process ever attempted by any branch of the chemical industry, from a raw material so limited in supply as to preclude all hope of making more than a fraction of the volume required. . . . The "total synthesis" of a complex substance [such as a steroid] from simple chemicals is something like the construction of a house, except that the chemist's task is very much more difficult because his building bricks, atoms, are too small to be seen. The molecules . . . must not only have the right number of the right kind of atoms, the atoms must also be arranged in a particular way, exactly as bricks make a specified house only when assembled in the specified way. Moreover, the more complex the substance to be synthesized and the more atomic bricks it contains, the more ways there are of putting it together wrong. . . . To synthesize something as complicated as a steroid, one must find, for each step of the way, a chemical reaction that places the right atom in the right spatial configuration; and the steps must be so ordered that later ones don't undo what was accomplished earlier.

The structure common to all steroids consists of four "rings" of carbon atoms. Some rings, however, share pairs of atoms with an adjacent ring. Thus, like Siamese twins, those rings are "bonded." It may be diagrammed like this:

C   D

A   B

Steroids always contain carbon atoms (each marked "C" in the rings above), some hydrogen, usually some oxygen, and sometimes one or two other elements. Thus, the carbon atoms, besides being locked to one another, are also connected to other kinds of atoms. In the basic four-ring structure shown above, there are twenty-eight hydrogen atoms jutting from the rings in various directions. Thousands of variants are known, and each has a different impact on human or animal life.

An extreme case of such clinging adornments is cholesterol, which dangles a long tail, or "side chain," containing eight carbon atoms and seventeen hydrogen atoms.

Some hormones are steroids, some are not. Steroid hormones are central to human life, for two sharply different reasons. Some steroid hormones are devoted to keeping the *species* alive, others to keeping the *individual* alive. Those in the business of sustaining the species from one generation to another do so by controlling the reproductive systems of both sexes. These are called sex hormones. Those that help sustain the individual go under various names: corticoids, adrenocortical hormones, or corticosteroids. One corticoid that became the object of the most active search was cortisone, because of its special value in treating rheumatoid arthritis. (The intensity of that search would soon hasten the discovery of the Pill.)

The sex hormones are divided into three principal kinds: androgens, estrogens, and progestogens.

Androgens are the masculine hormones. They originate mainly in the male's testes but also in other glands of *both* sexes. The most active androgen is testosterone. Androgens control the development and functioning of the male genital organs and promote such masculine characteristics as muscular strength, deep voice, and facial hair; they also largely account for the sexual urge of both sexes.

Estrogens are the feminine hormones, originating in the female chiefly in her ovaries. The most active estrogens are estradiol and estrone. The estrogens promote growth of the lining of the womb in preparation for pregnancy, the lining of the vagina, and development of the breasts.

Of the *progestogens,* the most influential is progesterone, the "pregnancy" hormone, produced in the woman in the corpus luteum (or "yellow body") that is formed in the ovary upon release of a matured egg. If the egg becomes fertilized, progesterone and estrogen act together to maintain the pregnancy. These hormones also regulate the menstrual cycle in the nonpregnant female.

If a way could be found to manufacture sex hormones from vegetation, Marker had begun to believe, the best source would be a variety of cholesterol called *sapogenin,* which occurs widely in wild plants, particularly roots. It has a structural similarity to steroids, but like all cholesterols, sapogenin presented the baffling obstacle of that side chain. When the human or animal body sets out to convert a cholesterol into specific steroids, such as progesterone, the side chain must be removed by chemical reactions. Our science-smart bodies know how to accomplish that removal with no difficulty. But doing so in the laboratory, at least from sapogenin, had so far defied the world's chemists.

Like Marker, Rockefeller pharmacologist Walter Jacobs had been sniffing the path of the Germans, particularly Tschesche and Hagedorn. They had been first to work out a sketch of the sapogenin molecule and to show that if the side chain were somehow snipped off, what would be left could be a long step toward duplicating the female hormones. But that was a theoretical blind alley, Marker had been told by his Rockefeller colleague Levene. Jacobs's conclusion that the side chain couldn't be removed was later confirmed more authoritatively by Louis F. Fieser of Harvard.

But is failure to be mistaken for impossibility? Marker repeated all of Fieser's experimental work and became convinced that the structure of the sapogenin molecule as described by Fieser and the German scientists was wrong and that the side chain could be removed under certain acidic conditions. But what were those conditions?

Marker and his graduate-student assistant, Ewald Rohrmann, labored through dozens upon dozens of manipulations, each one tediously slow to "cook" and analyze. They tried acids of various strengths. They tried temperatures not reported by Fieser, in vessels that would create different environments. Rohrmann, who habitually arrived at the lab at six in the morning, would set samples to heat, then next morning remove them and begin heating a new batch. Finally, their relatively crude spectrometer gave them color and wavelength readings that suggested they were arriving at what they thought they wanted. When they collected a large number of such samples, they sent them off to New York for analysis.

Two weeks later the results arrived. What they had produced was readily convertible with one step to a synthesis of human pregnanediol, which in turn was easily transmutable to chemically perfect progesterone. The successful trick was in heating the sarsaparilla-derived sapogenin (sarsasapogenin) with acetic anhydride at 200° F. in a sealed tube.

Marker submitted a three-paragraph communication to the editor of the *Journal of the American Chemical Society* (JACS) to announce his success. With the science world bubbling in steroid research, claims of discovery had to be filed with dispatch. He reported tersely that while sarsasapogenin "was inert in boiling with acetic anhydride for 24 hours, when heated in a sealed tube overnight at 200 degrees it gave a new product." His experiment illustrated how a minuscule change in laboratory procedure can profoundly change a chemical's impact on human life. The new procedure, a historic landmark in modern chemistry and pharmacology and still essentially unchanged today in the large-scale production methods of the multibillion-dollar steroid industry, has become known as the "Marker degradation."

"Until that time," Marker has recalled, "Fieser was refereeing [reviewing and passing judgment upon] all the papers I was submitting to the JACS. I think I had about forty of them on sterols. Arthur Lamb, the editor, was also at Harvard. Lamb got in touch with me and asked if I could visit Harvard. So I went over. Fieser and his wife, Mary, had invited me to dinner that night. When I continued to disagree with his finding, Fieser got so mad that he let his wife take me to dinner. He wouldn't show up. But Lamb said that if I was so convinced I was correct, he would publish it. He did.

"Soon there was an organic symposium where Fieser gave a paper on the sapogenins. I didn't go, but I was told that he repeated his position, then said, 'Others disagree with me.' "

Marker soon worked out a variation on removing the side chain, which led to the making of pure testosterone, the male hormone. It could then be converted into estrone, which required merely a single step to become estrogen.

Now certain of the route to making hormones from plants,

Marker undertook an extensive search for the particular variety of sapogenin that would yield progesterone most abundantly. At each scientific meeting he attended, Marker asked every botanist he met to send him plants that might contain sapogenins. Before long, Marker's laboratory overflowed, in fact reeked, with tropical greenery. A Japanese chemist had given him a plant called dioscorea, and Marker was struck by the ease of producing progesterone from its special variety of sapogenin, which its discoverers had named diosgenin.

Unsatisfied that he had yet found the best source, he set out on his own coast-to-coast search for American diosgenin-yielding plants. An early one he turned up was a species of yucca growing wild in North Carolina, called Beth root. Beth root had been a "trade secret" ingredient of Lydia Pinkham's Compound, a patent medicine widely sold early in the century for menstrual pain and disorders and almost as widely derided as a quack potion, perhaps because its generous proportion of alcohol explained adequately to some its talent for bringing comfort. Marker found that Beth root did indeed contain rich diosgenin, but the root itself was about the size of a thumb. He could not imagine harvesting more than a hundred pounds of it a year.

On November 5, 1941, about a month before Pearl Harbor, Dean Whitmore equipped Marker with a timely, if truth-stretching, letter to anyone it might concern: "This is to certify that the bearer, Professor R. E. Marker [is] working on some hormone projects intimately related to the National Defense. . . . The most hopeful indications come from some plants in Texas. . . . Since this work is of great importance not only to the National Defense but to medical science in general, we hope that you will be able, and find it proper, to cooperate with Professor Marker in every possible way in obtaining one ton of each of the following:" and he listed some varieties of agave and yucca.

The search took him to California, Texas, Arizona, and other places, usually assisted by a graduate student or retired botanist. In the course of discovering many new sapogenins, Marker grew fanciful in naming them. He called one *pennogenin* for Penn State; *kammogenin* for Dr. Oliver Kamm, research director of Parke-Davis; *markogenin* for himself; *rockogenin* for Dean Whitmore,

whose nickname was Rocky; *nologenin* and *fiesogenin* for fellow researchers Carl R. Noller and his Harvard critic, Fieser. One of
Marker's airy offerings that never made the written literature was
crapogenin; Arthur Lamb, the JACS editor, persuaded him to re-
name it *kappogenin.*

In Texas, spending a night at the home of a retired botanist,
Marker thumbed through an old regional botany book from his
host's library before going to bed. Looking up references to
dioscoreas, a major plant source of diosgenin, he came upon a
photo of an enormous tuber called *cabeza de negro.* Some grew, the
book said, to several hundred pounds. The caption said the pictured specimen was found in the Veracruz province of Mexico
where the road between Orizaba and Cordoba crosses a river
gorge.

"I called Whitmore and told him I thought I had found what I
was looking for and it grows in Mexico. He said, 'You don't want
to go to Mexico. There may be a war.' I said, 'Yes I do, but I don't
have enough money.' The embassy was advising all Americans to
stay out of Mexico. No one knew whether we were going to get
into the war or whether Mexico would be on our side or not.
Whitmore sent me the money. On the train the head porter gave
me the name of a hotel. It was a very nice hotel, but I later found
out it was where all the German spies were staying."

At the U.S. embassy Marker learned that the Mexican department of agriculture had stopped issuing plant-collecting permits.
He was advised to leave Mexico as soon as possible.

Marker left but returned in early January 1942, only a month
after Pearl Harbor. The embassy, still unable to get a permit,
helped Marker locate a Mexican botanist willing to go with him
to Veracruz, even without permits, to find the cabeza de negro.
The botanist, who understood no English, brought not only an
interpreter but his girlfriend as well—and her mother to serve as
chaperone. After three days on the road, the botanist grew nervous about anti-American jibes aimed at Marker and refused to go
farther. En route back to Mexico City, their truck broke down,
detaining them for two more nights. The embassy again urged
Marker to go home.

The next morning Marker boarded a bus alone to Veracruz

and, after an overnight ride, changed in Puebla for a rundown local bus. "It had pigs in the bottom, and the woman in the seat beside me had some chickens with her. I got to Orizaba at about daylight and found another bus to Córdoba. When we came to a stream of water, I asked the driver to let me off. At a little store there, I got the owner, a man named Alberto Moreno, to understand I wanted some cabeza de negro. He told me to come back *mañana*. Next day he found for me the first plants I collected."

Each tuber was about nine inches to a foot high. Such size, Marker learned, was the result of about twenty-five years of growth, thus making it impractical to cultivate as a crop, but the existing supply seemed, for his purposes, limitless. The tuber hada fuzzy-haired surface, appearing like the head—in Spanish, *cabeza*—of a black person, hence its name. A small portion stuck out of the ground, looking as if someone had been buried vertically. Cutting it open revealed a dense, heavy, chalky white meat, like a turnip's. The meat was used by local Mexicans as a kind of fish bait—actually a fish poison. Natives chopped it into fine pieces with machetes and threw it into the water, where it would close the throats of the fish, causing them to suffocate and rise to the top. Then the fishermen would scoop them out with a net.

Moreno placed the fifty-or-so-pound tubers into two bags and lifted them onto the open baggage rack of Marker's bus. When the bus reached Orizaba, both bags had disappeared. Marker gesticulated a protest to a policeman. The wall of language between them seemed impenetrable, but when Marker proffered the equivalent of ten dollars, miraculously the officer found one of the bags. Marker gestured that he had no more money. The policeman indicated he had no more root. This time Marker stashed his prize under his seat for the remaining ride to Mexico City and later slept with it in his Pullman berth to Pennsylvania.

Upon arrival home, Marker chopped and dried a portion of his treasure and, with alcohol, extracted a fine sapogenin. He took the remainder of the root to his financial supporter, Parke-Davis & Co. in Detroit, where he again performed his magic act of converting sapogenin to progesterone for the company president, a medical doctor named Alexander Lescohier, whom he had previously impressed by making thirty-five grams from the urine of pregnant women. Marker detailed to Lescohier the breadth and depth of his search for diosgenin, which now, because of its abundance in the wild, promised progesterone in unimaginable quantities. He went on to explain that the only practical place for reducing the huge mass of wild plant material into small bottles of hormone was in Mexico.

No, no, no, chided Lescohier. *Impossible!* Nothing complex, nothing precise, nothing scientific could be done in so backward a country. When Marker insisted that under careful supervision it could, Lescohier put forward incontrovertible proof: When he once was sailing off Acapulco, he came down with severe appendicitis, was carried off his boat and carted to a hospital where he got rough and incompetent treatment and he was sure he was going to die. That was all the evidence that he, as a medical doctor, needed for being sure that no chemist in Mexico could possibly carry out the delicate process Marker described. "Besides," he firmly assured Marker, "you have made progesterone from bull's urine, and that's enough for us. We're going to have a stable of bulls in here. We'll get a thousand bulls, if necessary. Science can't be done in Mexico."

Marker offered to set up and oversee production himself, pleading with Lescohier for a corner of a new building that Parke-Davis had just opened in Mexico City for a simple packaging operation for other products. Ten thousand dollars, he pleaded, would be sufficient to get into production. Lescohier waved that away, too.

For the remainder of 1942 Marker tried to convince other pharmaceutical houses. "CIBA was not interested," he says, "because they assumed I had turned the patent over to Parke-Davis who had been supporting me. I couldn't convince them that Parke-Davis had not applied for a patent and could not do so without me. I went to Merck and Schering-Plough and others. None of them were interested. They, too, thought nothing could be done in Mexico."

In July, stepping beyond the usual reserve of university scientists, Marker tried to advance his cause by consenting to an interview with a Tucson newspaper. Under the headline "Desert Plants to Furnish Medicine," Marker was quoted: "Practically all the medicine needed in this country can be found in desert plants . . . the same plants that are cursed now by cattlemen and farmers." The article reported him saying that native U.S. plants could make the country self-supporting medicinally, although it could take twenty-five years to do it. Marker declined to name the plants or medicines until "all problems of gathering and processing are completed."

Such articles brought no one running to sponsor him. By the end of the year Marker became convinced that the only way to make a reality of drawing hormones from Mexican roots was to do it himself:

"In October 1942, at the start of the dry season in Mexico, I drew out of the bank half of my meager savings and returned to Mexico. I arranged with Alberto Moreno, the store owner in Veracruz who had collected the first research plants for me, to collect about ten tons of cabeza de negro. With machetes he chopped the material like potato chips and dried it in the sun. I took that up to Mexico City and had it ground, then I found a man who had a crude extractor. He extracted the dried cabeza de negro with alcohol and evaporated it down to a syrup. I took the syrup to the United States."

In New York Marker went to a science acquaintance, Norman Applezweig, who had a laboratory. Marker proposed that Applezweig permit him to use his laboratory and, in addition, assume the cash costs of his work there. In return, Marker would give Applezweig one third of the progesterone he promised to produce, estimating they would get a little over two kilograms, about four and a half pounds. Marker had previously run a few batches at Penn State, producing about one kilogram at a time. Applezweig jumped at the offer. Actually, they got a little over three kilograms, worth about $240,000. This was the largest lot of progesterone that had ever existed in one place.

In September 1943, as abruptly as he had abandoned his previous sponsors, Marker sent his resignation to Penn State. Dean Whitmore countered with an offer to appoint Marker assistant dean of science, provided he would continue with his research there. But the single-minded Marker would not be diverted from his new goal.

Upon resigning from Penn State, Marker notified Parke-Davis that he was severing all his ties with them too and offered to sign patent applications if they were presented to him before he left Mexico. The company did nothing until the next year, in April 1944, when suddenly a Parke-Davis representative turned up with papers for Marker to sign. Embittered by their lack of confidence in him, Marker now refused to do so. From that day on, he refused to assign patent rights to Parke-Davis or anyone—including himself—and so made his discoveries available to all. He wanted, he said later, to "leave the field open to anyone who wished to produce in competition, to force the price of the various hormones down to a point where they would be available for medical purposes at reasonable prices." (Before long, Parke-Davis fired Lescohier, a step that Marker attributes to the disaster of its president's decision not to undertake steroid production in Mexico. "It was too late," says Marker, "for them to recover the lucrative hormone industry that soon developed in Mexico without them.")

When a scientist feels—*knows*—he is on the right path and every major drug company in the world, it seems, has turned him down, what is he to do? This man of blunt action and directness took blunt and direct action of the most extreme kind: Returning to Mexico City, he reached for the Yellow Pages of the Mexico City telephone book.

"I looked under 'Laboratorios' and found a listing for Laboratorios Hormona. Hormones. That sounded right. I took a taxi and went out to their address, and a German named Federico Lehmann was there." Marker drew out a bottle of a white crystalline powder and announced that it was pure progesterone. He said that he made it from something that grew in the tropics.

"Apparently he thought I was crazy or something. He excused himself. When he came back he said, 'Oh, I thought your name rang a bell. You are the Marker who has published all those papers.' Now he took me seriously."

The extraordinary story that follows has been told by Marker to five known interviewers, including this author, in the past few years. Each of these accounts is consistent with the others, but each contains details not appearing in the others. So the liberty is taken here of combining them into a single first-person narrative using his own language, edited for brevity, although the details were not all put together this way in the same place at the same time. To identify such reconstructed narrative, each paragraph of the passages, instead of opening with quotation marks, is marked by dashes, as follows:

MARKER:—Dr. Lehmann said that Dr. Emerik Somlo was the principal owner of the company, and he was in New York. Then he picked up the phone and called Dr. Somlo and told him to come back immediately. A day or two later I had a talk with Dr. Somlo.

Somlo and Marker talked about starting a new company. Marker learned that Somlo, a Hungarian, had trained in law at the Sorbonne in Paris and emigrated to Mexico in 1928 to represent for exports and imports several large German and Hungarian pharmaceutical houses. (Marker has since said to me, "Somlo didn't have a doctorate any more than I had. He gave it to himself. He was a lawyer and lawyers don't have doctorates. To sell his products, he called himself Dr. Somlo.") Soon Somlo decided that Mexico ought to be manufacturing its own pharmaceuticals. He placed an advertisement in a small chemical journal in Germany, and it was answered by Dr. Lehmann, who had degrees in both medicine and chemistry. In 1933 Somlo hired Lehmann and brought him to Mexico.

Several months after Marker walked in on Hormona, he mentioned to Lehmann for the first time that, in addition to the bottle of white powder he had originally displayed, he had also made several kilos of progesterone in the United States.

MARKER:—He was greatly surprised. He wanted to know what I had done with it. I told him that I still had it. When I got back to the States I got a phone call from Somlo. He wanted to know if I still had that progesterone. I told him I did. He asked me to meet him in New York in a few days. He took me to the Waldorf-Astoria for dinner to butter me up. He wanted to know if we

could work out something so that Hormona could sell the progesterone I had. He said, "We will set up our company in Mexico
valued at five hundred thousand pesos"—a little over one hundred thousand dollars in those days—"and we will make a deal
that you are going to have forty percent of the stock."

—Somlo had a small company in New York called Chemical
Specialties. I was to bring what I had to his man there. The receipt said—the man didn't know who I was—that I came in and
left two kilograms of what I claimed was progesterone. That was
the receipt I got for it. Somlo said that he would establish a price
of eighty dollars per gram, or one hundred sixty thousand dollars
for the two kilos. That would cover in full my payment of forty
thousand dollars for forty percent of this new company, and the
rest of the hundred sixty thousand dollars would go as income.
The stock would be split as follows: Somlo would get fifty-two
percent, Lehmann would get eight percent, and I would get forty
percent.

—Shortly after we started production of progesterone in
January 1944, Somlo told me he was applying for a charter for
our new company, and he suggested calling it Synthesis, S.A. I
told him I would prefer a name indicating it was Mexican, and we
agreed upon the name of Syntex, for synthesis and Mexico. Our
company was incorporated by that name in Mexico in March
1944.

—There was a vacant lot on the corner adjacent to Hormona,
and they put up some laboratories there for me to work in and a
place for the extractors. During the time I was producing progesterone, I received no salary. From time to time I would go to
Somlo and say, "I am short of money. I need to pay my hotel
bills." He'd give me a thousand dollars or so until I would come to
him the next time. No salary, nothing. After a year's time I had
spent all my money and I had to send my wife back to
Pennsylvania because it was cheaper for her to live at our own
home. In February or March, I went to him and asked about the
profits, because I knew there were substantial profits. I had made
at least thirty kilograms of progesterone. Some of it was going to
Argentina and was reshipped to Germany during the war. That

was another thing I objected to. I asked Somlo about the profits. He said, "What profits?" I asked if I could see the books. He said, "No, you wouldn't understand them anyway." I told him I would get someone who speaks Spanish to look over the books. Finally he got pretty mad and said, "There are no profits at all. I took them as salary." So I decided to leave. That was in May 1945.

—I got in touch with my friend [Applezweig] in New York whom I had given the one kilogram of progesterone, told him what had happened, and said I would like to set up some competition for Syntex. He said that he had made some nice profits on his kilogram and he would advance me the money. We set up a company at Texcoco, about thirty miles from Mexico City. We called it Botanica-Mex.

—In Orizaba I had a root collector who suddenly found that his roots would all be stolen during the night. So I went down to Orizaba to see him. We went to a place and had lunch. He got violently sick. I took him back to Orizaba, and he died that night. I don't know what happened. Then other very strange things happened like that. I had hired four of the women who had worked with me at Syntex. One of them—the only one who spoke English—was beaten up in an alley beside Sanborn's one night. She said she knew who did it: It was a friend of Somlo's. [Marker stated the name of a man employed in the oil industry before it was nationalized. "He is the one who did the beating. He kicked her in the legs for coming with me."]

—At about that time my lawyer got me two pistols. One was a Smith & Wesson, and I've forgotten what the other was. I had two watchmen. For their evening meal, one would stay on duty while the other would go out. They would alternate their sleeping. One of them left the place carrying his pistol with him, although I had given them orders not to take the pistols out. He was grabbed right outside of the door, and he was shot through the leg with his own pistol.

—Then Somlo wanted to know if I would have a meeting with him. We went to a good French restaurant on Reforma, which he paid for. He wanted to know whether I would come back with him. When he saw that I wasn't interested, Somlo said he wanted a process I had worked out at Penn State for making testosterone, which I hadn't published. Actually I didn't know at the time that after I left Penn State a patent was taken out on it by Parke-Davis in the name of two of their chemists. He said, "I'll give you a percentage of the profits from the testosterone." I said I wouldn't be interested. Somlo told me that he understood that at Texcoco I was having a lot of trouble with people, physically. I told him I was going to continue on. We talked until about one o'clock.

—Shortly after production started at Botanica-Mex, Somlo had a meeting with my lawyer and demanded that I turn over to him all correspondence between us, especially a letter stating that I was to have forty percent of the shares of Syntex. He said that if I did not do this, he would throw Syntex into bankruptcy and form a new company to produce hormones. He also threatened to have me jailed for stealing Syntex processes and eventually force me to leave Mexico. My lawyer advised me to comply with his request to avoid the possibility of my being forced to leave Mexico and to close Botanica-Mex. This I did, losing the two kilograms of progesterone I originally made and my shares of Syntex.

—Meanwhile, I began collecting new varieties of dioscoreas in various parts of Mexico. I collected about twenty. One of them was called "barbasco" by the natives. That's the one that's used at the present time. It had almost pure diosgenin. I couldn't use that earlier because I found it at a place called Tierra Blanca. The only way you could get it out of there during the war was to put it on a boat and take it over to Veracruz. When I broke off with Syntex and went on my own with Botanica-Mex, I started immediately on this barbasco root.

—My friend in the United States dropped the price of progesterone [as its sources became still more plentiful] from eighty dollars to ten dollars a gram and a few months later to five dollars a gram.

—Just about that time—in May 1946—a European company called Gedeon Richter asked me if I would merge Botanica-Mex into them. I decided to do so, and I turned the process over to them. They were going to operate it under a new name, Hormosynth. I remained for three years as a consultant for

Hormosynth. By that time the price of progesterone had been reduced to about two dollars a gram.

In 1949, SmithKline & French bought Hormosynth and changed its name to Diosynth, which eventually was bought again by a European drug company, Organon. It remains one of the major producers of hormones in Mexico. By that time, Russell Marker resolved that he would never again have anything to do with chemistry. To wipe away his painful memories, he destroyed his laboratory notes and letters.

Of course, there is another side to Marker's story, although its details are hard to come by. Emerik Somlo died a few years later in France, and Federico Lehmann, who lives in Europe, has steadfastly refused to talk about the conflict, feeling he was caught in the middle and wishing to leave the past to its slumber. His son, Pedro A. Lehmann, also a chemist, has written a history of the Mexican steroid industry and Russell Marker's seminal contribution to it. In 1991 he spoke at a symposium on the history of steroids at the American Chemical Society annual meeting in New York. Reviewing gingerly the conflict between the partners, he displayed on a projection screen a previously unshown document and commented: "It's been often said that Marker received no pay for all his work, but here I have evidence that he did." He showed a page taken from one of his father's notebooks listing the chemists who were working at the time, the number of grams of progesterone each delivered, and the total amount each was paid. "For Marker, I've underlined that he was paid $17,500, which was a very large amount at that time. This is 1943 or '44. It would be equivalent to one hundred seventy thousand dollars or more today."

He also displayed a "very threatening" letter from Marker, which said, "Remember, I know all." Lehmann says he asked both his father and Somlo what it meant, and "neither could remember." A possible interpretation, of course, is that the quizzical letter was not as "threatening" as it might sound but a stern reminder that he, Marker, was the only one who knew the process of turning a native fish poison into progesterone—or perhaps it was a reminder that the process was not patented, thus Marker, *or anyone,* was free to duplicate it anywhere.

Marker's last contribution to scientific literature—his 213th published article—before destroying his papers in 1949 was on a subject laden with irony. It was a review of possible new vegetable sources of the new wonder drug cortisone. What set the match to the explosive growth of the steroid industry in Mexico was, as Norman Applezweig, a steroid scientist-historian, described it, "the discovery that cortisone, an adrenal steroid, could alleviate the symptoms of rheumatoid arthritis. The demand for such therapy was so great that it sent chemists scurrying back to their laboratories and prompted the development of highly superior methods to synthesize and modify the structure of the natural hormones, creating a cornucopia of compounds for pharmacologic investigation and clinical use. *If it had not been for the impact of cortisone on steroid chemistry, the use of hormones for contraception would have seemed an impractical idea in 1950.* Instead, it became clear that hormones could be manufactured cheaply enough to provide fertility control for the entire world, probably with greater ease and lower costs than existing nonsystemic contraceptives." (Emphasis mine.)

Turning to their business significance, Applezweig added that natural and modified steroid hormones "represent a retail market of almost a billion dollars and are among the most important medical weapons yet devised. Our ability to produce these drugs in quantities and at costs which render them available can be traced directly to . . . the work of one man, Russell E. Marker."

The solution to the cortisone production problem underscored a point often overlooked in a big-money age. Big discoveries are made more by big minds than by big research budgets. And the leader in the race was a chemical manufacturer in presumably backward Mexico.

Almost a generation later, in 1969, Pedro Lehmann lured Marker out of hiding to be honored at a banquet sponsored by the Chemical Society of Mexico. Standing with Russell Marker near

Orizaba, at the site where the first specimen of cabeza de negro was unearthed, Lehmann asked whether serendipity had played a major role in the discovery. Marker glared at him and declared: "Serendipity? Hell, no!"

Nor had Marker, the classic "basic" researcher, been driven by a vision of an end product. He has asserted to me: "I was never interested in the *use* of the hormone, only in making it available. You just get curious, and you want to see how the end comes out. It's like playing chess. You become interested in the outcome. I didn't realize it could be used for birth control pills until I had quit completely."

7

A LEAP BEYOND NATURE

He who would do good to another must do it in Minute Particulars.
General Good is the plea of the scoundrel, hypocrite, and flatterer;
For Art and Science cannot exist but in
minutely organized Particulars.
—WILLIAM BLAKE, 1757–1827

At the starkly plain stucco-shell laboratory and office of Syntex in Mexico City, soon after Marker quit his partnership in May 1945, tremors of panic seized Emerik Somlo. He owned the only company in the world that had ever made precious progesterone from plant roots. But the one man in the company who knew how to make it had fled. Syntex held no patents on whatever the process was, so how long could its "knowledge" remain exclusive? Even worse, everyone in the chemical world knew that cortisone, the wondrous pure gold of cortisone—in profusion—lay just beyond their fingertips. Where there is abundant progesterone, now they knew, abundant cortisone dangles but a chemical step away.

THE PILL

Damn that Russell Marker! His last published paper reviewed all the possible new vegetable sources of cortisone—but not a word about how to *make* it. The prize of that new wonder drug made cheaply could fall into the waiting, eager lap of Syntex if only, if only . . .

Frantically, Somlo sent inquiries through a unique network that had formed throughout the Western Hemisphere—a web of scientists, mostly Jews, from Hungary, Austria, Germany, from wherever terror of Hitler had driven them across the ocean to whatever countries would accept them. A refugee in Cuba led Somlo to a young steroid chemist in a Havana drug firm. George Rosenkranz, a Hungarian like Somlo and self-described as "working at the glorious salary of a secretary at thirty-five dollars a week," seemed ideal for Somlo. Young Rosenkranz had earned his doctorate in Zurich in the lab of one of the great figures of early steroid chemistry, Leopold Ružicka. On his own he had followed Marker's publications and even succeeded in making both progesterone and testosterone from a sarsaparilla plant. Accepting Somlo's urgent summons to Mexico City for an interview, he wryly recalled the words of his illustrious mentor, Ružicka: "Rosenkranz, one last advice. Don't touch a steroid with a ten-foot pole. Everything that can be done has already been achieved."

"I couldn't fly directly from Cuba to Mexico because the war was still on and technically I was an enemy alien," Rosenkranz has recalled. So taking a roundabout route, "I arrived in Mexico on August sixth, 1945, the fateful date of the atom bomb at Hiroshima. I hurried to my appointment with Dr. Somlo and Dr. Lehmann for the strangest job interview I could imagine. Their first question was whether I knew how to run a certain piece of equipment. They provided me with a lab coat and led me to their lab so I could prove it.

"Not until later did I find out that Russell Marker himself had abandoned Syntex and that I was his replacement. It soon became obvious that the brilliant but secretive and suspicious Marker didn't want anybody to know the secret of his processes. Bottles of reagents and intermediates all carried strange code names. A hydrogenation catalyst he had used was labeled 'silver.' Solvents were identified by the workers by weight and smell. Well, all this cloak-and-dagger soap opera was too much for me. Instead of wasting time trying to reconstruct the fine details of Marker's process, I worked them out for myself. My initial staff consisted of eleven young women with high school education trained as laboratory assistants and one Mexican production chemist.

"After solving the problem of the lost art of manufacturing progesterone, I was asked to develop a method for the production of testosterone, which was in increasing demand. We never found any evidence that Marker had ever used at Syntex—maybe elsewhere but not at Syntex—a process to make testosterone." (Marker claims that he did.) "On paper, of course, the problem was as simple as the elimination of two carbon atoms. Well, you cannot imagine the gigantic effort invested in working out the ideal conditions for this seemingly very simple reaction. After a long period of experimentation—to me it appeared endless—we found the solution. Our last remaining industrial target was the synthesis of estrone from diosgenin. After long and patient experimentation, we finally succeeded."

The holy grail of synthetic cortisone urgently beckoned. But Rosenkranz, now vested as research director of Syntex, needed to find a Galahad. Through the refugee-scientist grapevine, he too located a candidate, a young Austrian just getting his career started in the United States after earning his Ph.D. at the University of Wisconsin with a thesis on chemically converting testosterone into estradiol.

Rosenkranz recalls, "After reading the brilliant work reflected in the publications of Carl Djerassi, I invited him in 1949 for a visit to Syntex. Apparently he fell in love with Mexico."

That is not the way Djerassi himself has reported it: "The idea of doing any 'serious' chemistry in Mexico seemed preposterous. But when I received an invitation to visit Mexico City with all expenses paid and no other advance commitment, I accepted. Rosenkranz made me a tempting offer. Having always worked alone or with one or two technicians, I was suddenly presented with the chance to head a research group that would attempt a practical synthesis of cortisone—and in a surprisingly well-equipped laboratory. Nevertheless, when I returned to my friends from Wisconsin and Harvard and told them about it, they thought I could not be serious."

In 1949 at age twenty-six, Carl Djerassi, yearning to try his hand at the hottest field of research in steroids, joined Syntex. He quickly synthesized cortisone through two different pathways. It was a breakthrough that drug houses everywhere sought. Next came the estrogens. Djerassi synthesized estrone and estradiol, again from diosgenin. By the early fifties Syntex became the major supplier of synthetic hormones to drug houses in Europe and America.

Later in 1950, Syntex set a brazen new goal: inventing a progestogen that was *better* than progesterone. For reasons not understood, progesterone itself was not effective when taken by mouth unless it was swallowed in very large doses. Therefore it had to be given by painful injection. Even by injection, to be effective, it had to be administered in much larger amounts than the other sex hormones. So what was needed was a "new progesterone" of increased potency, thus permitting a decrease in dose. Syntex set out to leap beyond nature to a "tailor-made" hormone.

The dare was not without precedent. Just before World War II, Hans H. Inhoffen of the Schering research laboratory in Berlin had introduced acetylene, a gas, into one ring of the estradiol molecule. Surprisingly, the converted molecule, unlike the original, remained stable when swallowed—and thus *could be taken orally*, for example, as a pill. Inhoffen then applied the procedure to testosterone, creating a compound with totally unexpected properties. Ethisterone, as it is now generically known, could not only be swallowed, but would act much like progesterone. This was the first evidence that an *orally effective progestogen* could be synthesized. It was precisely the hint that Djerassi needed in the summer of 1951.

In the scientific literature he found another lead: In April 1944—only three months after Somlo, Marker, and Lehmann had incorporated Syntex—a chemistry professor at the University of Pennsylvania, Max Ehrenstein, had read a short paper at the American Chemical Society meeting in Cleveland. Ehrenstein reported that to overcome the potency loss of progesterone when taken orally, he had devised a new molecule from the seeds of an African plant used by natives to poison their arrows. The extract, called strophanthidin, yielded a minuscule amount of a new compound that strongly resembled progesterone but lacked one of the carbon atoms extending from the rings. He injected his tiny amount into two female rabbits, whereupon he ran out of the potion. After absorbing the substance, *one* of the rabbits experienced the same changes to the lining of the womb that natural progesterone brought about.

Ehrenstein's announcement stirred no discernible excitement. Nobody raced to duplicate or apply his finding. But Djerassi, who had read Ehrenstein's report while at graduate school in Wisconsin, remembered it. The result struck him because it suggested that the absence of a carbon atom with three attached hydrogen atoms—a molecule known as carbon 19—was somehow a factor in the molecule's potency.

Now, six years later at Syntex, Djerassi had the idea of trying to modify the molecular structure of Russell Marker's synthetic white powder (derived from the Mexican yam) so it would resemble Ehrenstein's molecule (found in the seeds of the African plant). The rearrangement "took" with surprising ease. Then, injected into rabbits, sure enough, the new substance had four to eight times the potency of natural progesterone!

But how to make the "new progesterone" work by mouth? Djerassi played with linking Marker's powder to Inhoffen's testosteronelike steroid, ethisterone. What drew Djerassi to it was that in a strikingly important way, the synthesized male hormone seemed to reverse the behavior of progesterone. When injected, it *lost* potency, requiring a sixfold dose to do the same work. When swallowed, it behaved much like progesterone—but in full strength! Could it be that if he applied Ehrenstein's structural changes to this cousin of testosterone, he could have the best of both worlds? Would it do the work of progesterone? Would it too retain its potency when swallowed?

Djerassi designed seven separate chemical tracks to pursue, then assigned part of the pursuit to Luis Miramontes, a young Mexican chemistry student carrying out his bachelor's thesis in the Syntex laboratories. On October 15, 1951, the student found what appeared to be the molecule they were looking for.

For a certain test of its progestational activity, Djerassi mailed a sample immediately to a laboratory he trusted in Madison, Wisconsin. The compound proved to be highly active. Then the more severe test: How would it act in living systems? He sent a sample to Dr. Roy Hertz of the National Institutes of Health, who coincidentally had assisted Ehrenstein at the University of Pennsylvania. Hertz fed the samples to rabbits, guinea pigs, and monkeys with good results. Then he administered it to three women volunteers at the NIH clinical center "for the sole purpose of trying to control grave menstrual irregularities"—hemorrhagic monthly bleeding. The news that came back was triumphal. Hertz found that Djerassi's new substance, when taken by mouth, was eight times more powerful than natural progesterone.

On November 22, 1951, Djerassi applied for a patent for his new compound, to become known generically as *norestheisterone* or *norethindrone.*

Recalling the exquisite conquest years later, Djerassi was to say, still again, "Not in our wildest dreams did we imagine that eventually this substance" would become the active ingredient of an oral contraceptive.

Djerassi had not the faintest knowledge that, less than a year earlier, Margaret Sanger had dined with Gregory Pincus in New York to reveal her vision of a "perfect contraceptive," nor that both Sanger and Katharine McCormick would soon visit Pincus at Worcester to seal their pact to search for one. Yet the young chemist in Mexico City, with inscrutable timing, had just guaranteed the success of the dream.

Was his success a stroke of luck—stumbling into the right place at the right time? In some measure, almost every success is. Yet young Djerassi was cut differently. Perhaps the inexorability of his success is most clearly seen from the other end of the tunnel—with a backward glance at his extraordinary ambition, energy, and the life he eventually sculpted from them.

Both his Bulgarian father and Austrian mother were physicians, and Djerassi assumed that he would follow in their footsteps. Both parents being Jewish, they fled Vienna for Bulgaria after Hitler's takeover of Austria. Young Djerassi arrived in the United States with his mother in December 1939, at age sixteen. He recalls: "I didn't have any childhood chemistry sets. I never blew up our basement; I never had chemistry in high school—in fact, I never graduated."

Without a diploma he bluffed his way into enrolling for two semesters at Newark (New Jersey) Junior College, where a "superb" chemistry teacher, Nathan Washton, so dazzled him that Djerassi changed from a premed to a chemistry major; then another semester at Tarkio College in Missouri, finally charging through two semesters plus a summer at Kenyon College in Ohio, which awarded his bachelor's degree in 1942—after less than three college years. In only two years at the University of Wisconsin he whipped through to his Ph.D. in chemistry.

In the next twenty-five years Djerassi uncovered chemical secrets leading to antihistamines and an environmentally safe pesticide in addition to cortisone and the oral progesterone. After his career at Syntex, he began a second one that had long tugged at him, as an academic at Stanford University in California. He earned the National Medal of Science and the coveted Priestley Medal, the highest award of the American Chemical Society.

During his Syntex years, a distribution of stock to the company's early key employees had multiplied exponentially in value and made Djerassi enormously rich. He built a home on twelve hundred rapturous acres overlooking San Gregorio Beach a half hour's drive from Stanford, which he calls the SMIP Ranch ("Steroids Made It Possible" on some days, "Syntex Made It Possible" on others). Upon the death of an adult daughter, in her memory he transformed the ranch into a colony for artists and writers, whom he supports during their stays there. He also has invested millions in an art collection that includes works by Klee, Picasso, and Moore as well as pre-Columbian sculpture. Now, in spacious digs high on Russian Hill in San Francisco, he is given to hosting salons for personages of the arts, moving among his guests with a sprightly gimp, the result of a skiing accident that stiffened his knee.

"The pleasures of research," he grants, "are unbelievable, close to an orgasm when something works out. But research is mostly failure. So I always did more than other people. Even in graduate school I never went to the lab at night. Most graduate students do. But I took cello lessons instead."

Why, Djerassi always wondered, did most of the well-educated scientists he knew, chemists especially, turn themselves into "intellectual monogamists" with few interests outside their work? "Many of my colleagues never go to San Francisco. I see very few of them at the opera, and I don't think I've ever met a colleague at the Magic Theatre. I've promised a major collection of Paul Klees to the San Francisco Museum of Modern Art. I would suspect the vast majority of my colleagues don't know that. Klee is an intellectual painter, and I think an academic could appreciate Klee. But most of them couldn't be bothered."

In 1976, at fifty-three, he married his third wife, Diane Middlebrook, a Stanford notable in English who would later publish a best-selling biography of the poet Anne Sexton. Perhaps that was a predictor of his third career. He has since written a book of poems and two novels, the second of them published as *Cantor's Dilemma*, set in the intense politics and jealousies of academic science, as well as an autobiography and many articles on the politics and sociology of birth control.

Soon after Syntex and Djerassi patented norethindrone, a Chicago pharmaceutical manufacturer, G. D. Searle, and its chief research chemist, Frank B. Colton, registered a remarkably similar substance, which they called *norethynodrel*. Searle claims its research had progressed without knowledge of Syntex's work. Syntex had filed its patent application in November 1951 and published the results the following year. Searle did not file until August 1953.

Such competing claims in science are not uncommon. Sometimes researchers, half a world apart, learn of each other's similar work and engage in feverish races to be first to publish an

important discovery. Sometimes the competition drips with suspicion. Virtually never does a scientific achievement belong to a
single individual, nor are scientists, any more than other achiev-
ers, always selfless in allotting the glories of discovery.

For forty years Carl Djerassi was not to let go of a misgiving
that, in claiming a patent of its own, Searle was poaching. The
similarities between their two drugs raised exquisite legal issues.
Searle's norethynodrel was so similar to Syntex's norethindrone
that, according to Djerassi, "mild treatment of [Frank] Colton's
norethynodrel with hydrochloric acid or, in fact, even so weak an
acid as human gastric juice converts it to a large extent into
norethindrone." Which, Djerassi says, raises this enchanting
question: When the natural processes of the stomach "synthesize"
a patented compound, is that an infringement of an issued patent?
The similarity also brought into focus the interlocking dependen-
cies of drug companies, which often subjugate the ego interests of
scientists to their own profit interests. "I urged that we push this
issue to a legal resolution," says Djerassi, "but Parke-Davis [to
whom Syntex had licensed the use of norethindrone] did not
concur. Searle was selling an important anti–motion-sickness
drug, Dramamine, which contained Parke-Davis's antihistamine
Benadryl. Our norethindrone seemed in 1957 small potatoes over
which it was not worth fighting with a valued customer."

Frank Colton's version of the race is that his chemists at Searle
developed "some two hundred molecules" hoping to make a
progestogen "better than Mother Nature's." But his version intro-
duces still another potential claim to the confused authorship of
the "new improved" synthetic drug. Colton says that a chemist at
the University of Wisconsin, A. L. Wilds, and one of his students,
N. A. Nelson, had devised a better way to get rid of the carbon
atom 19. Wilds did not perfect and publish the improved process
until 1953. But in 1950, according to Colton, Wilds did tell
Djerassi, whose Ph.D. thesis Wilds had supervised. Also Wilds
described it to one of his former students, Dr. Jack Ralls, one of
Colton's coworkers at Searle. "So we [at Searle]," Colton has said,
"were following the same research line as Djerassi at the same
time." That hangs still another question in the air: Had both he

and Djerassi borrowed their "same research line" from Professor Wilds and his student, Jack Ralls, each merely extending it another step or two? That is the way of science far more commonly than "original" discoveries.

Like Djerassi, Colton too asserts that he had no idea he was creating a chemical for birth control.

In 1991, fully four decades after the Syntex patent application, Djerassi and Colton confronted each other face-to-face for the first time at a New York symposium of the American Chemical Society. The air misted with acid:

COLTON: After hearing Dr. Djerassi's story this morning, I asked myself, "What am I doing here?" Djerassi in the oral contraceptive field appears to have done it *all*.

DJERASSI: The way you reported the chronology was that your patent was in 1955, Syntex's patent in 1956. The fact of the matter is, of course, the dates of *issued* patents mean nothing. What counts is the filing date. Why did you not say "Syntex 1951" and "Searle 1953"? Why was your patent not filed until one and a half years after we made our first report? Why did you never publish the results of what you admit is *the* most important thing you have ever done—while we published our results immediately? To say today that the chemistry was concurrent seems to me to be nonsense.

COLTON: A very good point, Dr. Djerassi. We were negligent in publishing, and I will tell you why. Syntex was a universitylike atmosphere. The Mayo Clinic was a university-type atmosphere when I was there. The work we did was immediately published. But Searle was not a university-type atmosphere. Sometimes the more important the compound was, or was perceived to be by management, the more difficult it was to publish because of competition. There's another reason in my case: I did most of the chemistry myself. And to be honest, I don't like to write. I don't like to put everything aside for it. I wish in retrospect, however, we were a little more diligent in publishing.

DJERASSI: If that is true, why did you file the patent in 1953, a year and a half after we published?

COLTON: The patent filing was in the hands of the patent attorneys. Very often, instead of filing immediately, they would protect themselves and extend the life of the patent by disclosing the compound first, then later, filing on the intermediate use. There was no sinister plot. That's the way it was handled by our legal people.

One aspect of the *new progestins* (as the Djerassi and Colton compounds were soon to become known) made both of their competing inventors nervous: They were not dealing with new uses for old, known substances. These were new chemicals, not found in nature. While their subtle molecular changes appeared small, when introduced to the intricate machinery and mysteries of the human body those variations could cause pain, illness, even death. Much trying and testing lay ahead.

8

"In Science, Lizuska, Everything Is Possible"

Once when I was feeling depressed and thinking, "Oh God, it's all too much,"
I said, "Goody, let's go out and let's get a cart
and we'll sell lemons. It's just not worth it."
He said, "Stick with me, kid, and
you'll be wearing diamonds."
—ELIZABETH PINCUS TO PAUL VAUGHAN, 1969

When Gregory Pincus took his first step toward deliberate invention of the Pill, he knew nothing of Carl Djerassi's new progestin nor of Frank Colton's, neither of them yet published or patented. Indeed, Colton's new discovery and Pincus's new quest were like two huge ships sliding by each other in a fog, the two scientists unaware of their near miss. For Colton's employer, G. D. Searle, was the same company that since 1944 had been subsidizing much of Pincus's work at the Worcester Foundation. Yet neither Colton nor Pincus had heard of what the other was up to.

Pincus's main mission for Searle had been exactly the one that had brought Djerassi to Syntex: trying to find a new way to manufacture cortisone, lots of it, to satisfy a commercial market. One competitor, Merck & Company, had already succeeded in making cortisone in quantity but through a painstaking thirty-two stage process that continued to keep the cost high. One day Pincus had the arresting idea that he could "synthesize" Merck's process in a "natural" way. Instead of Merck's thirty-two stages, he would simply pump a serum through the adrenal gland of a huge sow. The animal's smart gland, he believed, would then insert the elusive and critical oxygen atom in exactly the right place for making cortisone. Similar tricks on nature had been performed before. The process is called *perfusion*. Skeptically, Searle let him go ahead. While Pincus coaxed his sow, an Upjohn Company biochemist took still another tack in the race for cortisone. He put yeast to work, starting a fermentation process not unlike the one that had been so successful in making penicillin and streptomycin. The stratagem worked. Upjohn relieved the world cortisone famine and thereby banked huge profits. For its years of investment in Pincus, G. D. Searle got the booby prize of a tax write-off.

Without apparent embarrassment, Pincus soon approached Searle's director of research, Albert L. Raymond, for new money to spend toward a new goal. Actually, two new goals: Pincus wanted a second crack at testing the perfusion process that had flunked in the cortisone adventure. This time he proposed that by flushing fluid through an animal gland he might develop a cheap, more effective progesterone. The shocker came next: With the open-faced sincerity of a child just given his first doctor kit, Pincus said that with the progesterone thus summoned he hoped to develop an oral medication—a simple pill—that would prevent human pregnancy.

The proposal struck Raymond as doubly bizarre. First, Raymond presumably knew (although Pincus didn't) that his own Frank Colton was at or near his goal, not only of a synthetic progesterone but one that would do its work when taken by mouth. But that second goal: Get Searle involved in a *pill* to *prevent pregnancy*? A pill with *Searle's name on it*? Had Pincus lost all sense of the practical world?

Raymond's reply to Pincus was prompt and blunt:

You haven't given us a thing to justify the half-million that we invested in you. . . . There is, to be fair, still some chance that . . . the perfusion process will prove useful. But to date your record as a contributor to the commerce of the Searle Company is a lamentable failure, replete with false leads, poor judgment, and assurances from you that were false. Yet you have the nerve to ask for more for research. You will get more only if a lucky chance gives us something originating from your group which will make us a profit. If I had unlimited funds I would undertake a large program in the steroid field, but I do not have such funds and the record to date does not justify a large program.

Before long, of course, as though by some inexorable intention of history, Pincus had the money. His new friend, Margaret Sanger, who had put him up to his radical new goal, saw to it that he got it from *her* old friend, Katharine McCormick.

The scolding from his patron at Searle smarted, of course, especially because it recalled other pains from Pincus's past. Since his earliest days in science, Gregory Goodwin Pincus had drawn recognition as a man of daring, far-reaching imagination—and somehow those qualities seemed to earn him distrust and often censure.

"Goody" Pincus was born in Woodbine, New Jersey, in 1903 to Russian-Jewish parents living in a philanthropy-created settlement of Jewish farmer-immigrants. Goody's father, Joseph Pincus, ran the local tract, lectured to settlers at similar colonies throughout the East, and for a time edited a Yiddish-language newspaper, *The Jewish Farmer.* As a boy Goody told his father he wanted to farm when he grew up so he could work with animals. There was no money in it, said Joseph Pincus, and he urged his son to go to college. After all, one of Goody's uncles had become dean of the agricultural college of Rutgers University. Besides, Goody was reputed, at least within the family, to have an IQ of 210. Was that to be wasted on chasing chickens?

At Cornell, Goody Pincus studied to become a pomologist, an

apple expert. His curiosity soon deepened into genetics and em-
bryology. One of his colleagues theorized that Pincus's fascina-
tion with genetics was sparked, in part, by learning to live with an
inherited defect: color blindness. Before earning his B.S. in 1924,
his expanding interests led him to become editor of the *Cornell
Literary Review*.

From Cornell, Pincus went to Harvard, where in 1927 he
simultaneously earned his M.S. and Sc.D. degrees. Perhaps
bearing out his colleague's theory, he studied patterns of inheritance in the coat coloration of rats. A fellowship from the
National Research Council afforded Pincus two years of postdoctoral research at Cambridge University and the Kaiser Wilhelm Institute. In 1930 he returned to Harvard as an instructor
in general physiology, studying further under William Castle,
a trailblazer in the new, high-fashion field of genetics. He
also collaborated on a series of research papers with William
Crozier who, at thirty-two, was Harvard's youngest associate
professor. Crozier, in turn, was a disciple of Jacques Loeb, who
had dazzled the scientific world in 1899 by showing that unfertil-
ized eggs of sea urchins, soaked in solutions saltier than seawater,
could be cultivated into viable embryos. His experiment seemed
to show that the sperm cell performed a task of mere physical
chemistry that could be accomplished by other means. "Theologians and animal watchers alike," commented James Reed, the
birth control historian, "would have to interrupt their musings
over the mysteries of life and take note that the artificial production and the control of living matter were within the grasp of
analytical science. . . . Loeb became a hero to a generation of
American experimental biologists, who were inspired by the
Loebian faith that the secrets of life are amenable to a sound research protocol."

Pincus's appointment in 1931 as assistant professor in Crozier's
department of general physiology "seemed to be the first landmark in an academic success story" during a time at Harvard, says
Reed, "when Jews had to be better than their competitors in order
to win equal consideration." In the next six years his delicate rab-
bit-egg work established Pincus as an authority on mammalian

sexual physiology. It earned him further support of the National Research Council and the Josiah Macy, Jr., Foundation.

That is when Pincus's imaginative research began bringing him notoriety as well as notability. In a startling experiment, he introduced rabbit sperm to their counterpart eggs in a test tube and coaxed them to fertilize, grow, and divide for a short period as though they were in a womb. Thus he produced the embryos of "fatherless" rabbits. His trouble began when he proudly announced it. The sensational publicity that followed, said Reed, "lent credibility to the charge that some of his senior colleagues wanted to get rid of a brilliant but controversial investigator whose work conjured up Frankenstein–*Brave New World* nightmares in the minds of many laymen." *The New York Times* ran an interview with Pincus under the headline "Rabbits Born in Glass; Haldane-Huxley Fantasy Made Real By Harvard Biologists."

"Pincus' research goals were large," Reed wrote,

and some of his colleagues believed that his experiments ranged too far beyond established knowledge. They found that they could not reproduce all of his results. Thus, Pincus' effort . . . exposed him to criticism from two sources. First, he aroused the hostility of the religious-minded in and out of science who resented his attempt to reduce the mysteries of conception and birth into mechanistic terms. Second, some experimental biologists, who sympathized with his goals, questioned the wisdom of trying to do too much too soon. The growth of knowledge depended on small steps, carefully planned. . . . Pincus, in contrast, designed experiments that were too complex to be carefully controlled or easily reproduced.

An article in *Collier's* titled "No Father to Guide Them" brought national prominence—or notoriety—to his "immaculate conceptions." The writer, J. D. Ratcliff, painted Pincus with subtle brushstrokes of antifeminism and anti-Semitism. It derided the "tricks" that biologists were playing on nature, labeling Pincus's forerunner in reproductive research, Jacques Loeb, as "the hugely famous Portuguese Jew." The article was adorned by a chilling photograph of Pincus, a cigarette dangling out of his mouth and a large white rabbit peering out from his arms with a pitiable look sure to inflame antivivisectionists. Pincus's work, the *Collier's* article went on, had "possibilities more thrilling than anything a detective story writer ever imagined: a world in which woman would be a dominant, self-sufficient entity, able to produce young without the aid of man. . . . Man's value would shrink. . . . The mythical land of the Amazons would then come to life. A world where woman would be self-sufficient; man's value precisely zero."

While Pincus at first enjoyed the controversy and his new fame, once he found himself pictured as a sexual manipulator and ethical deviant, this young, untenured researcher, with few of his colleagues willing to defend him, dutifully recited, to any reporter who gave him the chance, the obligatory vow of the "pure" scientific researcher: "I am not interested in the implications of the work"—only in the process of discovery for its own sake.

When Harvard celebrated its tercentenary in 1936, it cited Pincus's work as one of the university's outstanding scientific achievements in its history. The next year, when Pincus was thirty-two, Harvard further patted Pincus on the head with a paid leave of absence for the 1937–38 academic year and a research grant allowing him to return to Cambridge. Then upon his return Harvard denied him tenure, which meant he had to empty his lab and desk and get out. According to his Harvard colleague, Hudson Hoagland, the reasons "included some anti-Semitism and jealousy." Reed has added: "Pincus and his friends believed that the decision reflected prejudice against him as a student of Crozier and as a self-advertising Jew who published too soon and talked too much."

Neither of those observers mentioned that Harvard, then as now, only rarely granted tenure to young faculty, making exceptions only when a young professor's record of published research was truly exceptional. But Pincus's record *was* truly exceptional, as indicated by the special honors Harvard had bestowed upon him. Pincus was incredulous and crushed.

Pincus's suddenly wobbly career was given a steadying hand by his friend Hoagland, who by now had acted on his own antipathy for Harvard politics by accepting the chairmanship of the small biology department at Clark University in nearby Worcester, which had recently been put under a world spotlight by Sigmund Freud, who gave his first and only American lectures there.

Hoagland appointed Pincus a visiting professor of zoology, but his budget permitted him to offer neither a salary nor laboratory expenses. Together they scratched up two years of salary by getting contributions from Nathaniel Lord Rothschild, the third in the line of Rothschild barons, who had been Pincus's fellow investigator of the mechanisms of reproduction at Cambridge University, and from a wealthy New York business friend of Hoagland. The money enabled Pincus to move his family to Worcester.

Pincus continued to publish studies. By 1939 *The New York Times* depicted him and his work favorably for the first time in an article on "the first fatherless rabbit." His experiments, said the *Times*, would "provide much valuable knowledge that will eventually find clinical application for the birth of healthier human beings, the ultimate purpose of the experiments."

Other developments, meanwhile, were guiding Pincus's fortunes in yet invisible ways. One such influence was Russell Marker's Mexican adventure, which was forcing American drug companies into a race to improve upon Marker's methods. Within the new industry created by Marker's achievement, there emerged a sudden need for biologists with a working knowledge of endocrinology to test new experimental drugs on animals. Gregory Pincus, dispensable in the academic world of "pure research," was suddenly a hot property in the pharmaceutical industry. Pincus had never had any intention of going commercial, but his unstable and only quasi-official situation at Clark had become untenable. "It was at this time," Hoagland was to write, "we first seriously discussed a plan to set up a research institute completely divorced from control by any university or college. We welcomed the idea of freedom from bickering faculty meetings, futile committees, jealous colleagues, and teaching prescribed courses to often indifferent students."

They formed the Worcester Foundation for Experimental

Biology as a nonprofit Massachusetts corporation. Their purpose: to strengthen the links between practical medicine and the "new biology." While later, in the years following World War II, grantsmanship would become a way of life for American university researchers, in 1944 it was a high-risk way to go. Hoagland and Pincus dressed their letterhead with a distinguished board of directors that included internationally known scientists, a rabbi, and local businessmen who helped provide twenty-five thousand dollars for housing the infant organization in an old mansion. At first it was a two-man operation. To get by, Pincus became janitor of the Worcester Foundation's animal labs, and Hoagland mowed the lawn. Mrs. Hoagland kept the books. At least once Hoagland had to draw from family money to pay bills. Nevertheless, their knowledge of steroid research positioned them well. Grants soon came from the American Cancer Society, the G. D. Searle Company, and the federal government, which was just taking on the role of a major subsidizer of research. For their first year's budget they raised about $100,000.

Soon they attracted ambitious young scientists to their independent haven. One of these was Min-Chueh Chang, a thirty-two-year-old China-born graduate student who had been trapped in England by World War II and unable to return home. Pincus had met him at Cambridge, where Chang's early career was influenced by Pincus's book *The Eggs of Mammals*. "In our field *everyone* knew about him," Chang said years later of Pincus. "You must remember that until then no one knew mammals *had* eggs." Chang, thin, slightly stooped, with an exuberant smile, was paid so little he slept in a corner of a barn that had been converted to a lab, and he doubled as the foundation's night watchman. An inside joke that grew to an outside rumor in Worcester was that the foundation "kept a Chinaman chained in the basement."

In 1950 Chang won a one-thousand-dollar award for a paper on the fertilization of rabbit eggs. Another research associate, Oscar Hechter, won an award from the Endocrinological Society. Pincus, always the optimist, gleefully declared, "We don't have to worry about salaries anymore. The staff can live on its awards."

Later, almost immediately after Pincus accepted the assign-
ment by Sanger and McCormick, Chang was to conduct the day-
to-day lab work of the search for the "perfect contraceptive."

On April 25, 1951, according to a work-journal notation entered by Chang, the Worcester team took its first steps toward
creation of the Pill. Pincus had begun as any methodical scientist
would. He looked back to find what was known, or what was
known to be known. Ignorant of Colton's and Djerassi's recent
achievements, he directed Chang to duplicate and verify what
had been done eighteen years earlier (and largely ignored
throughout the rest of science) by Makepeace of the University
of Pennsylvania, who was first to inject rabbits with progesterone
and to find that their ovulation halted.

At first young Chang found his task tedious, unchallenging,
and out of his line. He was a specialist in sperm. He resented the
distraction of shooting progesterone into female rabbits day after
day, then pairing them with males for mating, and checking
whether the females became pregnant. It also meant "sacrificing"
many of the females (more plainly, killing them) to make sure
their egg production had been halted. But as the new work began
to engage him, he began to go at it with zest. "I like to feel the experiments through my hands," he once said. "If your technicians
do all the work, you lose the fun. Would you let someone else
play tennis or chess for you?"

Chang experimented with implanting sustained-release pellets
of progesterone under the skin of female rabbits and found that
he could thus inhibit ovulation for a long term (an approach that
became interesting to investigators of human birth control some
forty years later).

Chang soon did delegate much of the lab operation to a still
younger scientist, Anne Merrill, who describes some unexpected
aspects of her daily work:

"In those times, people didn't know how to talk about the
things they saw us do. We had these touring groups come in, like
the ladies group from church, because the Worcester Foundation
was always looking for support from the community. And we
might be sitting there weighing a rat's uterus or ovaries.

Sometimes we had taken out the ovaries so the rats wouldn't have any source of their own hormone, only what we gave them. Or sometimes, for Dr. Chang's other work, we'd be castrating a hundred rats. Rat testes have a blood pattern that makes them very beautiful. People would get interested and ask, 'What's that?' And we'd tell them, but they didn't know the word 'castrating.' When we helped them understand, we got treated with giggles or silence. I couldn't talk to anybody outside the lab because they were embarrassed even to hear the words."

For their experiments, Pincus and Chang were using what they thought was the state of the art, Marker's synthetic progesterone. With it they soon corroborated Makepeace. Progesterone could work as a contraceptive—at least, in small mammals—but with the same perplexing obstacle that had so recently confounded Djerassi: It required huge doses.

Then came one of those haphazard, inscrutable accidents that spangle the history of science. In 1952 at a scientific conference, Pincus ran into a Harvard gynecologist and researcher he had known two decades earlier. Dr. John Rock of Brookline, Massachusetts, only forty miles from Worcester, also was experimenting with the new synthetic progesterones. His goal was different—in fact, quite opposite—from that of Pincus. And it was their very oppositeness that magnetized the attention of each man toward the other's work.

While Pincus was injecting progesterone into fertile subjects to forestall their pregnancy, Rock was injecting progesterone and estrogen into *infertile* subjects with the goal of helping them *become* pregnant.

As though those mirror-image goals were not striking enough, their work contrasted in another way. Because his quest had no precedent, Pincus had to limit his experimentation, at least for the time being, to rabbits. But Rock was experimentally injecting women. As a person of extraordinarily lofty medical reputation, he could take on that risk following the considerable amount of animal experimentation with Marker's synthetic hormones. Rock was fastidious, however, in observing the rule of explaining to each woman volunteer "the experimental nature of the treatment and its unknown but probably harmless and only possibly helpful effects."

Few people in his field could match the reverence and trust commanded by this tall, straight rod of a man with a strong-jawed face, sparkling blue eyes, and silver mane. One male colleague described Rock addressing the American Gynecological Society: "He was such a stately, handsome man, impeccably dressed in a gray suit, dark tie and with that magnificent white hair. He looked like God himself standing there. No matter what he said, you'd believe it, but it was all true. He did far more to move gynecology ahead than any other single figure in the field. He was the most essentially kind man I've ever known, yet exhilarating and stimulating."

Rock's justification for the experimental risk was that his patients were afflicted in a most distressing way. With no discernible fault in their reproductive organs, their bodies simply refused to become or stay pregnant. Some ovulated on erratic schedules. Others, when they did successfully conceive, miscarried with rude regularity. Some may have had a subtle form of endometriosis that defied diagnosis.

Rock had developed a theory that the problem of some of these infertile women might be an underdevelopment, an immaturity, of their fallopian tubes or womb. If so, the influence of extra doses of progesterone and estrogen, the "hormones of pregnancy," might encourage those organs to mature. After all, wasn't that what those hormones did during a natural pregnancy? Furthermore, his years of practice with infertile women had led him to believe that by suspending ovulation for a few months—thus deluding the system into "thinking" it was pregnant—the system would "rest," then "rebound" with greater fertility. He tried what he called a "pseudo-pregnancy" on an accumulated sample of eighty "frustrated but valiantly adventuresome patients" who had been barren for at least two years. The volunteers, he reported in a professional paper, "were assured conception could not occur during the treatment. It couldn't because the hormones stopped ovulation."

Rock had to be exceptionally cautious in how he made that last assurance, scientific as it may be. It was all right to mention the blocking of conception as though it were an incidental, even an unwanted, side effect of his treatment. But certainly not as its *purpose:* There was that Massachusetts law banning dissemination of birth control information. Dr. Rock had no desire to go to jail for prescribing or even describing any form of contraceptive.

His experimental patients received daily doses of progesterone, bought from Syntex, starting with low, fifty-milligram doses, which gradually increased to extremely high doses of three hundred milligrams. They also received five- to thirty-milligram doses of estrogen. Because the hormones were given daily without letup, the women did not menstruate. They experienced all the classic markers of pregnancy, unpleasant symptoms included. Their breasts enlarged and became tender. Some complained of morning sickness and nausea. The "pseudo-pregnancy" was so much like the real thing that Rock often had difficulty convincing the women, each so hungry to conceive, that they were indeed not pregnant.

Within four months after Rock took each of his patients off the progesterone and estrogen shots, an astonishing number—thirteen of eighty previously infertile women—did indeed conceive. The evidence of the "Rock rebound" effect was clear. The reasons, however, remained less clear. Rock was never to be sure whether the "rebound" was caused by the state of "rest" or by the "maturing" of the reproductive system.

When the two men met by chance and exchanged experiences in 1952, Rock's description of his work astounded Pincus. All the drudgery of his animal tests had suddenly become a waste of time! It had all been preparation for a far more difficult and dangerous stage—tests on human beings—and Rock had already crossed that great divide. John Rock was already *testing a chemical contraceptive on women*—and *demonstrating that it works*. And, most delicious of ironies, under the protective cover of an opposite purpose, he was doing it in the heart of bluenose Massachusetts.

Yet Pincus still faced mountainous obstacles. He needed to field-test the drug on a large scale to prove both its safety and effectiveness when used deliberately as a contraceptive. Still other barriers loomed in every direction: religious opposition, the medical risks of long-term use, in some states the legal risks, and in all other states the more subtle but perhaps most daunting risk of all: a pervasive cultural antipathy toward the very idea of birth control. No matter where in the United States Pincus might locate his field test, when it became known, political harrassment would almost surely force him to shut it down.

Rock, in turn, was struck by Pincus's trials on animals using progesterone alone. Progesterone was known to produce fewer side effects than estrogen, and perhaps used alone it was sufficient for forestalling pregnancy in women as it was in animals. Urged by Pincus to do so, Rock decided to try progesterone alone on a new group of infertile women.

Early in their information exchange, Pincus also persuaded Rock to administer the progesterone for only twenty-one successive days, then stop it for a week. That would open a window for the menstrual flow, yet at the same time control its regularity, which had been a problem for many of Rock's patients. As a side benefit, Pincus pointed out, the women would be spared the false hope that they were pregnant. But Pincus may have had still another aim in mind. When he had sounded out G. D. Searle about their supplying synthetic progesterone for Rock's next experimental group, the company firmly warned Pincus that it wanted no part of any experiment that interfered with the menstrual cycle. In fact, Rock shared their fear. While he was willing to inject massive doses of progesterone for a few months to encourage the "rebound," he told Pincus that doing it for prolonged periods as a contraceptive was "unthinkable."

Pincus remained confident that each of those grim obstacles could be vaulted. One barrier, however, loomed, waiting to defeat him. No matter how successfully his clinical trials with women turned out, Pincus knew that as long as the synthesized hormone had to be injected—and in large doses—it simply wouldn't do for the "perfect" contraceptive that Sanger and McCormick had commissioned him to produce.

He knew, of course, that the pharmaceutical landscape was strewn with new varieties of synthesized progestogens, all fashioned from Marker's unpatented discovery and stirred by the race for cortisone. Hoping against doubt for a stroke of luck, Pincus sent out a call to his contacts at all drug companies doing hormone work. He asked for anything under development that might be a stronger progestogen, one that would allow smaller doses yet greater effectiveness. Most companies were happy to offer samples. Often the best customers for new drugs are developers of other new drugs. New compounds descended upon Pincus by the dozens. His old patron, G. D. Searle, sent him some two hundred varieties of progestogens. Chang and Merrill now were swallowed up in a new laboratory quest: shooting each of these new agents into rabbits and rats, searching for the strongest antiovulant.

Then a surprise burst upon Pincus that seemed too much to believe: Exactly the new variety of progesterone he was looking for had just been synthesized—not by one man, but by *two*, separately. One was Carl Djerassi of Syntex and the other, Frank Colton of G. D. Searle. Both claimed that these particular samples, intended as menstrual-cycle regulators, were *most effective when taken orally.* Neither sample was accompanied by the faintest hint that anyone had thought of it as a contraceptive. But unless Pincus was making some huge mistake, *the chemistry of the Pill he sought was already invented!*

Pincus and Chang tested and retested the two compounds on new platoons of rabbits and rats to verify their contraceptive effectiveness as well as their safety. That was an essential preface to clinical trials on women to test the new progestin as a deliberate contraceptive. The clinical trial must precede a large-scale field trial.

A *clinical trial* is usually designed and supervised by a medical doctor. A relatively small number of subjects, volunteers, are given a new drug. They are matched with a usually equal number of volunteers given a placebo, a do-nothing substitute. None of the subjects knows which they are getting, nor should the researchers conducting the trials know. Following a successful clinical trial, a *field trial* covers a large number of volunteers who take the drug and are monitored for its effectiveness, safety, and side effects.

To prepare for trials of the new progestin as a contraceptive,

Pincus needed a medical partner of high reputation, preferably a gynecologist. He considered two outstanding leaders of the birth control movement, Abraham Stone and Alan Guttmacher, and decided against them. Both were Jews, and Pincus felt that that might stigmatize the already unpopular cause. He was especially hesitant because opposition to the trials would surely come from Catholics and Christian fundamentalists. Such caution took other forms, too. The Worcester Foundation's 1955 annual report gingerly avoided specific description of work to control ovulation in animals. The 1956 report described experimental uses of steroids to help control the discomforts of menstruation. But contraception? Not a word.

The perfect choice for his medical partner, Pincus decided, was John Rock.

Margaret Sanger objected. She declared in a letter to McCormick, "[Rock] would not dare advance the cause of contraceptive research and remain a Catholic."

McCormick, who had a long-standing acquaintance with Rock, explained that he was a "reformed Catholic" whose "position is that religion has nothing to do with medicine or the practice of it and that if the Church does not interfere with him he will not interfere with it—whatever that may mean!" (Before long, Sanger revised her evaluation of Rock after watching him persuade other Catholics to the birth control cause. "Being a Good R.C. and as handsome as a god," she marveled, "he can just get away with anything.")

If he could "get away with anything," it was because his medical standing was coupled with his widely known devotion to his church.

In 1924, when Rock was about to marry Anna Thorndike, a "Catholic Brahmin" (a special Boston breed of elite that included the Kennedys), the bride's father suggested that Rock go see a priest who had been his boardinghouse mate during their college days. The priest had since emerged as the archbishop of Boston, William Cardinal O'Connell. The young couple visited the cardinal, who volunteered to perform the ceremony (the only other couple he had ever so honored had indeed been Joseph and Rose

Kennedy). The day before the wedding, Rock felt a crisis of religious conscience. He had performed several cesarean sections, delivering through surgical incision rather than vaginally, a practice forbidden at that time by Catholic doctrine. When he confessed it, his priest refused to give Rock absolution. That made him ineligible for the sacrament of marriage. His mother-in-law-to-be, alarmed that the next day's ceremony might be illicit, broke the secret to the cardinal. O'Connell laughed heartily and absolved the groom. Rather than comforting Rock, the act shook him. It was to trouble him for years. Were the rules of right and wrong changeable from priest to priest?

He filed the incident in a dark folder near the memory of another episode when he was much younger. During his high school days, he once recalled for a friend, he gradually became aware of "something," an organic expansion that he knew must be sinful even though it was involuntary. "It was quite a burden to me, and I knew it was wrong," he told the friend, "but it was uncontrollable. I remember picking up that 'how to go to confession' book, and I began to write down the dates and the number of times it occurred so as to be sure to remember. Finally a priest said to me in the confessional, 'Don't be so scrupulous, John.'"

Another memory stayed with Rock all his life and may have been the source of his strength when he took a stance independent of his church and its pope.

When he reached the age of fourteen, one Sunday morning he proudly wore his first pair of long pants.

As I walked out of church after the nine o'clock Mass, Father Finnick, a curate of my parish in Marlborough, Massachusetts, beckoned to me. Very shortly he was to drive to the Poor Farm to make his regular visit. Would I like to go with him? He had taught both our First Communion and our Confirmation classes. He was a saintly man, simple and quiet, but a good teacher. Although I had never had anything to do with him outside of class and the confessional, I liked him. . . .

I shall never forget the short slow ride in the small buggy down East Main Street to the Sudbury road, near the beginning of which was the small white building of the Marlborough Poor Farm, where a few very elderly men and women lived.

I don't remember how the conversation started, . . . [but] as we jogged along, . . . he said, "John, always stick to your conscience. Never let anyone else keep it for you." And, after but a moment's pause, he added, "And I mean *anyone* else."

He did not tell me more of it. I guess he knew that I had been indoctrinated with awareness of the voice of conscience within me, as all Catholics have. He told me to "keep" it.

Presumably, he was keeping it in 1931 when he signed a petition to repeal Massachusetts' anti–birth control law. Of fifteen of Boston's leading physicians who signed it, Rock was the only Catholic.

John Rock was scarcely a radical. He had once argued against the admission of women to Harvard Medical School. His daughters recalled their father's advice that women did not have what it takes to become doctors. In the 1940s, in his fifties, he hardened his independent views on birth control. Although not offering contraceptives to his own patients, he took risks with the Massachusetts law by teaching his med-school students how to prescribe them. In 1943 he took a public stand in favor of permitting Massachusetts physicians to give advice on medical birth control. Some years later, after the restrictive Massachusetts law was repealed, Rock made it known that he was fitting some of his patients with diaphragms, which so enraged some Catholic physicians that they called for his excommunication. In 1949 he coauthored a book called *Voluntary Parenthood*, explaining birth control methods for the general reader.

Rock agreed to join Pincus in partnership, and they further agreed to continue to cloak their clinical trials under the cover story of Rock's studies in fertility. By that time, Rock had come to terms with prescribing hormones as a form of birth control, but echoes of his earlier training lingered:

At first it bothered me a little bit, but not very much. By that time I had done my work with human embryos. It was not difficult for me to think that those few little cells that would eventually, if properly cultivated, grow into an adult were not really human. They were much like all the other cells in the body, and so it seemed to me that until the conceptus acquired visible characteristics of a human, that it wasn't human. The spiritual thing didn't bother me very much because I was dealing with tangibles and I wasn't very spiritual anyway by that time. But it did bother me, and I talked it over with many church dignitaries, and fortunately I found some very able ones who more or less agreed with me, and they were very helpful.

In 1954 Rock began his first, cautious tests of the new progestin as a blocker of ovulation. Although its true purpose was not mentioned, the tests were historic as the first human trials of an oral contraceptive.

The drug they used was the one developed by Colton for Searle, which had been named Enovid. Their stated reason for not using norethisterone, Djerassi's product at Syntex, was that in animal tests—perhaps because of the absence of estrogen, the hormone of feminine characteristics—it appeared to have caused an enlargement of the testicles in male rats. Therefore they feared the compound might have mild masculinizing effects. Even the slightest deepening of the voices of women, or growth of facial hair, would obviously destroy the compound's success as a contraceptive of choice. (In later uses, the masculinizing tendency never proved significant.) Although not the stated reason, Pincus also surely preferred Enovid because he had enjoyed long employment by Searle, not Syntex.

The new progestin was tried by Rock on fifty women volunteers. Like the eighty to whom he had previously given the original Marker synthesis of progesterone, he hoped to "rest" the reproductive system and get a fertility "rebound." This time, however, he made intensive examinations to verify that the compound stopped ovulation.

McCormick chafed over waiting out the tedium of trying to master "that 'ol' devil' the female reproductive system!" She asked Sanger, "How can we get a 'cage' of ovulating females to experiment with?" and hovered over Rock's work all through the winter of 1955, "freezing in Boston for the pill."

Rock, as a practicing physician, instinctively emphasized the safety of the volunteers over the laboratory researcher's concern for the outcome of the experiment. As a result, Rock's caution reached beyond normal practices of the time. A young Yale-trained obstetrician assisting Rock in the trials, Dr. Luigi Mastroianni, has recalled: "I don't really think I sensed the true significance of what was being done. The concept of informed consent that is so talked about now, and is a legal requirement of any research project involving human volunteers, didn't exist then. But Rock practiced it before it was ever defined. There were always long and large discussions of the risk factors. It didn't matter that Rock had no formal guidelines; he set his own, and they were high standards."

The work became known within the Rock team as the PPP—the Pincus Progesterone Project. But that was soon translated irreverently as the "pee-pee-pee" project to honor Mastroianni's endless task of checking daily urine samples from each of the fifty subjects.

The results were perfect. Not one of the fifty women ovulated. While requiring more proof than a trial with only fifty women, Pincus and Rock knew that they had identified an oral birth control pill.

Especially with as cautious a partner as Rock, Pincus was not about to run around shouting the triumphant news. Not yet. But he had to pour it out somewhere, to someone.

His wife, Elizabeth, who had a talent for distilling complex episodes into succinct summations, would never forget the moment her husband brought the news home. Using his intimate name for her, he said, "Lizuska, I've got it."

"What have you got?"

"I think we have a contraceptive pill."

"My God, why didn't you *tell* me?"

He replied that he was telling her now.

"Did you think you could *ever* get the pill?" she asked in awe.

Pincus replied grandly, "In science, Lizuska, everything is possible."

• • •

Sanger and McCormick itched to get the word out, to the scientific and medical communities at least, that a birth control pill was almost born. Pincus, too, wanted to stamp his name early on the new development. The perfect opportunity was the coming fifth conference of the International Planned Parenthood League scheduled for Tokyo in October 1955. The women asked Pincus to go to Tokyo and to float the news that was not yet quite news at that newsworthy gathering. Pincus jumped at the invitation and asked Rock to join him. Rock refused, feeling that any claims for the progestins as a contraceptive were premature. Furthermore, to present details of research at such a nonscientific meeting before publishing them in a medical journal broke the academic rules and would endanger his good name. The mere idea of chemical birth control was widely regarded among his colleagues as immoral and medically unethical. In addition, he knew that a fierce controversy with authorities of his church surely lay ahead. Why harden their opposition now, even before field trials, before he had data to defend his position?

Pincus bore no such burdens. He and Chang and their wives, at McCormick's expense, sailed off to Tokyo.

After 150 conferees heard a humdrum round of reports that held out little hope of progress in contraception, Pincus rose to burst his bombshell. He recited details of the absolutely reliable antiovulatory effect of the progestins on lab animals, then made clear reference to similar effectiveness on women in preliminary studies. (Back home in Massachusetts, Rock was greatly annoyed at what he felt was premature statement of results. The strain between them brought them close to a break for the first time during their partnership.)

Pincus also inserted daring predictions that were sure to win the news coverage he sought:

Much more investigation is, of course, needed, but they [the progestins] thus far are the most promising agents. We cannot on the basis of our observations thus far designate the ideal anti-fertility agent, nor the ideal mode of administration. But a foundation has been laid for the useful exploration of the problem on an objective basis. The delicately balanced sequential processes involved in normal mammalian reproduction are clearly attackable. Our objective is to disrupt them in such a way that no physiological cost to the organism is involved.

Then, making sure his point would not be missed, he concluded, "That objective will undoubtedly be attained by careful scientific investigation."

The truly big surprise of the conference, to Pincus at least, was that his bombshell fell like a dud. The only noticeable reaction was skepticism. After all, some delegates derided, the only thing new that Pincus had said was that these progestins worked (and worked effectively) when taken by mouth.

The history-making point had passed them by.

Unlike Rock, Pincus not only found the rules of the scientific world stuffy but at least once used a science forum for taunting his colleagues. A German researcher, Horst Witzel of the Schering company, recalled in 1992 Pincus giving a paper on his field-testing in Puerto Rico. It was delivered at a Canadian conference on hormones in the mid-fifties. Witzel's recollection of Pincus's speech: " 'You may be interested to know that I published these data'—and he gave them a sly raising of his voice—'in, believe it or not, the *Ladies' Home Journal!*' That was the first publication of the chemical news of the Pill. I think there was reluctance to accept scientific papers on this subject. It was a taboo, to put it mildly, to interfere with fertility."

A few months later, Rock felt ready to report on his clinical trials. He chose an audience of scientists involved in hormone research. No significant points were likely to be lost on them. The meeting was a subsequent conference in Canada, the annual Laurentian Conference on endocrinology, first organized years earlier by Pincus. Rock loaded his fellow scientists with clinical data, displaying slides of the appearance of vaginal and ovarian tissue as well as results of urine-analysis studies. Heavily on the side of caution, Rock said, "We are led to suspect that ovulation has been inhibited in at least a very high proportion of cases."

While continuing to avoid discussion of the drug as a deliberate contraceptive, he now wanted the participants to go home
aware that a major boundary had been crossed. He didn't have to
wait that long. At the closing session, a University of Georgia
professor finally said it out loud: "It seems to me we have antiovu-
lation."

"I didn't say it," Rock commented later with a grin, "but I allowed them to know."

Almost instantly the burst secret was the talk of the research
world and the drug industry.

"At last," Sanger wrote exultantly to McCormick, "the reports
. . . are now out . . . and the conspiracy of silence is broken."

In late 1955 Rock's moment of decision had come. "Was he
willing," asks his biographer, Loretta McLaughlin, "to be midwife
to the birth of a new era in birth control, to attend the delivery of
a tiny pill capable not only of taming the whirlwind of population
growth, but of unfettering human sexuality? It was an immeasur-
ably profound step, an historic benchmark of unparalleled significance. And one from which there would be no retreat."

In truth, there could be no retreat for Rock because his course
was already chosen. At conferences in Puerto Rico, Brookline,
and Worcester, Rock, Pincus, and their colleagues checked and
rechecked their own data. They culled reports from tentative trials with the Syntex pill carried out by Dr. Edward Tyler in Los
Angeles as well as by Searle's clinical staff. Together with Pincus,
Rock scrutinized the data to review their earlier evaluation of
which of the two finalist progestins—Searle's or Syntex's—had
fewer side effects. Finally, Rock decided: He picked Searle's [and
Colton's] SC-4642, Enovid, as the "pill of choice."

The time had arrived for Rock to come out of hiding—to shed
the safety of his long-used cover story of the "fertility drug" and
openly throw in his lot with the Pill.

BOOK II

9

"All These Trials Soon Be Over"

*Although the Pill appears to be safe, we don't know what its long-term effect will be.
You can run an automobile on the proving ground day and night
to equal ten years of wear, but there is no way
to telescope the living human cycle. It will be
two decades before we can sell
the Pill over the counter.*

—DR. CHRISTOPHER TIETZE, EXPERT ON MATERNAL HEALTH, 1959

Driven by the ruination of her own mother and the hovering memories of the Sadie Sachses, perhaps Margaret Sanger, having envisioned a "perfect contraceptive" so fervently and for so long, imagined that upon its simple announcement women the world over would seize it and be free.

But Gregory Pincus and certainly John Rock knew that first a long and intricate list of questions had to be posed and answered, and then the answers had to be tested, and only then might it be discovered that the critical questions had not yet been thought up.

In the Boston trials no women had taken the new progestin for longer than a few months. But if it was intended as a year-in, year-out medication for avoiding pregnancy, didn't tests now have to demonstrate that it was safe and effective for long periods? And didn't that mean tests had to be conducted where families tended to stay put for long periods of time so they could be tracked? And if quelling the outbursts of overpopulation was the Pill's ostensible—its stated—goal, didn't the tests have to be conducted where the birthrate was high (so that the new medicine could be shown to lower it)? Didn't they have to be where women had a clear interest in doing something—even something unusual—to help them space their babies? And didn't the locale of the tests have to permit close, almost day-to-day monitoring by medical doctors for accuracy of results and for unwanted side effects?

True, a couple of hundred of John Rock's middle-class patients in Boston, attuned to the latest in health care, had taken a pill once a day for twenty consecutive days, then stopped for, say, five or eight, then started the cycle again. But did that prove anything about penniless women who perhaps couldn't read, who might not understand what the day-counting was about, women in whose lives the disciplines of precision had never played a part? And weren't such deprived women, after all, the overpopulators, the targets of the fashionable new concern of the birth control movement? And when those women chose to have children after months or years on the medication, would those children be born healthy? Would the distribution of maleness and femaleness among those new babies be normal? What about *their* children?

Where could Pincus and Rock find a testing ground to suit that bewildering mix of questions? How could prolonged, elaborate testing be protected from the harassment of Victorian and religious snipers and subverters? One thing became clear to them: It would be best to conduct the field tests outside the United States.

These questions had occupied Pincus and Rock even before the Tokyo conference of October 1955, where Pincus let out the first hint of the new contraceptive. On March 5, 1954, Pincus had written to McCormick:

Mrs. Pincus and I recently returned from a trip to Puerto Rico where I was invited to give lectures to the Medical School and

Medical Association. I took the opportunity to confer with the Public Health physicians . . . who are concerned with the 67 birth control clinics . . . of the Public Health Service. I found these physicians to be quite efficient and knowledgeable, and . . . came to the conclusion that work could be done in Puerto Rico on a relatively large scale. These clinics have a sufficiently large patient clientele so that we could easily select 100 to 300 women who could cooperate with some intelligence in any studies that we would want to initiate. I have discussed the Puerto Rican situation with Dr. Rock, and he agrees that . . . we should attempt in Puerto Rico certain experiments which would be very difficult in this country.

In another letter to Mrs. McCormick on August 26, Pincus wrote, "It was decided best not to label the project as a study of contraception, but to call it the study of the physiology of progesterone in women."

A Puerto Rico Medical School faculty member named Edris Rice-Wray, who was also medical director of the Puerto Rico Family Planning Association and director of the Public Health Department's field-training center for nurses, had told Pincus she knew "just the place": a new housing project in a suburb of San Juan, called Rio Piedras, full of families loaded down with more babies than they wished they had. Those families had just been moved into the new project from a notorious slum called El Fangito, "the little mud hole." The superintendent of the new housing project was a "friend of family planning"—an essential qualification, but rare for an official in Catholic Puerto Rico. Although opposition to birth control was always vigorous in Puerto Rico, in 1937 the legislature had amended the laws so that distribution of information for the prevention of conception, previously a felony, was no longer a crime. To make the site even more ideal, Dr. Rice-Wray had on tap a former housing-authority social worker who knew the local people.

More than twenty years after Pincus's letter, Rice-Wray, a red-haired energy machine, recalled the world of Puerto Rican family planning in 1956:

"When I first came, the clinics weren't functioning very well, frankly, and they didn't call them 'family planning' clinics. That word was coined later. They called them Clinicus Pre Maternalis—'clinic before maternity'—because they didn't want to say what it was. So the nurse and I asked the women, 'How many children do you have?' and 'How many more do you plan to have?' and then they began talking: 'Oh, I have so many, and it's such a problem' and all that. And then we would say, 'Well, would you like to know about methods? We have a clinic for this.' 'Yes, yes, yes.' So I always had a little box behind me with the various methods in it, and I would show them. 'This is a condom, and this is a suppository, and this is a diaphragm, and I'll explain to you how you put it in. You have to be fitted.' And I'd explain how it goes in and all that. 'Now, how many of you would like to stay today and have your diaphragm fitted?' Well, some were menstruating and some were pregnant and some of them hadn't taken a bath that day or some silly thing. They lived in these little shacks, but even so, the Puerto Ricans are very clean people, and at the end of every day you just have to take a bath because you feel all perspiry. So they had some little place where they could put water in a can and sprinkle on their head, and while they were in this little place they could put the diaphragm in, you see. So they used them very well. Sooner or later we got them all.

"Well, I was just down here minding my own business, doing my work, when one day Gregory Pincus appeared. I don't think he knew me. At least I didn't know him. He had gone to the Family Planning Association to find out if there was a doctor associated with them, and I was the only doctor. So he asked me if I would do a study with this pill that he and Dr. Rock had developed. Well, I'd never heard of the Pill before, and I thought about it, and then I asked a lot of questions. He assured me that it was safe and that it was effective. I got a social worker, and she selected women who were mostly under forty with children."

Also Pincus enlisted Dr. Celso-Ramon Garcia, a Manhattan-born Spanish-American who had gone to Puerto Rico to set up a department of obstetrics and gynecology at the medical school.

One early visit by Dr. Rice-Wray's social worker was to the home of Maria Pinzon (not her real name), a cheery, buxom woman in her late thirties, who had delivered fourteen children. One had died, and Mrs. Pinzon, her husband, and their surviving thirteen filled a two-room house perched on a high ridge in the Puerto Rican barrio of Buena Vista with a sweeping view of the surrounding hills and the blue Caribbean. There they scratched a living out of skimpy patches of corn and tobacco and a bit of fruit.

"The social worker would say," Dr. Rice-Wray recalled, " 'I'm from the Family Planning Association'—like the Cancer Society or the infantile-paralysis campaign, so they could relate it to something they considered decent, you know. Then she said, 'We're very much interested in parents having the right to have the number of children they want when they want to have them. We have a pill that a woman can take twenty days a month and she doesn't get pregnant.' Well, they just couldn't get hold of it fast enough."

Not all women were perfect subjects: "There was one who had a husband who traveled, and she only took it when he got home. So, of course, she got pregnant. A few, their marriages broke up and things of that sort and they got pregnant. But I want to emphasize that nobody who took the Pill according to the instructions got pregnant."

Case notes told of one woman of thirty-two who had five children, three of them by a previous husband. She wanted no new ones because her second husband had just been admitted to a mental hospital for the third time. Another, though only thirty, had ten children, aged from sixteen years to ten months. "The husband," says the case note, "is very ill, drinks heavily and makes life difficult for the wife." She had to do odd jobs to support the family. Although she and her husband had daily intercourse he would not allow her to use any form of contraception. Only after a great deal of persuasion would he agree to her taking the new tablets.

Although a dutiful Catholic, Maria Pinzon did not shy away from the experiment on religious grounds, nor did many of her eligible neighbors.

"Among all Catholics," explained Dr. Garcia, himself a

Catholic, "there's a range of devotion from the fanatic to the devout to the marginal. But the vast majority of Catholics in Puerto
Rico were just plain practical in their attitudes. They felt they had
to choose between two evils. Maybe that wasn't the way they expressed it, but in their own way they were saying they had to
choose between the evil of birth control and the evil of not being
able to feed that extra mouth. The difficulty with the Pill was
they had to rationalize how they could live with a daily reminder
that they were continuously preventing pregnancy, and therefore, in the Church's eyes, were in a chronic state of sinfulness.
For a while, when the women went to confession, if the priest no-
ticed they hadn't been pregnant for a while, he'd openly ask
them, 'Are you on the Pill?' They came around by themselves to
the idea that they were refraining from ovulation"—rather than
committing the sin of sexual intercourse while deliberately preventing conception. "Despite admonitions by Catholic social
workers, warnings from the pulpit, and outright condemnation
over TV and in the Catholic press, the women just decided that
the worst thing they could do was bring another baby into the
world that they couldn't afford. To develop family problems was
the greatest evil. They feared they might come to hate the child.

"Then, the Catholic Social Workers Guild put on a TV program and made a lot of statements about how dangerous the Pill
was. About ten percent of the volunteers dropped out.

"All the dropouts, all, soon became pregnant. After that we
never had a problem enlisting volunteers. They came flocking,
not just the indigent population, but schoolteachers, the social
workers themselves, everybody."

Most of the women arrived at the clinic dressed in their best,
many surprised when told to undress for a more intimate examination than they had ever had.

"Some of these women looked like great-grandmothers, all
wizened up and gaunt," recalled Anne Merrill, Pincus's laboratory
associate from Worcester, who flew regularly to San Juan. "I'd
look at their records, and the woman would be thirty-four, you
know, with ten children.

"After a while, some women for one reason or another decided
they wanted another baby. So they just stopped the pills. I can re-

member the terror that hit us: Are we going to get all girls, or all boys? What's going to happen? What a relief to find out that we got a fifty-fifty proportion. Then we started examining the babies that were arriving, and they were fine."

Soon after the start, the unsympathetic Puerto Rico Secretary of Health threatened to throw the project off track when he accused Rice-Wray of implying the Health Department's endorsement of the Pill. Rice-Wray replied that her work for Pincus was on her own time and separate from her public-health duties. He responded that her work could not be divided into parts. In December 1956 she decided to take a job with the World Health Organization, leaving Puerto Rico for Mexico. That, of course, threatened the progress of the field trial. Before leaving, however, Rice-Wray provided Pincus with her data on 221 subject women.

"Enovid," she reported, "gives one hundred percent protection against pregnancy in 10-milligram doses taken for twenty days of each month. . . . However, it causes too many side reactions to be acceptable generally."

Pincus was shaken. *Too many side reactions to be acceptable!*

Rice-Wray went on to specify: Thirty-eight women out of 221, or 17 percent, had unpleasant reactions. Of them, twenty-nine felt dizziness, twenty-six "felt sick," eighteen reported headaches, and seventeen actually vomited. Nine reported stomach pains, seven "felt weak," and one had diarrhea. Many reported more than one reaction. Rice-Wray cited others who dropped out and why they did so: Two decided to become sterilized. The husband of one hung himself, which she presumed was from desperation over poverty. Several patients found they were pregnant before starting the tablet. Nine dropped out on the insistence of their husbands, and six because doctors warned against the trials. In three instances, a woman's priest ordered her to stay out of the sinful business of birth control.

Yet those instances of nausea and other discomforts did not discourage other women. The secretary of the Family Planning Association wrote Pincus: "We have more cases than what we can take for our study. Continuously they are ringing this office asking for the pill."

One consolation was that the "failures" provided valuable information. Most dropouts who had taken the Pill for several months quickly became pregnant when they stopped, thus demonstrating that it did not adversely affect a return to normal function of the reproductive system.

The prevalence of side reactions did not square with Rock's findings in Boston. A bold idea struck Pincus. Perhaps these women were having "morning sickness" from their systems' being fooled by the Pill into "thinking" they were pregnant. For one sample of the women he added to the daily regimen a simple antacid. Indeed the antacid "cured" 80 percent of the complaints. Pincus wondered further, however, whether some of the unpleasant reactions might be caused by his fastidious warnings about possible side effects. With Dr. Manuel E. Paniagua, Rice-Wray's replacement at the Family Planning Association, he worked out a new study. Its subjects were volunteers who had been using conventional contraceptives. They were told they would be given Enovid to see if they were "suited" for it, but were to *carry on using their accustomed devices* "just in case" Enovid didn't work for them. The subjects were read the usual warnings of possible side effects. The pill they received, however, was not Enovid but a placebo. Seventeen percent of the women—exactly the same proportion as in the true Enovid sample—reported unpleasant aftermaths, confirming Pincus's suspicion that they were probably the result of suggestion, not of the drug itself. That conclusion was further corroborated when another group was given genuine Enovid tablets to accompany their conventional contraceptive. This time no warning of side effects was read to them. The actual incidence of side effects dropped to a little over 6 percent.

In still another test group, half the side-effect complainers got antacids and half placebos. For every ten antacid takers who found relief, seven placebo takers also felt better. Perhaps most important, after a few months of taking the Pill, all complaints declined.

The dosage of these first field-trial pills was ten milligrams of progestin, about 1/2800 of an ounce. That was a "best guess" quantity, chosen because Rock and Pincus knew it was enough to assure freedom from pregnancy. In their watchfulness, however, they discovered an error—an inexcusable bungle—that could have destroyed the entire experiment. They accidentally found that one batch of the Enovid had been contaminated at the Searle factory by a minuscule amount of synthetic estrogen. Would that mistake give them an impure reading of the consequences of long-term use of progestin? Could it throw the whole result of the field trial into question?

Pincus and Rock swallowed hard and decided to push on—with a new shipment of properly pure progestin. Their next readings surprised them even more. On the pure progestin, women reported a higher incidence of intermittent breakthrough bleeding during their menstrual cycles rather than at the end. That puzzling finding led the researchers to check back into previous batches of Enovid that had worked so effectively. Those original batches too had contained unwanted traces of estrogen. So the happy accident saved months, perhaps years, of trial and error. It had accounted, in fact, for the choice of Searle's Enovid over the Syntex version of progestin, which had not performed quite as well. Thereupon Searle redesigned the ten-milligram Enovid tablet to include its production error, permanently adding a minuscule 1.5 percent of estrogen.

(A more detailed account of side effects of the Pill and how they almost sidetracked its further development and acceptability by the women of the world is contained in a later chapter on side effects, "Pill Kills!")

Edris Rice-Wray's new job in Mexico City did not end her work for Pincus but rather expanded it when she helped Pincus lay groundwork for extending the field trials to Haiti, an island country of a quite different culture. In March 1957, Pincus explained some of the reasons in a letter to McCormick:

The idea of birth control is completely foreign, and probably unknown to about 90 percent of the inhabitants of Haiti. Furthermore, among the peasants who constitute this 90 percent, children are regarded as capital goods. There is no compulsory schooling, and the children are put to work at the age of five.

Furthermore, because of a degree of polygamy, women are pretty much in the same category, and the ignorant peasants for whom these women and children work would certainly not want to see any reduction in the number of offspring. In and about the cities, there appears to be a small lower middle class which has some appreciation of the idea of planned parenthood. . . .

The situation in Haiti is complicated by the political climate at present. They have, as you know, no president; they will be holding elections in April. Among the candidates for the presidency, the one who seems superficially to have the best chance is a man who is an ardent Catholic and who would certainly oppose any official sponsorship of the work. For this reason, we attempted to have the project considered as a straightforward scientific research project which should be independent of political considerations. . . . Dr. Pierre-Noel, who is the man who invited me to Haiti and who is now the acting director of the Public Health Service, is responsible for this suggestion. He is a very earnest and sincere man, and unless he becomes a political football himself, will probably see to it that the committee is maintained and encouraged.

In Puerto Rico. . . . Again, we were unable to find any complications that could be attributed to the medication, and for the most part, everything has gone along uneventfully with these women.

On more immediate details that burdened their progress, Rice-Wray soon wrote Pincus:

Dear Gregory,

I have just returned from a three week stay in Haiti. . . . I feel that you should get started as soon as possible. . . . Fifty patients have been interviewed and are anxiously waiting for the pill. . . . All the docters [sic] feel that if we don't get started we shall lose the interest which has been built up. If we get started, we can have fifty more within a couple months. The residents will get them from the Maternity ward. I talked to the residents and they say they have yet to find a patient who is opposed to using a contraceptive. They have 25 deliveries a day.

Dr. [Felix] Laraque [the Haiti project's new medical director] and I made a tentative budget which came to approximately $2000.00 a year. He will write you and give a breakdown. You will notice that salaries are very low in Haiti. By the way this worker is not a social worker at all. Just a docters secretary who has worked in a docters office and learned to be a docters helper. I was wondering if you would like to consider giving each ten dollars more a month than that in the budget. This would make them feel that they had a good job, which they would not like to drop for another. . . . Also would you like to consider a small stipend for Dr. Laraque such as $300.00 per year. . . . Medical practice is bad right now. Noone pays. He will do it anyway and is anxious to start. I got a small stipend in P.R. . . . and I was glad to get it. . . .

Let me know what you think and please don't broadcast it that I am taking any responsibility for this project. Officially, I can't be involved in Birth control activities until the member nations [of the World Health Organization] give us the green light. (there are a few who threaten to withdraw if we include B.C. activities). Also I am there to work on my job and not do something else. . . . Excuse this lousy typing but I can't do this in the office.

Almost immediately, on March 11, 1957, Pincus wrote to Dr. Laraque stating in meticulous detail how the Puerto Rican trials were conducted as a guide to what he expected in Haiti. Pincus's instructions are one indication of why some have called the Pill "the most carefully researched drug in history":

I should like to stress again the procedure used with the patients. Each patient is given two vials, containing twenty tablets each, of the medication. . . . The patient is instructed to take one tablet a day, beginning on the fifth day of the menstrual cycle, until twenty tablets are exhausted. A withdrawal bleeding will follow the taking of the last tablet, within one to three days ordinarily. If any interval bleeding occurs while the patient is on medication, she is instructed to take two tablets on the day on which she notices the bleeding and to continue this for three or four days, when she may reduce the dose to one tablet. In order to complete the taking of medication to the twentieth day, this will involve the removal of three or four tablets from the second vial. These tablets are replaced by the nurse when she visits the patient. In practically every instance, the doubling of the dose promptly arrests the bleeding, but occasionally a true escape occurs in spite of this. The patient is instructed to continue the medication despite this escape, and day one of the succeeding cycle is considered to be the first day of bleeding following cessation of medication. . . .

As was evident in our discussion in Port-au-Prince, it is essential that the nurse or social worker visit with each subject at least once a month in order: (1) to supply an additional vial of the tablets, (2) to record the menstrual history, (3) to record the degree of regularity or irregularity of tablet taking and particularly to note the number of days on which each patient has failed to take the tablets, (4) to refer to the physician in charge any case exhibiting an unusual history, and (5) to record any side reactions that may have occurred, effects on menstrual flow, and so on. We feel that each patient should have a thorough pelvic examination before being admitted to the study. This will enable the physician in charge to exclude from the study those having obvious abnormalities.

It will be advisable to have as a control group for this study a group of women not taking the medication, selected on the same basis as those admitted to the study. These women should be visited also at intervals to obtain health records and particularly to obtain records of fertility. In our control group in Puerto Rico, a visit is made at least once every six months.

By this time in Puerto Rico and Haiti combined, the Pincus team had closely followed 25,421 monthly cycles. The failure rate among them was 1.7 per 100 woman-years. Meanwhile in Mexico City, Dr. Rice-Wray had launched a new field trial and reported a failure rate of only 0.6. To compare, a 1961 study in Puerto Rico of unwanted pregnancies while using other methods showed a failure rate for the diaphragm of 33.6 per 100 woman-years, 42.3 for vaginal suppositories, 36.1 for vaginal creams and jellies, and 28.3 for the condom. In locales of higher literacy, those older methods usually yielded better results, but the figures for Puerto Rico tended to highlight the superior effectiveness of the Pill. Even more supportive of the Pill, all the statistics showed that if women followed instructions without letup, the pregnancy rate was zero.

Before long, the test program would be extended from the original one hundred Puerto Rican women to a full field trial, involving more than twenty thousand in three countries. As these initial clinical trials expanded, Pincus and Rice-Wray tested other companies' new progestins in addition to Enovid.

One scientist-businessman who kept a close watch on the trials was James Balog, a research chemist at Merck and Company. "The length of time for each trial," Balog recalls, "was usually six months. When the six months were up and Edris turned in the cycle information to the drug companies, they'd say that's enough. So Edris would wonder, 'What do I do with my ladies now? Do I tell them go out and get pregnant? Now I have to find some money to buy pills so I can keep them contraceived.' That was her real interest to begin with, not the testing as much as helping these women. I'd say, 'Why don't those drug people keep giving you those pills free? They cost them pennies. You're giving them extremely valuable stuff.' I talked to the drug companies myself and said, 'You guys are crazy to let these women on these trials get away. Right now you need six months. Someday you're going to need a year, someday you're going to need the second generation. You're going to need to know what happened to the health of the *daughters* of the women who were on the Pill and what happened to *their* babies.' I told this to every drug company I visited that had an oral contraceptive. I would tell them they were crazy to let these people back into the jungles without keeping them in the continuous stream of data. Edris was the only one who had *offspring* of women who were on the Pill. They had been married at twelve, had a baby at twelve-and-a-half or thirteen, and *they* could have babies. So she already had the second generation—she had *daughters.* And these drug people were dropping her, and I told them they were nuts."

But Balog's company, Merck, was not a supplier of pills. What drove his concern?

"Well, I had a business motive to be informed of all that was going on. But basically, I felt that the Pill was pretty important stuff. I'm not a Roman Catholic, I'm Russian Orthodox, but the philosophies on birth control are not very different. I felt that the Pill might be the theological way out. It's not abortive because there's no ovum, you're not interfering with the will of God by putting a spermicidal jelly on that poor little sperm who's trying to find this wonderful little ovum and do his thing. There's no fertilized ovum. If you believe the soul is created when the sperm unites with the ovum, *there is no fertilized ovum*, therefore *no soul*. This reasoning is my own, and so I had this notion that this might become a good theological way out."

As the atomic bomb had done a few years before it, the Pill stirred a number of its scientist-creators into ruminations beyond science about the meanings to society of what they had wrought.

One of these was Étienne-Émile Baulieu, the young Frenchman who would one day become famous for promoting the abortion pill, RU 486, whom Pincus had invited to give a talk at the Worcester Foundation: "Pincus suggested that I stop off in Puerto Rico on my way home. . . . I was reluctant. It troubled me slightly that Hispanic women from a poor outlying territory were chosen for clinical trials (although the first human experiments had been done in Boston). . . . Pincus urged me simply to take a look, and I thought I detected a twinkle in his eye. . . . No sooner on the ground in San Juan, I saw the answer. One look at the clinical data suggested how the Pill could change women's existence. From that Puerto Rican experience, I realized the impact on human life that the product of birth control research could have. I was hooked, although I didn't know it yet."

These field trials, for all their investigative thoroughness and their immense breadth in numbers—twenty thousand women in three countries—did not put the Pill in the clear. It was yet to suffer a major institutional attack by the Catholic church—to the surprise of no one. But a far more surprising challenge was to come from champions of women's health, in the name of defending the very birth-burdened women whom Sanger, McCormick, Pincus, and Rock thought they were rescuing.

Those new kinds of trials—political trials—lay far ahead in a raucous decade that the emerging life of the Pill was yet to endure.

10

A RACE FOR THE PILL

We recognized right from the start this was going to be as controversial as hell,
and we moved very cautiously. If anyone had told us
the Pill was going to be discussed at bridge parties
and across dinner tables, well, frankly,
we would have disbelieved it.
—A SPOKESMAN FOR G. D. SEARLE COMPANY, 1969

Pincus's longtime sponsor, G. D. Searle Company, had spurned support of his work on a birth control pill and later wanted no part of using its drug to interfere with menstruation. Now, soon after the start of the Haiti trials, Searle appeared to be on the verge of an earthquake of a decision. Its vice president for research and development was participating in the first meeting between Pincus and Dr. Laraque, his medical director in Haiti. The subject of the meeting was the full field test for Enovid.

Could other drug companies afford to keep pretending that oral contraception didn't exist? Was it time to reexamine their skittishness? Or were the hazards of leaping in still too terrifying?

An awkward Alphonse-and-Gaston ballet among huge companies in the ensuing months ranks among historic pinnacles for industrial indecision, confusion, misjudgment, misreading of the marketplace, and reversals of commitment.

Following the apparent success of the Enovid trials, Parke-Davis, which had supported Russell Marker, then had turned its back on him, was now urged by Warren Nelson of the Population Council to take a second look. The company conferred with Pincus, then with its own soul, and settled for dropping in the poorbox a good-will contribution of five thousand dollars for the Worcester Foundation, reiterating its policy of staying away from contraception.

The Upjohn Company inquired about the possibility of including one of its own progestins in the Puerto Rican trials, but after talking with Pincus, they, too, backed away.

The Ortho Pharmaceutical Corporation, the industry's most successful makers of diaphragms and vaginal jellies, inquired gingerly into prospects for the Pill. Many expected Ortho would jump in, especially since its research director, Carl Hartman, was on the medical advisory committee of the Planned Parenthood Federation of America. Hartman still doubted, however, that the marketplace would accept an oral contraceptive. But if Searle entered the market and made a success of it, he avowed, Ortho would take another look.

Charles Pfizer and Company paid Syntex for an option to market norethindrone. Then Pfizer's president, an active Roman Catholic, decided not to exercise the option. He wanted nothing to do with any drug that even potentially could be used for birth control.

Meanwhile the new synthetic hormones plunged inexorably forward down other paths. As early as 1951, Syntex had received a telephone inquiry from the Upjohn Company. Could Syntex deliver ten tons of progesterone in the following twelve months? The world had never seen a hormone batch close to such size. Upjohn wanted its mammoth amount not for use as a sex hormone but as a step toward conversion to cortisone. Despite its original control of the precious new drug, Syntex had been going broke. Somlo, its owner, had no money for expansion. He had no sales force, and if he had had one, he had no final product to sell to consumers. He had only a chemical ingredient saleable to other, far more powerful drug companies. That invited the drug giants to dictate strangling terms, just as Upjohn was now doing. Before the formation of Syntex, world output of progesterone had been only a few pounds a year, and in 1951 less than one ton. Pre-Marker progesterone had been priced at eighty dollars a gram. Late in 1945, Syntex sold it at eighteen dollars a gram. The price dropped within a few years to three dollars and continued to fall. Syntex delivered its ten tons to Upjohn at forty-eight cents a gram. Said a Syntex executive ruefully, "Through the mid-fifties we were busy fools selling vast quantities of raw materials for other people to make a profit out of." At one point Searle looked into buying Syntex to remove a competitor, then decided not to bother.

In 1956, a New York investor, Charles Allen, who had parlayed millions he had made in the scrap-materials business into an immensely successful holding company, ignored the contrary advice of his lawyer and most of his family and bought Syntex for a price estimated at two and a half million dollars. For that price, the buyer acquired Syntex's patents and the talents of Carl Djerassi and George Rosenkranz. Rosenkranz soon became its chief executive.

Even before its norethindrone was approved for sale as a gynecological medicine in the United States, Syntex had begun producing it commercially in 1956, licensing it to Schering A. G. in Germany to sell in most countries of the world as Primolut-N. Also, E. R. Squibb and Sons used it in a pregnancy test it marketed as Gestest.

And eventually, of all possible petitioners, along came Parke-Davis asking Syntex for an exclusive license to market norethindrone. It was potentially Syntex's biggest deal of all. Parke-Davis intended to apply to the U.S. Food and Drug Administration for approval of the drug for treatment of threatened or habitual miscarriage, dysfunctional uterine bleeding, painful menstruation, and infertility. These conditions formerly had been treated by more radical and expensive therapies, including hysterectomy. Parke-Davis, one of the most conservative of all drug companies, appeared quite unembarrassed that it had turned its back on Marker, forcing him into the founding of Syntex, then had refused to consider chemical manufacture in Mexico. But the new Syntex, under its New York financier-owners, demanded and got a fixed percentage of what would be Parke-Davis's retail sales price. The deal truly locked the two former antagonists into an unaffectionate partnership by giving Syntex a major financial stake in Parke-Davis's commercial success with norethindrone.

About the same time that Parke-Davis applied for Food and Drug Administration (FDA) approval, G. D. Searle asked the same for its progestin, norethynodrel.

The year 1957 produced a great leap forward toward the Pill, although few realized it at the time. That was the year the FDA approved the Syntex-Parke-Davis drug, to be marketed as Norlutin, as well as Searle's Enovid. The approval was limited, of course, to treatment of disorders.

Katharine McCormick, who recently had complained that FDA approval was "as slow as molasses," now cannily grasped the meaning of the limited licensure. "Of course this use of the oral contraceptive for menstrual disorders," she wrote, "is leading inevitably to its use against pregnancy—and to me, this stepping stone of gradual approach to the pregnancy problem via the menstrual one is a very happy and fortunate course of procedure."

Even though the word was out that the new drugs would work as birth control pills, few watchers of the drug industry expected that the FDA would soon—if ever—approve them for birth control, nor even that consumers wanted them to. James Balog, the science-trained financial analyst, recalls: "The issue among drug people was whether any woman would take a pill every day for twenty-one days to prevent the chance she might get pregnant. They believed nobody's going to do that, not when they're not sick—and they're not sick! This was a prevention drug—prevention as a social activity as opposed to prevention of cancer or something. Secondly, it was expensive. It would cost about ten dollars a month, at least at that time, and that was a lot then.

Thirdly, you had the religious problem. The chairman of Parke-Davis was a Roman Catholic, and so was the chief executive of Merck. Neither one would go for it. Finally, there was a lot of side-effect talk. Disrupting the female cycle for twenty-one days, who knows what will happen? Then other legal problems. As late as 1964 or thereabouts, when I did my first report to Merck, our lawyers advised us that we couldn't mail this report into Massachusetts or Connecticut. Disseminating information on birth control was punishable."

Indeed in the late fifties, after the FDA approved the compound as a gynecological medicine, fully seventeen states had laws limiting the sale, distribution, or advertising of contraceptives. Connecticut outdid all the others by making it a crime "to *use* any drug, medicinal article, or instrument for the purpose of preventing conception." Ironically, in Massachusetts, the state where the contraceptive pill was born, anti–birth control laws were the most restrictive, and that state would remain the last to repeal them, in March of 1972. While it was not illegal to *use* contraceptives, Massachusetts law made it a felony to "exhibit, sell, prescribe, provide, or give out information about them."

For years newspapers had splattered evidences of the trouble awaiting any drug company that would dare market a birth control pill. Why would a pharmaceutical company of rapidly growing respect, like G. D. Searle, for example, voluntarily expose itself to incidents such as these?:

• As far back as February 1952, in Poughkeepsie, New York, a city of forty thousand, St. Francis Hospital, a two-hundred-bed institution run by Roman Catholic Franciscan nuns, found itself on the front page of *The New York Times* when it demanded that seven of its staff physicians sever their connections with the local Planned Parenthood Association or resign. Four of the physicians, all non-Catholics, refused to do so. A fifth quit Planned Parenthood, saying he would stay on staff until his four hospital patients were discharged, then would leave the hospital. The controversy pitted Protestant clergy against Catholic clergy. Monsignor Michael P. O'Shea, the dean of the area's Roman

Catholics, tried to encourage calm: "This is not a fight between Catholic and Protestant," he reassured the community, then reawakened the opposition, adding, "This is a point on which Catholic, Protestant and Jew should all be agreed, since it is the Bible which expressly forbids birth control."

A committee of Protestant and Jewish clergy retorted that the hospital's "attempt to police the thoughts and personal actions of individuals . . . is un-American." The Planned Parenthood physicians, in the end, were not forced out of the hospital, and their privileges were quietly renewed a year later.

• In 1953, animosity burst upon the front pages again when the Planned Parenthood Federation's Mothers Health Centers applied for membership on the Welfare and Health Council of New York City, a powerful federation of 391 private and public social-service agencies. The application was opposed by fifty-three agencies comprising Catholic Charities for the Archdiocese of New York, which said they would walk out rather than work with Planned Parenthood people. Fearful that the important Health Council would be wrecked, its board voted fifteen to five against admitting the family planners. Interfaith cooperation stalled for six months as a bitter struggle raged across headlines: "Parenthood Dispute Reported Spreading," "Catholics Boycott Welfare Conference," "Welfare Council Faces Vote Fight," "Roman Catholic Church Accused of Obstructing Parenthood Group," "Planned Parenthood Backers Win in Welfare Council Vote; Catholic Agencies Are Expected to Withdraw Membership," and a climax on *The New York Times*'s first page: CATHOLICS QUIT WELFARE UNIT IN PARENTHOOD CONTROVERSY. At last, fifteen months after the battle broke out, the Catholic groups relented: "Welfare Council Hints End of Rift."

• As late as 1958, a year after FDA approval of Enovid, a conflict erupted when a diabetic Protestant mother of three appeared at Kings County Hospital in Brooklyn, presenting her doctor's prescription for a diaphragm. Another pregnancy, he had told her, might cause her death. A hospital staff physician, Dr. Lewis Hellman, refused to fit the woman, saying his orders had come down from Dr. Morris A. Jacobs, New York City's commissioner of hospitals. Dr. Jacobs was Jewish and professed no personal objection to birth control, but he said the city hospitals had "had a policy" against providing contraceptives. Mayor Robert F. Wagner, a Catholic, turned the matter over to his legal aides for study and left for a vacation.

The New York archdiocese of the Roman Catholic Church, backing the hospital, quoted Pope Pius XI on the church's position as first designed by Saint Augustine in the year 400 A.D.: "Every attempt on the part of the married couple during the conjugal act or during the development of its natural consequences to deprive it of its inherent power and to hinder the procreation of a new life is immoral. No indication or need can change an action that is intrinsically immoral into an action that is moral and lawful." The archdiocesan office added, lest the point be lost, that the pope's statement "holds good today just as much as it did yesterday. It will hold tomorrow and always, for it is not a man-made precept, but the expression of the natural and divine law."

Soon the controversy broke into a war of accusations among the Protestant Council of the City of New York, the United Lutheran Church in America, the New York Congregational Church Association, the Presbytery of New York, the New York Board of Rabbis, the American Jewish Congress, the National Council of Catholic Women and Men, the Catholic Physicians' Guild, and the American Civil Liberties Union. The vehemence finally settled down to this definition of a central issue, which continued, in New York and elsewhere, to defy resolution: Should Catholics have the power to prevent non-Catholic doctors from prescribing contraceptives for non-Catholic patients?

These conflagrations did not appear to trouble the new Wall Street owners of Syntex. Marvyn Carton, a lawyer and member of its pragmatic board of directors, recalls: "We had paid for a research study. What the researchers did was count how many condoms were sold. They analyzed and extrapolated the numbers in different ways, and they came up with the conclusion that, yes, there was a fifty-million-dollar market for this oral drug worldwide. What they didn't fully realize was that while this product functionally did the same thing as the physical and mechanical contraceptives that existed at that time, *esthetically* it was *a different world*. Even though fifty million seemed attractive, when you have something that changes the approach functionally and esthetically, then all bets are off. We didn't underestimate its potential. We didn't know *what* to estimate."

Wasn't Charles Allen, the new principal owner of Syntex, inhibited by the religious question? "I don't think he even knew what his religion was," Carton replied.

As speculation about the new drug's birth control possibilities persisted, in April 1958, soon after the FDA approval, the magazine *Fortune* forecast for its business readers:

> There is a vast difference between dispensing the drug as a safe means of inducing temporary sterility for therapeutic purposes, and dispensing it for "habitual use" as a standard contraceptive. It will take perhaps five years of research to satisfy any drug firm that it is ready to apply to the Food and Drug Administration for permission to so label its product; and it would probably take five years after that before the FDA—which says it may well require clinical data on "thousands of women," not 500—would approve such applications.

Gregory Pincus did not doubt, however, that widespread oral contraception was at hand—and soon. As early as January 1957, Pincus, whose good name at Searle had been restored by the Pill trials, pleaded with the company's president that a cheap and reliable supply of plant steroids was essential if Searle was to think seriously of marketing a birth control pill. Exploring possible Mexican sources of sapogenins had been the reason for Pincus's 1951 visit to Russell Marker in Mexico. Searle bought two small Mexican companies, Root Chemicals and Productos Esteroides, both started by former Syntex employees.

At Searle's suburban Chicago headquarters, decision makers found themselves gripped by the traditional competing pulls of temptation and fear. By late 1959, only two years after its approval, fully half a million women in America alone were swallowing Enovid daily for their menstrual disorders. Were those disorders all real? Or were many convenient and imagined, with the collusion of doctors? Many of these users (and surely their doctors) were fully aware that the pill was a contraceptive. So women *would* voluntarily take a chemical that would block conception. How many were taking it *for that reason?* How many more would take it if it were approved and openly marketed *for that reason?*

If the cash registers ringing up a half million Enovid pills a day were already promising to swell G. D. Searle from a small, ambitious company into a large, rich one, what if those pills sold to untold *multi*millions every day—year in, year out, perhaps for generations? Couldn't that marketing trick be pulled off quietly, discreetly? After all, wasn't Goodyear Rubber fattening itself by $150 million a year marketing condoms, while concealing its fine, distinguished name from the package of the shameful product? Didn't a vigorous market in diaphragms hum along almost silently, with scarcely a whisper passed that would embarrass huge, upright Johnson & Johnson, whose Ortho subsidiary manufactured them; nor did they bring a blush to the doctors who quietly prescribed them, nor to the druggists who passed them out under their palms? Did the middle- and upper-class women who purchased them surrender a trace of their accustomed dignity?

But on the other hand, the company that dared market this wholly new kind of thing, this Pill, might find itself swiftly ruined.

Searle decided to go out on what someone called "the longest limb in pharmaceutical history." They decided to file an application to license Enovid—exactly the same pill, the same formula already approved for disorders—as a contraceptive. The application was based on field trials with 897 women who had been "on the Pill" for 10,427 cycles.

The company's "corporate policy counselor," James W. Irwin, has recalled: "We were going into absolutely unexplored ground in terms of public opinion. My fear was that this would provoke an avalanche of letters. So I went to see various people. I sat down with a lot of my friends at the *Saturday Evening Post.* They decided to do two major pieces about the Pill. I said, 'Look, I want to warn you. This is controversial. You may get a sizeable protest.' But there was none. I told the same thing to *Reader's Digest.* And there was none.

"We were overly cautious. All my experience told me that you could not do this without getting your teeth knocked out—or some of them. And we didn't lose any teeth. We had underestimated the receptivity of the product. We got quite a surprise."

Searle's chief competitor was not so bold. "Parke-Davis [Syntex's licensee] got cold feet just as it became obvious that oral contraception was about to be converted into practical reality," according to Carl Djerassi. "They refused even to consider marketing norethindrone as an oral contraceptive. Thus, while Searle was proceeding full steam ahead toward FDA approval, Syntex suddenly had to find an alternative marketing outlet or be deprived of any financial benefit from a field in which, relatively speaking, it had committed a larger proportion of its resources than had anybody else. Parke-Davis's apprehension about a possible Catholic-inspired boycott of some of its other products was perhaps not unreasonable, but such a consumer backlash never developed."

Syntex signed a contract giving the Ortho division of Johnson & Johnson the marketing rights to norethindrone. But that breakthrough soon led to a monumental snag. Parke-Davis refused to release the results of its studies on norethindrone with monkeys, which Ortho would need for its FDA application. "I can only speculate," said Djerassi, "on the reason for this refusal: 'If I dare not taste this tempting cake, why should I help someone else do so?' As a result, Ortho had to repeat a number of studies that had already been carried out and thus was unable to receive FDA approval to market Syntex's norethindrone for contraceptive use until 1962 [under the trade name Ortho-Novum], two years after Searle first hit the market."

As the Searle application meandered among the bureaucratic desks of the Food and Drug Administration, it brought disquietude and trembling. That potent drug for perfectly healthy women? For unprecedented use as oral chemical birth control?

After long, ominous delays and silences, on a bitterly cold day in late December 1959, Rock and Searle's medical research director, Dr. Irwin C. Winter (I.C., or Icy Winter, as he was known) descended upon the Washington bastion of the FDA for the hearing. The FDA occupied a barrackslike wooden structure built as a "temporary" almost half a century earlier during World War I. For more than an hour and a half the visitors stood in a barren vestibule waiting for their hearing officer. Finally, as Rock has recalled, "in came a nondescript, thirty-year-old . . . Can you imagine, the FDA gave *him* the job of deciding. I was furious." Rock soon discovered that the expert in charge of the application, Dr. Pasquale DeFelice, was a staff physician at Georgetown University Medical Center and part-time FDA reviewer.

Rock was "livid," according to Winter. DeFelice asked questions about the Pill's possible relation to cancer as well as probable "moral and religious objections, saying the Catholic church would never approve of it," Winter has recalled. "I can still see Rock standing there, his face composed, his eyes riveted on DeFelice and then saying in a voice that would congeal your soul, 'Young man, don't you sell *my* church short!' "

DeFelice, himself a Catholic, recounts a strikingly different experience:

There I was, a thirty-five-year-old, qualified but not yet board-certified OB-GYN man. Standing before me was John Rock, *the light of the obstetrical world!* Yes, there was *some* discussion of the morals of the pill. . . . But I personally was not concerned about the morality of it. . . . I tell you very frankly, I'm a philosophical Catholic. I felt the Church was really not right about birth control and I never could understand why the popes didn't back down on it. . . . But I had my job to look out for, you know what I mean. I've only met about three doctors in my entire life who I would trust with anything. Rock was one of them. He still is. . . . They, the Searle people, were very anxious to get the pill on the market. The important things to us at the FDA were quality control, purity of product and side effects. Actually, Searle wanted us to license lower doses of Enovid right away, 5 and 2.5 milligram doses as well as the 10. But, it wasn't because they were worried about the side effects. It was because the 10-milligram pills cost them a lot of money. They were worried about costs.

But I knew what was going to happen once we licensed it. I knew that birth control pills would be flying out the windows. Everybody and her sister would be taking it. You know something: I was stupid. I should be a millionaire. I should have bought Searle stock, but I didn't. Somehow I thought I shouldn't since I was the person who approved it. But I often wondered why I never got an award for okaying the Pill. It changed the whole economy of the United States.

Bringing the encounter to a conclusion, DeFelice said, "I'll go over it all again, and you'll hear from me."

Rock's recollection: "I stood up and I grabbed him by his jacket lapels and I said, 'No, you'll decide right now.' He said, 'Oh, all right.' I don't think he knew the significance of what he was doing."

Of course, it was not sanctioned by a minor official on the spot, as Rock's telling of it implies. But five months later, on May 11, 1960, the Pill was formally approved. The "perfect contraceptive" was born.

The FDA's caution justified itself as early as the next year. From the "blunderbuss" of its first ten-milligram Pill, Searle soon got approval to reduce the progestin dosage to five milligrams, and further dosage reductions soon followed. Moreover, progestins emerged in the early 1960s that were effective at dosages of two milligrams or less.

The estrogen content was also reduced. In Germany, Schering followed Searle's Enovid almost immediately with its pill, called Anovlar. "From the beginning Anovlar was better than the Pincus pill," claimed Dr. Friedmund Neumann, a Schering researcher. "The estrogen content of the Pincus pill was approximately three and a half times higher than ours—one hundred fifty micrograms. Anovlar's was only fifty micrograms—very low at that time. Based on our experience with other hormones right after the war, we knew by instinct not to go above that. Years later, when publications came out about side effects [see chapter 19], it was recommended that a doctor should not prescribe pills containing more than fifty micrograms."

"In 1964," chided Djerassi, "Parke-Davis finally woke up to the facts of life and decided to enter the contraceptive market after all," obtaining a license from Syntex. So while Searle initially had the oral contraceptive market to itself, by the mid-1960s Syntex had gained the major share of the U.S. market, first through its two licensees, Ortho and Parke-Davis, and later through establishing a sales force and marketing its own brand.

While Syntex was losing two years, Searle played its advantage. The company's sales, having totaled $37 million in 1960, just before FDA approval of the Pill, soared to $89 million in 1965. In 1964 alone its net profits rocketed beyond $24 million. But once Syntex fully entered the race, its stockholders profited at a rate virtually unprecedented in the history of the stock market. Between 1960 and 1966, when its sales multiplied from $7 million to $60 million, its earnings per share of stock ballooned from three cents to two dollars. To avoid certain costly tax hazards, most of the shares had been sold or given to key employees of Syntex, research director Carl Djerassi, as noted, among them. The shares, originally valued at two dollars each when issued, were soon traded at eight to ten dollars each. Before long the price reached three hundred dollars a share, then split three for one. After that, stockholders enjoyed six additional two-for-one splits. An original single share thus fissioned by 1993 into 192 shares, exploding the value of an original two-dollar share to eight thousand dollars.

By this time, Russell Marker, whose original discovery of treasure in a lowly yam had made all those riches possible, was long gone from the scene of his handiwork, gone from chemistry, and, for all any of his former colleagues knew, dead broke.

By the end of 1961 an estimated 408,000 American women were taking the Pill. In 1962 the figure was 1,187,000, and in 1963, 2.3 million and rising.

Dr. Sheldon J. Segal of the Population Council's Bio-Medical Division estimated in 1965 that one of every four married American women under the age of forty-five had used or was using an oral contraceptive and that the actual users at that time numbered some 3.8 million. His deadpan conclusion: "American women are interested in oral contraception."

Norman Applezweig estimated in May 1967 that 12,612,000 women throughout the world were taking the Pill. By 1984 estimates ranged from 50 to 80 million women worldwide.

Soon after American women, by their acceptance of oral birth control, set fire to the industry's timidities and doubts, G. D. Searle, the company that had called the idea "unthinkable," began to describe itself as "the inventors of the Pill." Searle officials even considered a plan to copyright the words "The Pill" as a trade name. "We toyed with the idea, but it was never tried," said James Irwin. "After all, if you could patent the word 'Coke' for Coca-Cola, why not 'The Pill'?"

11

THE PILL CHANGES LIVES

*F*or several years I've stayed at home while you had all the fun,
And every year that came by another baby come.
There's gonna be some changes made right here on nursery hill
You've set this chicken your last time 'cause now I've got the Pill.
—"THE PILL," LORETTA LYNN, 1973

The world-changing news of 1960 was not that the FDA approved the Pill as a contraceptive. The fact was that the women of the world had not waited for the FDA. Millions were already "on the Pill."

Searle's package for Enovid, approved originally for gynecological disorders in 1957, was required by the FDA to contain a "warning"—to many, a clear invitation: This medicine would cause the women who used it to stop ovulating, which could have the effect of "possible contraceptive activity." By late 1959, sales of both Enovid and Parke-Davis's Norlutin surged beyond the happiest expectations of their makers. "It was *amazing*," wrote Loretta McLaughlin, Rock's biographer, with feigned wonderment, "that so many women suddenly had menstrual disorders requiring treatment with the Pill, women who had never seemed to have menstrual problems before." Searle's Dr. I. C. Winter commented with suppressed glee, "It was like a free ad."

In Germany, to accompany Anovlar, its version of the Pill licensed from Syntex, Schering put out a leaflet similar in language for doctors in February 1961. Ostensibly describing a cycle regulator that incidentally had the "harmless" side effect of preventing pregnancy, it read:

> ANOVLAR—A specific therapy to suppress ovulation and yet retain a regular menstrual bleeding. . . . This treatment is harmless, and the effect only lasts while the patient is taking tablets. . . . Effect of suppression of ovulation: Without ovulation, conception is not possible and provided the tablets are taken strictly as directed the patient may be assured that pregnancy will not occur while undergoing treatment. This effect only lasts while the patient takes the tablets and normal prospects of conception will return as soon as treatment is suspended. A pregnancy that occurs because the patient fails to take the tablets regularly will develop normally.

"This was to me a moral joke," the wife of a professor at a leading American Catholic university has recalled. Ironically, her husband's research focused on population and fertility rates. "The joke was that among Catholics, birth control, and later the Pill, was not licit, not permissible. However, if you had an irregular cycle—and, by definition, *every* woman has an irregular cycle—you could use the Pill to 'regularize' it. Now, who was kidding whom? There were Catholic doctors who didn't ask too many questions, and women quickly found out which ones to go to. Some doctors would say absolutely no. So if you really wanted to take the Pill, you would find out who one of the liberal Catholic doctors was and switch to him. There was a whole circuit, an enormous exchange of information of who you would ask, of who you would go to. Women would call me up on the phone and ask, 'Is this okay? Is it moral?' I was in the odd position, because of my husband's work, of being better informed than any parish priest about the Pill. It was so much a part of our lives, I used to joke that we had it for breakfast, lunch, and dinner.

"These women wanted to do something, but their big question was how does it work—from the moral point of view? Does it destroy the fetus? That would be a real moral problem. Or does it just prevent conception? That presented a fuzzier problem. At that time, nobody knew for sure."

According to a 1967 survey by Charles F. Westoff of Princeton University and Dr. Norman E. Ryder of the University of Wisconsin, fully 53 percent of American Catholic couples were regularly using some form of birth control other than church-approved rhythm. Francis C. Mason, who shared with two Catholic partners in Norwood, Massachusetts, one of New England's largest group practices in obstetrics and gynecology, asserted: "I don't practice medicine as a Catholic. If a woman asks me for medical advice, I give her medical advice."

In Spain and Italy, two almost completely Catholic countries, "menstrual regulation" became a widespread euphemism for using the Pill for contraception. In Ireland, 93 percent Catholic, a standing joke was that the Irish had the "highest incidence of cyclical irregularity in the world." In Japan, where the Pill was illegal for contraception—and remains so (the birth control method of choice continues to be abortion)—the subterfuge of "irregularity" was to persist for more than thirty years.

This breach in old moral defenses against birth control gaped threateningly at the Catholic hierarchy. "Certainly the drugs cannot be used as a disguised form of contraception," wrote Catholic theologian Hugh J. O'Connell sternly in a pamphlet, *The New Sterility-Fertility Pills*, published under the auspices of the archbishop of St. Louis, Joseph Cardinal Ritter. "Unless there is a condition present which the drugs are intended to cure, the woman cannot take the drugs through the whole span from the fifth to the twenty-fifth day of the menstrual cycle, for she is not regularizing her cycle, but is totally inhibiting ovulation. She has no fertile period at all."

Many besides Catholics were troubled by the moral perplexities of the Pill, and their questions were not always drawn from theology.

"I remember talking to my mother about it after church one day," said Horst Witzel, in 1961 a young researcher in steroids and the Pill at Schering's laboratories in Berlin. He would later become chairman of that huge company. "It was at lunch at her home, and she said to me, knowing that I was one of the guys responsible for this, 'What you are doing will open the box of Pandora. Think about it. The chance of becoming pregnant has always kept a girl on the right path. What's now going to happen when nobody is afraid of becoming pregnant? Now they will live it up. Now they will do anything they want.'

"She was not too educated in the natural sciences, but we had always talked quite freely in our family about the birds and the bees and all the facts of life. I reminded her of a lady she knew who had just had her eleventh baby and who had only wanted two. Now, I said, any of her daughters would be able to live quite differently—able to have two babies if that's all she wants, and have them when she *wants* them. 'Don't you think that is an aspect of the Pill you should think about?' I asked my mother. It developed into quite an argument. My father, an engineer, was sitting there amused, smoking his pipe. Nobody was talking at that time about emancipation of women. And we certainly were not yet thinking of the unmarried woman. But my mother was. She said, 'You will give the chance to young girls to have sex just for pleasure, without thinking of the consequences.'

"Also I remember that when young people went for dinner, or had friends over, ten minutes wouldn't pass before this topic came on the table and everybody was commenting on whether it was safe, and how can we get it. Most of them were married. I was married. The girls would whisper to one another: 'Have you already seen your doctor? Do you have your prescription?' This was the number-one topic because it was such a revolution. I remember how friends we had known for many, many years would, for the first time in a small group, disclose their contraceptive techniques. Talking out loud. This had been taboo for many, many years, and now one of the guys was able to say, 'So now you take the Pill. What had you been using all those years?' By the way, the diaphragm had never caught on in Europe. The main methods were coitus interruptus and condoms. And rhythm."

The Pill's acceptance in Germany stumbled at first because of what Witzel terms a "tragic misnomer." The medication began to be called the "antibaby pill," apparently the creative headline shorthand of a journalist. The repellent term kept appearing in German newspapers and magazines and still does today.

In the first five years after the FDA approved the Pill as a contraceptive, it became a staple in the lives of a majority of young American married women. And the younger the wife, the more likely she was to use the Pill. In a nationwide survey in 1966 of fifty-six hundred women, Westoff and Ryder found that 56 percent of married women under twenty were using or had used the Pill. The more educated the woman, the more likely she was to use it. Among non-Catholic married college graduates under twenty-five, an astonishing 81 percent were Pill users. Of all American women using any form of contraception, 24 percent of white women were on the Pill, the percentage of black women slightly less. Twenty-seven percent of all Protestant women used the Pill, 22 percent of Jews, 18 percent of Roman Catholics. The researchers cautioned that the figure for Catholics may have been low because some Catholics said they used the Pill for reasons "other than contraception."

Even more surprisingly, a growing number of Catholic priests were advising parishioners to follow their own consciences regarding Pill use. Another survey by Ryder and Westoff found that 21 percent of Catholic wives under forty-five were Pill users. This compares with 29 percent of non-Catholic wives, regardless of educational level. In matching use of the Pill against family size, they concluded that Catholics were more likely to use contraceptives to terminate childbearing, whereas non-Catholics used them to space their families.

Another study concluded in 1965 that married women who used the Pill had sex up to 39 percent more often than users of other means of contraception.

Less than ten years later, in 1974, the impact of the Pill on sex within marriage had not only lasted, but deepened. In another broad-sample survey, of five thousand married women under age forty-five, Dr. Westoff found that married couples of the seventies were having sexual intercourse more often than corresponding couples of the sixties. Coital frequency was highest among couples using contraception, especially the Pill, the IUD, and vasectomy. In 1970, Westoff's subject couples reported their average coital frequency in a four-week period as 8.2 times. For 1965, their average had been 6.8. Women using the Pill reported an average of ten sexual acts in a four-week period, 25 percent higher than the average for women using other contraceptive means. Those findings left Dr. Westoff both surprised and suspicious: "I didn't believe the whole thing, and bent over backwards to show the increase reflective of greater willingness of women to talk about their sexual behavior rather than a change in behavior." But every test he applied to the data, Dr. Westoff said, supported the accuracy of the findings.

The most remarkable societal impact of the Pill often was explained in elemental ways—and not always from terror of pregnancy. A Chicago mother in her late twenties told a writer for the *Saturday Evening Post* in 1965, "Oh, I know I've put on a little weight since I started on the Pill, but I think it's just from contentment. I used to worry a lot about having another baby, and that kept me thinner, but I never have to worry anymore. . . . I'd prefer to wait for the next until the youngest is at least two years old. And now I know I *can* wait."

Dr. Seymour Sholder, a Chicago gynecologist, detected another freedom: "Frigidity is often caused by fear of pregnancy. With that gone, so too is the frigidity."

In Chicago in 1965, the number of women seeking contraception at Planned Parenthood centers increased 47 percent over the previous year, with "a large majority" choosing the Pill. Nationwide, of 320,000 Planned Parenthood clients, 150,000 were "on the Pill." While clinic clients often gave up using diaphragms and other devices after brief trials, a mid-sixties study of the Chicago centers showed that more than 70 percent of Pill users were still taking them after two and a half years.

Impoverished women who visited Planned Parenthood centers, upon learning they could get birth control materials at little or no cost, were found to choose the Pill substantially more often than other options. In Wolfe County, Kentucky, where eight dollars of every nine dollars of income was federal money, the county's only doctor, Dr. Paul Maddox, began recommending and dispensing the Pill when all other methods of controlling the birthrate left him feeling hopeless. In just three years the birthrate dropped down to the national average from one that had been 50 percent higher.

In England, too, the Pill changed the relationship between family-planning clinics and the women who, often timorously, came to them. "In the fifties when I started," recalls Sylvia Ponsonby, a clinic volunteer, "all the people who came to the clinics were sort of lower-middle-class people. But, of course, the Pill changed all that very significantly. Once the Pill arrived, working-class people began coming. That was very striking. They weren't sent by doctors or health visitors or anybody else. It was that they knew somebody who used the clinic, so they came and tried it themselves. This is how the word was spread.

"I remember the rather dramatic way I saw that. At the clinic in Battersea where I worked, we had a press release announcing that the Pill was going to be available for the first time on such and such a day. As a lay worker, I used to go and get the clinic set up before the doctors and nurses would arrive. When I got there on that first day of the Pill, I found a Roman Catholic priest standing outside the door with a woman who looked about fifty. He asked if he could come in, and inside he said quietly that this member of his congregation had had ten children and only half of them were still living, that she had pneumonia every winter, that she had bad legs, that she'd deserted her family twice, and please could she have the Pill? I was impressed with this really courageous act of a Roman Catholic priest to bring one of his congregation to the clinic. He left her with us and went away. She became our first patient who was given the Pill. And she was a very good one. She returned regularly, then began sending her children round sometimes to fetch her pills. One day she sent a little thing about this high that said, 'I've come for Mum's pills.' So I got out her card and she was written up for them, so I gave them to him, and he had six-and-eightpence clutched in his hand, and I took his six-and-eight and said to him, 'Do you know why your mum gets these pills?' And he said, 'Yes, stop her having any more kids, don't they?' And I thought, really, we've begun to win the battle if children grow up knowing that mothers don't have to have more kids if they don't want them.

"After that, I didn't see that first mother for a long time, five years, I think. I was dispensing one day, and I saw her name on a card and looked up. I wouldn't have known her. She looked young. She looked pretty. She was well dressed. She looked *well*, and I said to her, 'Do you remember when you first came to this clinic? I was here then, and you came with Father so-and-so.' And she said yes, that day had changed her life. It was really most moving. It showed what this new contraceptive could do for people."

In December 1961, long before any demand developed in the United States to subsidize birth control, the British government announced that the National Health Service would supply women with a month's supply of the Pill at a token cost of two shillings (twenty-eight U.S. cents). The government would pay the rest, estimated at $2.08 a month.

Across the range of classes and pursuits, lives changed. The Pill enabled not only the planning of a family but the planning of a life through the timing of a career. In Washington, D.C., Karen Hastie Williams had inherited a sense of precision from her father, Judge William H. Hastie, whom President Harry S Truman had appointed as the first black member of the United States Courts of Appeals. "I'd been married a year and a half to Wesley, a lawyer," said Mrs. Williams, "when I decided to go to law school myself, and we began to think about the best time to have children. We decided that the sequence should be law school first, have a family second, and then go into the career." After finishing law school at Catholic University, in Washington, D.C., she was appointed to a year's clerkship in the U.S. Court of Appeals for the District of Columbia, which she interrupted for six weeks in the spring of 1974 to have her first child, Amanda. Following a clerkship for Supreme Court Justice Thurgood Marshall, the young mother joined a private Washington firm, taking five weeks in the spring of 1976 to have a son, Wesley. Soon Mrs. Williams became chief counsel of the Senate Committee on the Budget.

"Modern woman," declared the playwright Clare Booth Luce, "is at last as free as a man is free, to dispose of her own body, to earn her living, to pursue the improvement of her mind, to try a successful career."

"The career of someone like Margaret Thatcher," says Madeline Simms, a London family-planning activist, "would simply not have happened fifty years earlier. She would never have got to the stage of becoming a minister. She couldn't have done it because she would constantly have been worried in her earlier years of Parliament as to when the next baby was going to turn up. Almost the only women members of Parliament in the prewar period were single women or widows. Or the very rich."

Tom Davidson, in his forties, a former president of the Young Presidents Organization (YPO), whose members are heads of American companies, has observed a "significant shift in the family demographics of YPO. My age group grew up in the world of the Pill, where there has been much greater control over fertility and reproduction. Ten years ago the average forty-five-year-old YPOer had an average twenty-two-year-old son or daughter. Today the average forty-five-year-old YPOer has an average thirteen- or fourteen-year-old. The age group that is coming through YPO now tends to get married later. They just have opted to have kids later in life, whereas my generation got married in college or just after finishing college and had kids within a couple of years."

The Pill was no panacea for women planning careers. Two scholars, Deborah Swiss and Judith Walker, among many others who studied the subject, surveyed 902 women—594 of them mothers—who had graduated from Harvard University's professional schools between 1971 to 1981 and published their findings in a book, *Women and the Work/Family Dilemma*. "Many women thought that if they could do the hard stuff and prove themselves, the battle would be won," wrote Swiss and Walker. They found, however, that 25 percent of mothers with Harvard MBA's had already left the workplace before 1993, many feeling they were forced out of the best jobs once they had children. Of all the women surveyed, 85 percent felt their careers would be harmed by reducing their work hours or refusing to work evenings and weekends. A majority of the women said their advancement would be slowed or halted by taking time off for a child's illness or turning down travel assignments.

Before the Pill was many years old, it was showered with credit and drenched with blame for all manner of changes not easily proved—or disproved. In 1968, Harold L. Graham, vice president of Pan American World Airways told the American Marketing Association that "the development of the Pill has done a great deal for the air transportation business. By delaying the family during the early marriage years when both husband and wife are working, we have a combination of disposable income and the desire to travel. It is not that we advertise birth control. It just happens that it helps business." The Pill invited fanciful cause-and-effect speculations by sociologists. In a 1978 issue of *Human Behavior*, Robert T. Michael of Stanford University unearthed a statistic that during the first five years of marriage, child-free couples had twice the likelihood of divorce as those with young children. "Thus it stands likely," Michael leaped to conclude, "that the availability of the Pill and the IUD would be a major cause of divorce."

In June 1990, the *Ladies' Home Journal* celebrated the Pill's thirtieth birthday with a declaration that the Pill "transformed our lives like nothing before or since. . . . It's easy to forget how truly liberating the Pill seemed to be in 1960. Nothing else in this century—perhaps not even winning the right to vote—made such an immediate difference in women's lives. . . . It spurred sexual frankness and experimentation. It allowed women to think seriously about careers because they could postpone childbirth. And it sparked the feminist and pro-choice movements. Once women felt they were in charge of their own bodies, they began to question the authority of their husbands, their fathers, their bosses, their doctors and their churches. As Founding Feminist Betty Friedan has said: 'In the mysterious way of history, there was this convergence of technology that occurred just as women were ready to explode into personhood.' "

12

TWO DECADES, TWO WOMEN

*My mother used to say, Delia, if S-E-X ever rears its ugly head,
close your eyes before you see the rest of it.*
—ALAN AYCKBOURN, *BEDROOM FARCE* (ACT 2), 1978

After World War II, when Our Boys by the millions returned to the embraces of Rosie the Riveter and their Girls Next Door, the birthrate rocketed as never before in the history of the world. And by the time their progeny, the baby boomers, arrived at *their* teens, the look of the world was transformed by sheer numbers. By the mid-sixties, 40 percent of the American population was age twenty-four or under.

It was as though the sudden wave of youngsters had no older brothers or sisters. Before they boomed onto the landscape in 1947, '48, and '49, the number of births had declined steeply for five, six, seven wartime years, and before that the Great Depression had squeezed out a birthrate as low as any in America's history. So, as the phrase-laden sociologist might ask, to whom were the postwar babies—arriving at their teens in the sixties—to look for role models?

They had more role models, in a sense, than any generation in history. They had one another. They shaped themselves after themselves. Rejecting their fathers' three-piece suits and mothers' party gowns, they adopted a conformity of their own: the dressed-down uniform of the laborer and the farmer—blue jeans. *Only* jeans. *Only* blue. Carelessly (but carefully) patched at the knee.

When a protocol-defying singing quartet from Liverpool accompanied their hit songs with hair that draped their ears, (neatly, the way only girls had previously worn it), long hair for boys sprouted across the land. Parents gasped, not knowing what to make of it. Someone invented the word *unisex*, and the word became magic when painted on the windows of barbers and clothing shops. Boys with sensitive eyes for the cool conversed with delicate gestures of the hands; girls hunched their shoulders with artful toughness and adopted the barreling walk of athletes, never, ever swiveling demurely. (That was only for Elvis.) In a crowded street, you could scarcely pick the males from the females except by the width of hips. Boys made bead necklaces and wore, of all daring things, *earrings*. Girls—*nice* girls—smoked. Together, boys and girls donned finger-painted costumes, decorated themselves with flowers, and conducted dance-and-song-and-play events in the park called "happenings." These baffled not only parents but the police and the mayor, who were helpless to stamp out a rebellion that was so utterly harmless—at least, it appeared to have no clear enemy target. No one could understand what in the world these kids were up to. What were they trying to *say?* The kids, of course, couldn't understand what their elders were so upset about. But they didn't need to understand their elders, or to be understood by them. There were so many of them, they needed the understanding only of one another.

The two generations achieved almost total disconnect. So, in a sense, the Pill was born twice. Or more aptly, once, but as

unidentical twins. For parents on the one hand and for their
ripening progeny on the other, the Pill was a different happening.

"Members of the younger generation are making it clear, in
dress and music, deeds and words," wrote Marshall McLuhan and
George B. Leonard in 1967, "just how unequivocally they reject
their elders' sexual world. It is tempting to treat the extremes as
fads; perhaps many of them are. But beneath the external symptoms, deep transforming forces are at work."

There was no escape from the baffling omnipresent sexuality
of the unisex generation. Spilling into the streets, it overflowed
into the language and around the burning ears of their elders.

"Sex is the politics of the Sixties," wrote a suddenly bewildered
social critic, David Boroff, who went on to describe the night
Norman Mailer gave a reading at the Poetry Center in New York
City.

He started out tamely enough reading an excerpt from a magazine piece; then, in an atmosphere of mounting excitement, he
began to read short verse selections, some of which could be regarded as obscene. Showing a mock solicitude for the squares in
the audience, he said he would "semaphore" if a selection he was
about to read would jar their sensibilities. He was wagging his
arms before reading his seventh or eighth selection when the curtain came down. The management of the Poetry Center decided
they had done enough for one evening to advance the cause of
freedom. For a while, it looked as if there might be trouble. Many
of the young people in the audience for whom Mailer oozes
charisma were hurling imprecations at the weak-kneed guardians
of public morality. It was only after Mailer came out, in duffel
coat and mischievous smile, and asked them to leave to show
their "discipline" that the crowd grumblingly dispersed.

In November 1967, *Newsweek* declared that the Western world's
morals and manners had "changed more dramatically in the past
year than in the preceding fifty." Under the headline "Anything
Goes," the magazine particularized the new era that had brought
the Pill:

The teen-age narrator of Norman Mailer's *Why Are We in Vietnam?* peppers the reader with a stream of profanity unparalleled in American letters. On stage in *The Beard*, Billy the Kid and Jean Harlow assault each other with salvos of four-letter words, then end their sexual dual in an explicit act of oral intercourse. In *Scuba Duba*, playwright Bruce Jay Friedman unleashes a bare-breasted floozie before sellout crowds. In *America Hurrah*, giant dolls copulate on stage. And at Joseph Papp's exciting new theater, teen-agers in a musical called *Hair* sing gutter profanity with the cherubic straightforwardness of choir boys.

Even Mailer himself struggled to keep his footing: "We're in a time that's divorced from the past. There's utterly no tradition anymore. It's a time when our nervous systems are being remade."

Some seemed to maintain balance better than others. Ed Sanders, the lead singer of a rock group called The Fugs (a name drawn indeed from the "unparalleled stream of profanity" of an earlier Mailer novel) expounded: "You have to get to the point where people aren't shocked anymore. It's not being jaded—it's when people know sex is not a threat to them and they accept it."

Where had all this suddenly come from? What did it mean, really *mean*? Were these kids just talking sex, or were they *doing* it? Few parents were immune from terror over what was going on in their children's beds. Perhaps worse, few had any idea what was going on in their heads.

Just as there were suddenly two Americas—the baby boomers and the not-to-be-trusted over-thirties—there were actually two revolutions. They were not only simultaneous but, to many, so intermingled as to seem the same. One was the obscenity revolution. The other came to be called the "sexual revolution." Between the two, the revolution in sexual standards was the less—well, the less *revolutionary*.

To get a sense of that swift change, we need to understand the ideas and feelings of young people of that time, to enter the scenes they remember. Following are reminiscences of two women who "got on the Pill" a few years apart, although the interval might seem like a century.

In 1962, two years after the Pill was accredited by government, Wanda Stein (only the name is fictitious) enrolled at Jackson College, then the Tufts University branch for undergraduate women, near Boston.

"I must have heard about the Pill in high school," she told me, "but it wouldn't have meant anything to me. It wasn't something that was talked about, for sure. I recall in my freshman year of college someone in my dorm, another freshman, standing out in the hall ironing her shorty nightgown and announcing she was going to sleep over with her boyfriend for the weekend. It shocked us. That anyone would *say* it.

"We had curfews. You could never go away for the weekend unless you had the permission of your parents. The girls who slept around sure were quiet about it."

Was it to be assumed that most first-year women were virgins?

"I don't think I assumed anything. It wasn't an issue at all. There had been the girls who got pregnant in high school, but of course they were—'Why were they doing this? What was their problem?' It wasn't that you assumed everybody was a virgin. There were two kinds of girls, at least in high school in the late fifties and early sixties. The good girls and the other kind.

"My father was a businessman and my mother a housewife, very traditional, very conservative. I subscribed to their attitudes at the time, but it became increasingly difficult the older I got because clearly the sexual mores were beginning to change. Toward the mid-sixties, I began having a lot of difficulty with two different sets of messages. One message was the old-fashioned one, follow a straight line and be a virgin when you get married. The other, of course, was what I knew was happening among many of my peers.

"Most girls were not taking the Pill. I mean, you had to go to a doctor. That cost a fair amount of money. Most of us weren't in a position to go to a doctor, either financially or otherwise—just not up to discussing this.

"At the end of my sophomore year, I met the man who was to become my husband. He was an undergraduate. Eventually we became engaged. When we graduated, he was to go to medical school and I to Yale for a master's. Before we graduated from college, we started a sex relationship, but it was very limited. Kind of frightening because neither of us were—I was not about to go to a doctor. He would, like—withdraw. The oldie but goody.

"A few months before we got married I went on the Pill. Now here is the really incredible thing. At the same time I am going through this long engagement, my college roommate is sleeping with several boys at the same time, which I don't know about. She was a beautiful girl, Catholic and extremely strict. The whole thing was very strange. She gets herself pregnant in September or October, the first year in grad school, and she doesn't admit anything to me, and she doesn't admit she missed her periods or anything. But she is telling me how bad she feels, how sick she feels. So by about the time she is four months pregnant, I say, 'Grace, I really think you better get yourself to a gynecologist.' The thing that really came out was that she didn't know who the father was. That blew me away. It was beyond my comprehension. I had been engaged for goddamn ever, and I was afraid to have sex with my fiancé, and she was sleeping with at least two guys at once, sometimes the same weekend. Anyway, we kind of hid her in big shirts so she wouldn't be kicked out of school, so she could get through classes and take her finals. She was five months pregnant. She did not want to do an abortion, out of the question. She took her finals, and her father shipped her off to a home. She was having her baby the same month I got married. She didn't come to my wedding. I remember once that she was railing about how *I* was getting married and *she* was going to have a baby. She gave up the baby for adoption."

"It's sort of a cliché by now, but it's really true," says Tina Mariello. "We really felt that we invented sex."

In the fall of 1971, when the Pill had just turned its first decade and Wanda Stein was five years out of Jackson College, Tina carried her suitcases for the first time into a woman's dorm at the University of Michigan. Like every other freshman on her floor, Tina was scared.

Today, after a dozen years as a magazine researcher and editor, she is a mother of two, approaching her fortieth birthday.

Squinting nostalgically, she recalls how the apprehensions that engulfed her on that first day in college were not to endure:

"We were totally convinced that we were doing things differently than anyone had ever done them before—*drugs, sex, and rock and roll*. Especially the drugs and sex. *We* were the ones who were breaking free and *doing* it. Yes, I know now that a whole lot more premarital sex had gone on than I thought. But I do believe we changed things.

"Our whole idea was that everything was going to be better. That's all nonsense to me now, but we were in this incredibly lucky window of opportunity of drugs and sex. I mean, the truth is we had a hell of a good time. It was a small period when there was good birth control and there wasn't AIDS yet and you could do drugs on a nice recreational level. It was very tied in together. The same people who smoked dope were the same people who had sex, with very few exceptions. In my dorm, mostly freshmen and sophomores, I'd say seventy-five percent of the kids took drugs. I can think of some people doing drugs and not having sex because they didn't have boyfriends. But they were *willing* to have sex. Unless they were, you know, straight, *straight, so* straight."

Tina, raised by Catholic parents in a Midwestern suburb, learned in college to design her sexuality from the way others were designing theirs. She shared with them the centuries-old fear of uninvited pregnancy but also their knowledge of the magical new Pill that promised to obliterate that fear.

"When I first moved onto that dorm floor, sex was a complete *obsession*. It dominated *everything*. Every girl who had a boyfriend was fending off sex. Many of us were trying to decide whether to or not, and there was the constant fear. My first semester was the very first time the dorm was allowed to have open houses for male visitors on football home-game afternoons—for a couple of hours before the game and after the game. Doors had to remain open, and the rules actually did say things like at least one foot on the floor. By the time I moved off campus, in those few years we had reached the point of guys being able to stay in the dorms till eleven.

"My roommate Becca, who was my friend from high school,

got pregnant on the first night we were there as freshmen. *Literally* the first night. It was a boyfriend from back home whom she had not slept with until then. He was older and had an apartment with some other guys. That first night, I went to bed and looked over, and her bed was empty.

"The whole first semester for all of us was dominated by *her* situation. I mean the gradual suspicion that maybe . . . That fear of getting pregnant was just the most *terrifying* thing. The hardest part was the shame—before parents and the rest of the world. You know, even though we could intellectually rationalize that away, it was still *so* strong. Everything around you culturally was telling you to have sex. This was what everyone was doing. At any given time, half the girls on the wing were afraid about being pregnant. Getting pregnant was still, *still* felt so completely unacceptable. The shame of it, the moral kind of shame, even though everyone knew that was ridiculous. The embarrassment of being stupid enough to let it happen, let it utterly change your life and mess everything up. And then the other kind of terror: that it was, *is*, something happening inside your body and you *can't* do anything about it and you *can't* get away from it. It's like once you had the sex and weren't one hundred percent sure that you took care of things, and you're nervous about that, and you might have to wait two whole weeks before your period was supposed to start. You were constantly going into the bathroom and checking. Constantly thinking maybe you felt something. It was *just* terrible!

"I remember one afternoon sitting in my dorm room with the door open, studying, and suddenly heard the bathroom door down the hall *bang* open, and this girl Kelly screaming at the top of her lungs, 'Jesus, Mary, and Joseph, I got my period!' And I just sort of calmly stuck my head out and said, 'Great, Kelly,' and went back to studying. She was a sophomore when I was a freshman. She was a character. She and her roommate used to boast about which sports team they had been through. Pretty religiously, they slept with somebody on every team. They were not typical. Most girls had a boyfriend.

"For Becca, the time came and her period didn't start. Then, of course, there was always this idea that your period wouldn't start because you're stressed out. She was a freshman and a very emotional person, and one Saturday night she freaked out. She went to the student health center and they let her spend the night there and gave her a pregnancy test, and the next morning they told her that it was positive, that she was pregnant. She came back to the dorm, and of course we were all *really* upset. In a week or two it became obvious. She was a smoker, and suddenly the smell of cigarettes made her sick.

"Her first idea was 'I'm going to have this baby.' Her boyfriend didn't want her to have it, and he came up with this plan for her to go to Oklahoma and live with a family there and give birth and place the baby for adoption. At the time it was like, What kind of scum is he, what a ludicrous idea, and then, slowly but surely, finding someone to do an abortion started to become the obvious answer. She had to go to New York. It seems to me she got pregnant at the end of August and she had the abortion in November. She didn't want to tell her parents, so everybody among us who had any money at all scraped it together. I think I put in fifty dollars and everybody put in a little bit, and we put together the three hundred fifty dollars that it would cost. Then Tom told her that he was terminating the relationship. Actually, before that she talked about killing herself. One night I was in the shower and she comes to the shower and said, 'Could you come back to the room as soon as you can?' I did, and she was sitting there in the room blotting blood with Kleenexes. She had slit her wrists, then just ran to the shower and told me to come back because she needed help. The first thing she said was, 'Go call Tom and tell him.' I didn't quite understand that at the time. She was trying to tell Tom, 'You are hurting me so badly.' Of course, what she really wanted was for him to say, 'Okay, let's get married.' I think she thought she wanted to marry him. That was the closest way you could make everything go away or be right.

"That was the only abortion that I was personally involved with, but I believe that at least half of the women I know today have had abortions.

"During that freshman year, I knew there was a high likelihood that within the next few months I was going to start having sex with my boyfriend. He was a high school boyfriend at another college, a year ahead of me. I remember at one point saying to him, 'Couldn't we just wait until I'm finished with my freshman year?' It was just like I couldn't deal with all of this at one time. But knowing that I was going to at some point, I knew that I had to do something about birth control.

"I guess I was afraid of the sex itself to some extent. Just the scary unknownness of it, but much much more, the fear of getting pregnant. So I knew the smart thing to do would be to start using birth control. But something about it seemed wrong. Wasn't it all supposed to come about romantically and spontaneously, not as if you were expecting it? You weren't supposed to *plan* for it. Being prepared was too clinical. Yet part of me really knew that I should. You would think that the being afraid would overwhelm the other part, but it just didn't. I mean, the idea of going to the Planned Parenthood clinic before I was having sex—I just couldn't *do* it. I was just going to have to try to be careful, take my chances and do it afterwards. We had to have sex first, and *then* I would go.

"The night Joey and I first had sex I came home and looked in the mirror. Literally, I remember looking and saying, 'This is a person who is not a virgin anymore. This is a woman who is having sex.' It was just a tremendously big deal to me.

"Joey and I had sex when we were at my home one weekend, and then on the drive back to school we spent a good part of the drive back discussing birth control. I had already discussed it with my older sister. She and her boyfriend were of the group that did the 'everything *but*' routine. They slept together all the time and did everything but. And I remember telling her that my boyfriend and I finally had sex. It was something I needed to share with her right away. She said, 'What are you going to do?'

"We discussed the Pill and IUD. I had enough of the Catholic still in me that the IUD bothered me as an abortive device. So on that drive back, my boyfriend and I decided I was going to go to Planned Parenthood and get on the Pill. That was very romantic, that discussion. Because we were taking care of things together. I really liked that he was interested, and I needed this because it was all so scary. So I told my sister I had decided I was going to go to Planned Parenthood. Wednesday night was clinic night, where you could go without an appointment. The girls who were secretive just went out on Wednesday evening and didn't tell you where they were going. My sister said, 'Where are you going to put the Pills? When you go home, how are you going to have Mom and Dad not see them?' My father was a practicing Catholic only on the surface, but Mother was serious. It would have really bothered her that we were having sex. I said, 'I'll keep them in my purse. Mom never goes into our purses.'

"So I went the next Wednesday night. The clinic was very close to campus. It was a house, a converted residence. Going there I had a complete mixture of feelings. I felt so grown up. I was taking charge of my life in a way that I had never, ever done before. This was like making a decision—where to go to college was just tame, nothing compared to this. I felt really proud of myself, but there was still some layer of guilt and fright. This little fear that maybe this was wrong. There were maybe a dozen chairs, almost all filled. I didn't know anybody else there that night but felt some sense of kinship with the others. There was a nurse receptionist, and you signed in. I'd never had a pelvic exam before, and you just sat there and waited until your name was called.

"The pills were pale yellow. Some people had them in round cases and some in oblong cases, from different manufacturers. I got the oblong case. That's what they prescribed.

"My roommate Becca had gone on the Pill after she had the abortion. The night she took her first Pill, it was as though she'd suddenly become Kelly. She went out in the hallway and sang at the top of her lungs, 'Ta-da, I'm taking my Pill tonight.' I felt very superior, sorry for her through that whole thing. She was clearly a mess. By that time I had become good friends with the girl across the hall, and we were both equally disgusted at Becca's display. But I remember that the night that I was to take my first Pill I went into my friend's room and parodied Becca, 'Ta-da, I'm taking my first Pill.'

THE PILL

"I remember my parents coming up for a fun weekend when this dark fear just lurked over me every hour. I was a day or two late with my period. They went home Sunday evening, that typical kind of a Sunday-evening feel, and I was very depressed. I fell asleep on my bed. It wasn't bedtime, but I just had to escape. When I woke up, my period had started. It was such a relief, and I just had to go to sleep again. Once I got on the Pill, all that was over. I felt *completely* secure. Because I took it right. I mean I always took it the way I was supposed to.

"The following year Joey and I were sort of taking a break from each other. We decided we were going to pretend the other didn't exist. For a while I got involved with a Southern Baptist, an interesting guy from a very, very small town. He was a graduate student in art but a small-town Baptist, and he couldn't quite reconcile those two things. The first time we slept together—what do you call it?—I can't remember the term for it, but the guy pulls out before he comes. We discussed nothing ahead of time, and he did that, and I said to him afterwards, 'You didn't have to do *that*. I'm on the Pill.' I remember that he was like, Gulp, swallow. On the *Pill! That* I remember.

"Then some years later my boyfriend and I broke up for the first time, *really* broke up. I went off the Pill immediately. There was something wrong about being on it if I didn't have a boyfriend."

Twenty years after her college days, what does Tina now want for her two small children? Asked, she grins sheepishly and turns pensive. "I suppose it's typical that we think that whatever our own experience was, that was the right one. I still have this vague feeling, though, that it really would be nice if kids wouldn't have sex in high school, the way they do today. If they just would *wait* until they're freshmen in college, like I did. Wouldn't it be nice if high school kids really got to feel that they are just too young?"

13

WHAT SEXUAL REVOLUTION?

The first time I ever had sex, it was in the Cuban missile crisis. And I was still a virgin. And I remember thinking, "Oh, if this war is going to end it all, I want to have sex before it does, so I know what it's like." I was going to call up my then boyfriend because I didn't want to die without having had it.
—INTERVIEW WITH A FORTY-NINE-YEAR-OLD AMERICAN WOMAN,
NAME WITHHELD

When the bombs were dropping [on London in World War II], all my girlfriends used to say, "You know, you can't die a virgin." I was the only virgin in Chelsea, or so they told me. You felt you could die in a moment, and he could die in a moment, and England could be invaded and taken over any moment, and nothing really mattered apart from making life as pleasant as possible for your fighting men—and yourself. It's the greatest thing in the world to be in love and have sex for the first time, particularly against the background of war. Oh, absolutely worth the risk, yes.

I remember the first night of the blitz. We were in our shelter in the garden in the Fulham Road, sitting by candlelight, and heard the bombs for the first time. The first thing I thought to myself was, Tomorrow I'm going to go and get seduced. I'm going to lose my virginity, and, you know, the next day I went off and lost my virginity. The day after that I went back, and the house where I'd lost it was a hole in the ground. There was the bed in which I'd been done, and it was hanging out over the street. Then, luckily, my boyfriend came hurtling down the road carrying his guitar and saying, "Don't worry, I sneaked out for some fags and I missed the bomb." So we went on from there. It was that excitement and tension and never knowing when anyone's going to be killed, even the day after you've made love to them for the first time, that made it all so exciting and worthwhile.
—INTERVIEW WITH JOAN WYNDHAM

W hen humans experience the major and the unexpected, they crave an explanation of cause. The
world is struck by a calamitous weather system, or a
new style fad, or a sudden plunge in church attendance, and we demand of science or of journalism a
precise answer to *Why?* And so we have demanded, What *caused*
this confounding sexual revolution? Wasn't it surely this new,
radical, awesome Pill?

Before the sexual revolution can be pinned on the Pill, there is
a prior question: *Did* a revolution indeed take place?

Well, something surely did.

At Columbia University on April 16, 1968, the judicial council
of its sister school, Barnard College, listened gravely as one of its
sophomores admitted that for two years she had been living out
of wedlock with a Columbia undergraduate, violating school regulations. The scandal made the front pages and *Life* magazine. But
Linda LeClair told the council unrepentantly: "I have disregarded
a regulation which I believe to be unjust."

The council made its political assertion too, wrist-slapping
Miss LeClair by suspending her from the Barnard snack bar.

"Terrific!" the accused cheered gleefully. Perhaps that verdict,
widely reported, marked the moment at which the morality of an
era changed.

At sedate Vassar College, president Sarah Gibson Blanding declared, unremarkably for 1962, that sex relations without the benefit of marriage constitute "offensive and vulgar behavior." What
was remarkable for that year, however, was that a student responded and was quoted (unnamed) in the New York *Herald
Tribune*: "If Vassar is to become the Poughkeepsie Victorian
Seminary for Young Virgins, then the change of policy had better
be made explicit in admissions catalogues."

At Boston College, a citadel of Catholic education, the dean of
students, Father Edward Hanrahan, reassured parents with an uneasy guess that these couples living together off campus "have
less sex than they did before they started living together."

Life, the ever-reliable chronicler of popular ways, noted that a
"neo-Biblical" phrase had entered the language: to *have* sex.

In 1968, posters around New York announced a course called "Are You Premarital?" offered by New York Theological Seminary, which *The New York Times* described as "once a bastion of Protestant conservatism." The poster listed a score of topics to be discussed, among them: "Should I feel guilty when I go to bed with someone?" "Is virginity a meaningless state?" "What is the positive value of sex before marriage?" "I want to know more about masturbation." Dore Schary, author of the hit play *Sunrise at Campobello*, commented: "Sex has gone as public as AT&T."

And how did the Pill figure into this sea change? "Many people believe that the foundations of contemporary sexual morality may be threatened," said a stalwart defender of those foundations, the *Saturday Evening Post*, in 1966. "College health officers have shocked parents across the country by publicly reporting that coeds come to them for prescriptions for Pills." Some physicians felt an obligation to dissuade those students. One, at a Midwestern college, said he responded when asked for the Pill, "How old are you?"

"Twenty-one."

"You have a particular man in mind?"

"Well, yes, I do."

"Have you ever stopped to think," asked the doctor, "that you might someday want to marry a man who holds virginity in high regard?"

Her answer, the like of which the doctor had never heard from a college woman: "Yes, but I'm not at all sure I want to marry a man like that."

A major rumble broke out at Brown University after the campus health director prescribed the Pill for two unmarried students at Pembroke College, Brown's undergraduate school for women. Newspaper reporters descended upon the director, who pointed out that both students were over twenty-one and had been carefully questioned. Brown's president came to his defense. So did the Pembroke student newspaper, which asserted that the college ought to be "geared to safety and efficiency and not to the ordering of the personal lives of its students, or to the legislating of chastity."

Dr. Freda Kehm of Chicago's Association for Family Living had other apprehensions: that the Pill might not only encourage promiscuity but might be used by college women as a predatory weapon: "The girl tells the boy she is using the Pill when she really is not. Then she has him trapped."

Reports exploded at random like land mines: An Austin, Texas, gynecologist admitted to prescribing the Pill "without qualms" for eight to ten coeds a month. "I would rather be asked for the Pill than for an abortion," he said. A druggist near the campus of the University of California at Los Angeles began giving a 10 percent discount on the Pill when bought by students.

Under the barrage, one of the first casualties was the age-old "in loco parentis" role of colleges—the duty to chaperone the lives of students. John U. Monro, Harvard dean, made peace with the new rule permitting men of the all-male college to entertain women in their dorms. But he lamented that it had "come to be a license to use the college rooms for wild parties and sexual intercourse." Economics professor John Kenneth Galbraith, Harvard's intellectual laureate who had scarcely unpacked after returning from his tour as President Kennedy's ambassador to India, sided instantly with the newly liberated students in a letter to the *Harvard Crimson*: "The responsibility of the university to its students is to provide the best teaching. . . . Rules need only reflect the special requirements of the academic community—the quiet, good order and opportunity for undisturbed sleep that facilitates reflection and study. No effort need be made or should be made to protect individuals from the consequences of their own errors, indiscretions or passions."

Across the sea Malcolm Muggeridge, the British journalist, curmudgeon, and former editor of *Punch*, took an opposite stand. At sixty-five, Muggeridge had been elected by students of Edinburgh University to be rector, their spokesman on the university's governing body. When students demanded that females be supplied with the Pill free of charge, Muggeridge quit in "contempt." His indignant explanation: "I am supposed to be the spearhead of progress, flattered and paid for by their admiring seniors, an elite who will happily and audaciously carry the torch of progress into

the glorious future opening before them. How sad; how macabre
and funny it is that all they put forward should be a demand for
pot and the Pill."

Sturdy old walls of tradition continued to crumble. As if admitting women across the thresholds of Harvard dorm rooms was
not enough, colleges across the land opened their dorm buildings
for *shared tenancy* by men and women. In 1969 *Look* magazine displayed a full-page photo of more than a dozen solemn-faced male
and female students in formal pose, all costumed in bathrobes and
nightgowns. The caption, by author Betty Rollin, read: "First they
wanted classes together—a trickle of girls at Yale, a smidgen of
boys at Sarah Lawrence. And what now? They're *living* together,
that's what now—not shadily in dank off-campus pads, but on
campus, in dormitories, even in fraternity houses. The kids in the
photo above, for instance, are not models in a nightie ad. They
are (*all*) members of Lambda Nu Fraternity of Stanford University. Well! For some people, this new kind of integration is
even more threatening than the old black-and-white kind. Fearful
queries about ruined study habits and promiscuous sex are rising
like war whoops."

In the serious journals, some behavior-watchers resisted being
swept away by all the bedlam. The world had not turned promiscuous, they insisted, at least little more so than it had ever been.
One such deportment monitor, Stanford psychologist Nevitt
Sanford, tabulated reports of twelve years of sexual activity at
three colleges: an Eastern one for women, a large Western state
university, and a private Western coed school. His conclusion,
reported in the *National Education Association Journal* in 1965, flew
directly in the face of the media hubbub: Young unmarrieds had
not changed their sexual ways substantially since their grandparents had courted! Sanford reported "no revolutionary change in
the status of premarital intercourse since the 1920s." He found
that only 20 to 30 percent of women in his samples were not virgins at the time of their graduation. Earlier studies, according to
Sanford, indicated that this was about the same as the percentage
of graduating virgins forty years earlier.

Yet something new had crept in to change the ways of college women—the Pill. Richard Moy, head of the student health service of the University of Chicago, reported in the mid-sixties: "Many girls have [the Pill] when they come to school. Their family doctors at home have prescribed them. Or they borrow from each other or use the prescription of a married sister or they put on an engagement ring and get them as part of preparation for marriage. It's not a very formidable task to obtain the Pills." A West Coast doctor agreed: "There is certainly a lot of Pill swapping, like sugar and eggs."

Scholars of sex behavior rushed updated reports to reconcile these conflicting reports of what this new breed of college students was up to. Ira L. Reiss of the University of Iowa, Joseph Katz of Stanford's Institute for the Study of Human Problems, William Simon of the University of Houston as well as Drs. William H. Masters and Virginia E. Johnson, coauthors of *Human Sexual Response*, all pushed forward from where Alfred Kinsey, the pioneer sex researcher of Indiana University in the 1950s, had begun. *Life* summarized their combined findings:

"Promiscuity was not invented by the current college generation, and the evidence indicates that it is no more common than it ever was. There has been much speculation that the Pill has accelerated the willingness to engage in sex. The studies all refute this. . . . 'The presence of the Pill does not make people decide to have sex,' says Dr. John Gagnon. 'It is after they decide to have sex that they go get the Pill.' "

What was occurring, Isadore Rubin, a clinical psychologist and widely published social critic, argued in 1966, was not a revolution but an evolution that had been ongoing for generations. "How much of the post-Kinsey sex revolution is based on actual changes in behavior," Rubin asked, "and how much is based on simply increased freedom to *talk* about that behavior—and to admit to it without apparent shame or guilt?"

Buttressing Rubin's skepticism, in January 1965, Mervin B. Freedman, assistant dean of undergraduate education at Stanford and a research associate at the Institute for the Study of Human Problems, had published a study of American college women. Freedman kept track of a sampling of students at an Eastern women's college through their four undergraduate years, interviewing them repeatedly. He found that more than three out of four unmarried college women remained virgins; that when premarital intercourse did take place, the partner was usually her future husband; and that promiscuity among the women was relatively rare. "The Puritan heritage," Freedman concluded, "has by no means passed from the scene. . . . It is probable that the incidence of nonvirginity among college women has increased . . . little since the 1930s."

During the same year, 1965, a survey in Great Britain by Michael Schofield for the Central Council for Health Education studied the sexual behavior of 1,873 males and females between fifteen and nineteen years of age. One boy in three and one girl in six, the survey found, had experienced sexual intercourse at least once. But only 2 percent of girls had had more than one partner during the entire preceding year. The figure for boys was, unsurprisingly, larger—almost 12 percent. These numbers did not match a growing alarm sounded in the popular magazines, tabloids, and on television that promiscuity among the young had sharply increased from that of previous generations.

Ira L. Reiss of the University of Iowa (later, at the University of Minnesota) produced further evidence that a new birth control technique cannot be readily tied to increased sex activity. Although the diaphragm first came into use in the 1880s and the condom had been perfected even earlier, "these methods produced no immediate radical changes in the extent of premarital sex. Sexual standards and behavior seem much more closely related to social structure and cultural and religious values than to the availability of contraceptive techniques. If the opposite were true, we should find higher rates of premarital intercourse among the better-educated and higher-income groups, where contraceptive knowledge is more widespread and devices more readily afforded. Yet according to available data, the premarital intercourse rate is greater among the less educated and less affluent."

"Other evidence also supports the view," Rubin further added, "that sex behavior cannot be explained simply on the basis of the availability of contraceptives. The British study of teenagers, like many other studies, shows that large numbers of young people who know about contraceptive techniques make no use of them in their premarital relations. Over and over again pregnant young girls reported that they considered it immoral to take any precautions. They could justify their conduct to themselves only by feeling that they had been swept away by passion over which they had no control. . . . If it is true, as almost all authoritative research seems to conclude, that there has been no revolutionary change in the sexual conduct of unmarried girls in the years after Kinsey, it makes even more striking the one moment in American history when a 'sex revolution' did take place. Girls who became sexually mature in the years up to 1920 were not likely to have had premarital intercourse; only fourteen percent of these women informed Kinsey's researchers that they had not been virgins when they were married. But after World War I a sharp and dramatic change occurred. Among girls who became sexually mature during the period from 1920 through 1950, 36 percent—roughly four out of every eleven women—experienced premarital intercourse. It is a sobering thought that if there has been a sex revolution in recent times it was made, in fact, not by today's teenagers but by their grandmothers who are now past sixty years of age."

Lewis Frank, executive director of the Information Center on Population Problems, scoffed at the idea that the Pill could be blamed for what many in the sixties were calling a declining moral standard: "A lot of nonsense. Promiscuity and the fruits of promiscuity have always been with us. There are 125,000 to 300,000 illegitimate children born each year. Many youngsters, using Saran wrap and Seven-Up douches, are practicing contraception on the level of the ancients. We get the results of that in foundling homes and shotgun marriages." (In making a significant point, Frank may have been using language loosely, meaning to refer to careless sex, not promiscuity. Studies have consistently shown that promiscuity—indiscriminate sex with many partners—is not a typical choice of teenagers.)

As late as 1968, Ira Reiss reported that "perhaps one or two percent" of premarital sexual events occurred because the Pill permitted a feeling of greater safety. All the rest, he suggested, would have taken place anyway, promoted by a courtship system that had been evolving for a hundred years in the United States permitting young people to choose their own marriage partners, and which therefore encouraged choice of when as well as with whom to share sex.

But Reiss added a prescient comment: "The impact of the Pill well might be more powerful in the next generation." It had taken, he pointed out, a full generation for the diaphragm to find acceptance and to influence sex behavior.

Whether or not a revolution was taking place, and to whatever its cause was attributable, by 1967, according to a survey of the 315-member American College Health Association, fully 45 percent of the nation's college health services were prescribing the Pill for students.

Surveys of college students, however, failed to disclose all that was happening among young Americans. One just had to look at pregnancy figures to perceive some kind of upheaval of sexual norms in the lives of teenagers four years younger and more.

In 1966 the Connecticut Department of Health estimated that among thirteen-year-old girls in the state, one out of every six would become pregnant out of wedlock before she was twenty. Not much comfort was to be found in the stereotype: that these were daughters of welfare mothers or of the otherwise dispossessed. The figures would not support that political cliché. According to *U.S. News & World Report* in 1966, especially on the East Coast, "high school girls of the middle and upper income classes join a steady traffic reported among college girls who fly to Puerto Rico for legalized abortions."

Was the newly introduced Pill a factor in speeding—or slowing—that traffic? Then as now, a complex mystery has beclouded the very young person's decision to use or not to use the Pill. As Dr. Faith Spicer pointed out regarding teenagers and the Pill, early in its history as well as later: "We found at the London Youth Advisory Center that they would come to us for a bit and then stop coming. Why? Sometimes because they'd quarrel with their boyfriend. One kid I remember threw the whole lot down the loo. Then they didn't come back to the center because how did they know if they were going to have another boyfriend? It takes several weeks, you know, to get established on the Pill. If you're dithering around about whether your boyfriend's going to be with you or not, you don't take the Pill. So there are all sorts of subtle reasons why people don't take the Pill. It's not the panacea."

A seventeen-year-old from North Carolina, Charles W., told a *New York Times* writer that he somewhat resented the Pill even though he guessed it was 90 percent effective. "You know, a woman can leave her dude now and go with somebody else. She's on the Pill." Lillian H., a Philadelphia seventeen-year-old, took umbrage at that, even while reluctantly agreeing: "Just because a woman can take the Pill she can make love to anybody, but that doesn't mean she *will*. But anybody who takes the Pill helps her man and herself too."

"Teenage girls have a tremendous amount of mythology as far as fertility and birth control are concerned," said Dr. Alwyn T. Cohall, a thirty-one-year-old pediatrician at St. Luke's–Roosevelt Hospital Center who worked at high school and neighborhood clinics in New York City. "There are still girls who think you can't get pregnant the first time you have sex, or that you can't get pregnant having sex standing up, or if you douche with Coca-Cola right after having sex that you will prevent yourself from getting pregnant."

As the Pill moved in its earliest years from the hands of its first user, the married woman, to her younger sister, and then to her ripening daughter, by the late sixties the Pill's use by teenagers threw some parents off balance. A Long Island mother was quoted by Jane Brody in *The New York Times:* "My pills seem to disappear twice as fast as they used to. I wonder if Joanie [her teenage daughter] is snitching a few. I hope she's at least taking them according to schedule." Some doctors, Brody reported, "balk when faced with an unwed teenager not accompanied by a parent. But others are inclined to give her the Pill because they fear that she will turn to a less effective form of contraception, if any at all, and risk an illegitimate pregnancy."

• • •

By 1980, the twentieth anniversary of the Pill, the age of its "average" user had skidded steeply. It was now more likely to appear in the purse of the unmarried fifteen- to twenty-four-year-old than of her older sister or mother.

If a truly discernible revolution took place as a companion to the Pill, it was among the second generation of Pill users. Unlike their baby boomer elders, they adopted it as a rite of passage, like a driver's license. The *Ladies' Home Journal* quoted a twenty-two-year-old editorial assistant: "Most of my friends started the Pill when they began going steady. It's part of a commitment you make."

The second Pill generation did not limit itself to Manhattan trainees. Indeed, there is evidence that it hit them last. To locate the advance messengers of this second generation, better to look outside the urban centers: for example, a town of seven thousand in north central Pennsylvania that revolves around a single factory. The town is real, although its name here, Plainfield, is not. "Beverly" graduated from Plainfield High in 1982 in a class of 102 students.

"On the day we graduated, out of the fifty or so girls, ten were pregnant or had already had children. One was married. Several married right after high school." Beverly, now a bank branch manager in another city, thirtyish and a mother of two, pulls out those numbers with no air of reporting a notable change from earlier times. "The first time someone in my class got pregnant was ninth grade. Parents would just say it was unfortunate—you know, concerned that she was so young. When people became pregnant, they had the children. I mean, that was just part of the culture. When we hit eleventh grade, it became more real. I mean, around eleventh and twelfth grade is when people started having children. Sometimes they married the father afterwards, sometimes not. Part of the reason for not getting married before having the baby was to collect insurance. As minors, they were still on their parents' insurance policies. If they married, they would lose the insurance. The father probably wasn't working yet."

At Plainfield High, the Pill plumbed new meanings of family planning.

"There got to be a sense of urgency about getting married as we entered our senior year. A close friend of mine, Loretta, had been taking the Pill since she was fifteen. She was a cheerleader and became our homecoming queen. Her boyfriend, Chuck, was an athlete, very liked. Her mother had become pregnant at sixteen, so she'd tell Loretta, 'You don't want to be pregnant so young.' I don't think her mother ever advised her never to sleep with Chuck, but just to protect herself. Then in our senior year she stopped taking the Pill and at seventeen became pregnant. She said the reason was that she'd heard it wasn't good to take it for more than a few years at a time. The real reason was graduation was coming and to hold on to Chuck.

"The main method of birth control in our senior class was the Pill—I'd say twenty were on it out of the fifty graduating girls, most of the ones in long-term relationships. Condoms were probably second. Maybe five were virgins at graduation. Most of the people who took the Pill got it from going to the doctor with their mother. Or certain girls would go with the mother of another girl if they were afraid of discussing it with their own mother. They probably went to St. Johns, a town a half hour away, to buy the prescription. To buy it in our own town, everyone in the store would know you, and the druggist would know you, and everything was talked about."

Beverly's classmates, sexually active when they were fifteen or sixteen in 1979, may well have been the children of parents in their young thirties who were finishing high school when the Pill had come on the scene.

"I think that most of the people I knew on the Pill had somehow gleaned from their parents that the Pill was the thing you should do until you were close to graduation. As soon as we started driving, a lot of the pregnancy happened in cars. Almost all these girls were at mass every Sunday. At least sixty-five percent of the town was Catholic, probably higher. I don't know if they confessed the sex or taking the Pill, but they didn't seem to have any concern about it. We had an older, conservative priest.

I'm not sure what they were willing to tell him and what they weren't.

"When I look back on it, being a mother now, I don't feel any scandal about it. But I just wonder how they *did* it. How someone has a child at sixteen and is able to cope with all of it."

That rural factory town provided an early sighting of the second-Pill-generation shift, the locale of which was soon to change. Near Kansas City, Andrea Warren, a writer, lives in a more-than-comfortable suburb of 100,000 with amply budgeted public schools. When her daughter entered her midteens in the late eighties, what Warren found in the new culture of teenagers' families recently led her and her husband, Jay Wiedenkeller, a sex therapist and teacher of the sociology of sex, to write a book called *Everybody's Doing It: How to Survive Your Teenagers' Sex Life (and Help Them Survive It, Too)*:

"We found parents who rent hotel rooms for their kids after the prom. We hear about parents hosting boy-girl slumber parties at their homes. Supposedly the parents are present, but they just discreetly disappear or go upstairs to bed and never check to see what's going on. My assumption is that they don't want to know.

"I did an interview a couple of weeks ago with a nice middle-class kid in my town who became sexually active at age eleven. She got pregnant and had an abortion. She told me she never liked it and it always hurt, but 'doing it' meant she could have a boyfriend, or the other girls thought it was neat, and it was just expected. She said she became active because she was the only one among her friends who had not yet had sex.

"These girls get the Pill very easily. At the clinic they're a dollar a pack. You see a doctor, you have an exam, and you're on the Pill. From what my kids and other kids tell me, most parents just drop out when their kids turn thirteen or fourteen. They put blinders on. Like, I don't want to know, don't tell me, just take care of things. The kids don't learn anything from their parents. I've asked these kids where they get their sex education. They get it from each other or from the movies, and they get a lot of it wrong.

"My husband, who counsels teenagers a lot, says that as he gets older it's harder and harder to stay in touch with what it was like to be thirteen, fourteen, seventeen, eighteen. You forget how strong those feelings were, how great those desires were, and we seem to think as adults that kids really can put the brakes on. The way *we* were controlled was through *fear:* You'll get pregnant, you'll get VD, you'll end up in hell. I remember my mother pointing out to me a young woman who lived across the tracks. Her family was poor as church mice, and she had to drop out of high school to get married because she was pregnant, and then she had a scad of kids. It was very clear. I'd look at that girl and say, 'Boy, I don't want that to be me.' So that's how we were controlled: fear.

"I became a teenager in 1960. When I was the age my daughters are now, there was no birth control. I mean, guys my age couldn't buy condoms in the drugstore. For those who knew about it, birth control meant withdrawal or rhythm.

"Then when I got to college, suddenly there was the Pill. So my generation went from being raised just like our mothers to—suddenly—here's all this freedom and free love and the psychedelic Volkswagens, sleep with anyone you want to, three different guys in a night if you want to, and it was very confusing. I swear to God, that's why so many of us have married three or four times. I mean, we really got screwed up. So now, to keep our kids from getting screwed up, we think we're raising them with all this communication that we didn't have with our own parents. Well, maybe we've succeeded. Maybe our kids are not screwed up. They're out there having sex early, and loving it, and *we're* still confused. And maybe our screwed-up generation always will be."

Two Johns Hopkins University professors, Melvin Zelnik and John F. Kantner, tracked the transition between the first two stages of life with the Pill in a 1972 survey for the Presidential Commission on Population Growth and the American Future. "The picture is not one of rampant sexuality," the report said, but the researchers were struck by "the pervasiveness of chance-taking." More than three fourths of the experienced young women had never used contraceptives, or used them only sometimes. That finding of the willingness to take risks in early sexual experience, rediscovered in survey after survey, is perhaps the most decisive evidence that the sexual revolution, such as it was, was not principally driven by the sudden availability of the Pill but was feeding on a momentum of its own.

The decision to begin sex activity is governed far more by ethnic and cultural influences than by the availability of the Pill or any other contraceptive choice. At each age interval, the presidential commission learned, nearly twice as many young black women as whites had begun their sex lives. Averaging the entire fifteen-to-nineteen-year-old group, the black proportion was 53.5 percent compared to 23.4 percent for whites. Yet the study found that "it is the white nonvirgins who have sex more frequently and are the more promiscuous." While promiscuity was not high in either group, 16 percent of sexually experienced white women had had relations with four or more partners, while among black women the figure was less than 11 percent. In both groups about 60 percent reported only a single partner and said they intended to marry him.

The sharpest difference between the race groups was in the incidence of pregnancy, probably reflecting a difference in cultural acceptance of children born out of wedlock. Ten percent of the white young women had been or were pregnant. Among the blacks, the figure was 41 percent.

Seven years later, in 1979, the same researchers, Zelnik and Kantner, reported a dramatic rise in premarital sexual activity among urban teenagers. While previously sexual activity among teenage girls in metropolitan areas had been lower than the national average, they now found that it had risen from 30 percent in 1971 to 50 percent in 1979. The increase was mostly attributable to teenage whites.

They also found a decrease in first-choice reliance on the Pill due to a "general nervousness, which is pervasive about this method." Between 1976 and 1979, use of the Pill among teenagers declined from 32.8 percent to 19.4 percent. In the same period, the practice of withdrawal doubled.

Like almost everything else in American popular culture, the shift to an ethic of a second Pill generation made its way across the Atlantic, and not just to the high hot spots of European capitals.

It also turned up in unexpected places, for example, alive and in full flourish among young teenagers of the most obedient of Catholic lands, Ireland. Maxine Brady, born nine years after the FDA's approval of the Pill and a year after the Vatican officially banned it, was in 1992 the president of the Union of Students in Ireland (representing fully three quarters of the nation's college enrollment of 145,000). In the shadow of staid Trinity College in the heart of Dublin, Brady, in her midtwenties, talked to me of her personal witness to new degrees of sexual activity in the lives of young teenagers.

"When I was in secondary school, maybe thirteen, fourteen years old, most of my peers were sexually active," she recollected. She heard her schoolmates recite the same myths that flitted among early-teen Americans. "All of them had this thing that you couldn't get pregnant if it was your first time, you couldn't get pregnant if you were standing up, you couldn't get pregnant if you douched afterwards. Obviously all of us were curious about sex, and we used to go to parties. Couples would go into bedrooms, and there would be maybe six couples in one room, all of them having sex. I think I was the only one of my group who wasn't sexually active at that age, but I was the only one who had actually read books and knew what was happening. I used to have to advise them about when the most fertile time was and which contraception to use. Quite a few people I went to school with were pregnant by the time they were sixteen and had to leave school and bring up their children. Out of thirty women in my class and the class above mine, there were at least nine pregnant by the time they were sixteen. Several people I went to school with got engaged after they were pregnant, had a diamond ring on their fingers, by the time they were sixteen. One friend left school when she was sixteen, came back during exams, and I saw her about a year ago, and she now has three children and still is not married."

Did the Pill make a difference for them? Were any of her school friends on it?

"No, none of them. I think they felt that to be on the Pill, in some odd way, was a way of saying you're sexually active, some kind of commitment to it. Some were under the impression that they could get one month's supply of the Pill and take it, and that would protect them against pregnancy. Some believed if they took a tablet before going out with their boyfriend, they would be okay that night. They took that risk if they could get one Pill, because we weren't taught how the Pill works. We weren't taught about condoms. Most of the people I ran about with didn't really bother to protect themselves in any way whatsoever. They just didn't think about it. You used to have girls who would feel so guilty and scared afterwards, they would actually break out with some form of vaginitis.

"The church? I think a lot of young people at colleges have given up on the church and its teachings. They may go at Easter. They may go at Christmas. I guess most of them go during weekends home with their parents."

Just as the interview was ending with Tina Mariello, to whom the new sexual liberties of her University of Michigan days had been "a hell of a good time," she said, "This may not exactly be on the subject, but I don't want to think I left out this chunk."

"This chunk" soon went to the heart of the ages-old search of women everywhere, and indeed to Margaret Sanger's driving concept: a yearning to disconnect the sexual appetite from the ever-looming consequence of pregnancy.

"I told you that my boyfriend and I broke up, but we could never make those breakups stick. So I used other methods. One time in the middle of making love, suddenly—'Oh, my God, I forgot!' I was really scared because I had never done *nothing* before.

"I didn't get my period. That was the complete low point of my life. This was still during that time when you couldn't get a pregnancy test until so many weeks after you missed your period. So I went through a fantasy of 'All right, if I'm pregnant, I'm going to have it and raise it.' My boyfriend just couldn't deal with it. He said he would have to leave town because he couldn't possibly bear to be in the same place with me and a child of *his*.

"The thought of waiting for a pregnancy test, then waiting until I could have an abortion was complete torture. All I could think of was to make this thing go away as quickly as possible.

That feeling: *This is your body, you can't get away from this, no escape.* So, without even telling my boyfriend I made an appointment with a doctor in town who did menstrual extractions—really, abortions before you know whether you're pregnant or not. I have no regrets about that at all. My only feeling of incompleteness was this weird question, *Was I pregnant or not?* I know now there was a good chance I wasn't, because of just getting off the Pill. But I would like to know: Was I ever pregnant, or wasn't I?"

This "weird question" grew into one of nagging importance later. A life-governing disappointment to Tina was that several years of marriage brought no children. After a few years of infertility treatments, with all their attending hopes and setbacks, Tina and her husband adopted a child, then a second.

"Infertility and reproductive technology have become just a minefield of values for me. I said earlier that sometimes when I think of sex and reproduction, I wonder: *Why are these two things linked?* Maybe I felt that quandary more than others, having thought I was pregnant when I didn't want to be and, later, not being pregnant when I wanted to be. Because when you live in the infertile world, you get to the point where they aren't linked at all. The two have nothing to do with each other. Most of the time, having sex is one thing. Procreation is entirely another thing. How is it that they are still so linked together?

"Everything in the culture, in advertising, in the media—*everything* tells people to have sex, yet we clearly have not been able to separate sex from pregnancy. To make a birth control mistake is so easy, but the results, the *consequence,* is this *gigantic* thing. *You're pregnant!* The entrance of another human being that you're responsible for! The proportions have changed. Sex has become *nothing*—but *pregnancy* is *huge.*

"This out-of-proportion connection between sex and pregnancy—the little act and the big consequences. I guess maybe sex is going to have to be brought back into proportion, become important again, have greater meaning. If the act has these huge, terrible consequences, people are going to be in trouble if they do it too casually.

"But we've gone so far toward making sex casual that I don't know how far we can back up."

BOOK III

14

A LIFE OF CHOICE:
CREATING HER OWN RULES

In so far as the generation of offspring is impeded,
it is a vice against nature which happens
in every carnal act from which
generation cannot follow.
—SAINT THOMAS AQUINAS

In the privacy of the love bed, the Pill burst as a gift of liberation. In the timeless war between the genders, it contributed to a zone of peace. In the silent, sweeping, rising tide of the world's population, it offered a protective dike.

In still another region of human activity, the Pill attracted an attention that approached a fearsome awe—a prolonged search for its deepest meanings; for how it helped define the meaning of life itself. That place was the Roman Catholic Church.

The study of the Pill by a surprised and unprepared church reached into and transcended almost all the imaginable ways in which this innocent chemical tablet was about to change the world. Therefore our search to understand the Pill must—*must*—pause and regard in detail what happened inside the Vatican when a hierarchy of sage men—all men—circled around this speck of an inert object, gazed solemnly at it, asked themselves what it was, what it meant, and what they were to make of it. Their scrutiny focused a brilliant light on how the Pill challenged age-old beliefs and traditions, not only of Catholics but of all humanity.

For its best telling, that story might be shaped to begin somewhere else—with a woman, a most private woman and her private thoughts.

In the first days of the 1960s, Anne Greene Biezanek, an English medical doctor, was in her midthirties. Tall, handsome, and blond, she was also keenly cerebral, intense, and had a tongue like a whip. A convert to Catholicism, she gave herself to her faith with an utterness one is likely to observe only in those who accept that faith as an intellectual choice.

This daughter of an Oxford-educated Quaker had chosen the faith so wholly that Thomas Aquinas himself would have taken notice and perhaps faintly raised an attentive eyebrow; so completely as to cause the foundation of St. Peter's Basilica to quiver uneasily, which indeed it soon did. Her devotion would also thrust Dr. Biezanek into the harsh glare of a sudden, unsought world notoriety.

When only nineteen, this young woman of radiant mind and beauty had achieved her bachelor of science degree at Aberdeen University in Scotland and immediately entered medical school. At about the same time, she searched for a religion. Not one for halfhearted undertakings, she buried herself in theological readings. The fine points of Catholic tradition and teachings she thus absorbed were far more complete than the instructions she was routinely receiving from her priests. She became not only a true believer but a learned one, prepared to defend her new beliefs—chapter, verse, and innuendo—from any assault.

While still a student she married a Polish ex–army officer, Jan Biezanek, a mechanically obedient Catholic, who in civilian life had been a lawyer and judge advocate, a profession he was not licensed to continue in Scotland, where he had come to live. He took employment as a transoceanic ship's steward, which meant long absences from home and consequent marital deprivations.

Anne birthed two children before she took her medical degree. To help support their new family, she went to work as registrar of a mental hospital. Part of her compensation was a nearby house, which the Biezaneks could not otherwise have afforded. After a third and fourth child as well as a miscarriage, the young wife was soon to write, "I sought relief in my religion and my husband. . . . And the closer we got, the very communion of matrimony, that which binds and comforts a couple in adversity, could only lead to yet another baby." While continuing in her demanding job, the young doctor already felt pushed beyond her limits.

In desperation, she went to the hospital's Catholic chaplain. When he impassively repeated the "rules," she protested that the church's moral law had trapped her into her present fix, and that something was wrong with it. If she had another child, she could manage either the job or the brood, but surely not both. And if she and her sexually demanding husband followed moral law, she was sure to have another, and another after that. The chaplain told her simply that she did not understand, and he recommended that she take on a spiritual director, an older and wiser priest. But he warned that enlisting a spiritual director required obedience to his counsel, no matter how burdensome. The devout young doctor did so, then began to urge her new advisor that, in a case like hers, contraception was not only urgent but morally justified. He ruled that out and instructed her to go on in both her job and married life and to trust the design of providence. She pointed out the danger to her health and to her competence at work if her husband's embraces should lead to another pregnancy. "He replied that that danger was entirely subordinate to my duty to love my husband," Dr. Biezanek has written, *"and as long as I did that, God would look after me."*

Before long Dr. Biezanek had a fifth child, a second miscarriage, and a physical and emotional collapse. "I had had faith and had submitted to my husband as directed to do and, as my training had quite correctly led me to suspect, the resultant pregnan-

cies left me unfitted for any further medical work. . . . I left with
the solemn conviction that I would never practice again." She
gave up her job, house and all. With her five children and husband she moved in with her willing but bewildered parents. In six
months she was pregnant yet again. After her sixth birth, she
committed herself to a mental hospital in Edinburgh.

Her therapists pressed questions that challenged her to reconcile the voices warring within her: those of her religious convictions and of her self-identity as a wife and mother, not to mention
as a doctor. "My problem was this: How could my own marriage
survive without the bond of sexual union that my husband demanded? My psychiatrists were assuring me that it could not.
How could I survive if more babies were to come? I knew I could
not. . . . At this point I began to suppose, with an agonized dismay, that I must be the only Catholic with a problem of an order
as great as this."

Under the pressure of her illness (if her plight was to be called
by that name), a new clarity shone on her dilemma: "The Catholic wife wants to be a loving wife, yet knows she must limit her
pregnancies. If she uses contraceptives, she is called wicked by
her parish priest. If she follows the advice of her priest and re-
frains from sexual intercourse, she is called cold by her husband.
If she doesn't take steps, she is called mad by society at large."

In the quiet of the hospital she pondered the most challenging
question of all, which she was later to unfold in a remarkable
memoir, published in England: "Why should the needs of your
soul force you to disrupt the lives of your whole family?" While
tormented by the question, she received no visits from her spiritual director or pastor. That, she wrote, "was the worst part of the
suffering that I then endured. I was standing out for what I had
been taught to believe was a fundamental Catholic doctrine and
hardly a Catholic I knew seemed interested. . . . Until the reasoning that underlay my complaint is understood, the reason for the
bitterness that is about to tear the whole Roman Catholic body
will not be understood. . . . For this reason did I, in these bleak
days, begin to see my personal dilemma as one that was bringing
judgment day itself to the intellectuals of my church."

After five weeks in the hospital she returned to her family. "At the time we were living with my parents in Scotland, living there with my six children, and then, oh God: Jan was sailing from Liverpool and we slept in a downstairs sitting room. We couldn't go to bed until everybody else went to bed. And because I had six children, I was really very tired and he'd been at sea and we used to want to go to bed. But, of course, he would want sex, which was understandable. So I had to tell them all again I was pregnant."

Anne's non-Catholic parents found the new pregnancy so inexplicable that she felt impelled to leave their home, board out her children, and spend her time trying to borrow money from relatives, medical societies, and Catholic charities. She knew that what she needed above all things was her own home. The chaplain at the hospital where she had worked arranged a guarantee of a mortgage on a redbrick house in Wallasey that looked across the River Mersey to the giant loading cranes of Liverpool. The simple house, set halfway up a hill from the shoreline, was soon to become famous.

Now the mere thought of another pregnancy was a nightmare.

My husband's brief and unpredictable appearances in the home became in themselves a source of torment. As I wished the children out of my way, so did I wish him. When his presence was added to my broken nights and chronic sleeplessness, I used to feel, and say, that if he as much as touched me I would take a running jump at the window. . . . Thus was I driven down the suddenly fashionable, Roman Catholic line of thought that maintained that the solution to all marital problems of my type lay in the abolition of sex. My husband had to be banished from my presence, into a room of his own. Everything in me that attracted him to me and me to him had to be suppressed. All this I attempted, and heaven is my witness. . . . Hate became the order of the day. In such an atmosphere, even prayer, or shall we rather say, above all things, prayer, becomes the most dangerous activity. When you pray, you must needs let your defenses down, you must lay yourself open to the influence of a loving and generous

Spirit. The next thing would be that you find yourself betrayed into actually kissing your husband good-night, and from then on. . . .

Dr. Biezanek told her parents that she felt that she must now start taking the Pill and that they ought to be the first people to know. They responded with an "explosion of relief and joy."

Her husband's response was less jubilant. Dr. Biezanek, today in her vigorous sixties, recreated for me that stressful time. On a Sunday afternoon a few miles north of Liverpool, she and I sat at a picnic table atop a windy knoll of a country pub, couples and children frolicking about. "My husband thought it was fine that I should take the Pill as long as I kept my struggles with my faith to myself," she recalled, underlining her words with strikingly large, strong hands. "It was my business, not his. In a properly run Polish Catholic society, you know, the women discuss this amongst themselves. The women know what to do, and the men are not troubled by it."

With absolute confidence in the rectitude of her position, Anne Biezanek proceeded. She has written of it:

First I had to obtain the Pill and this entailed a consultation with my family physician. I was covered with bruises the night I went to see him. I said, "I've been trying to work on abstinence. But my husband doesn't like abstinence. He only comes home now and then, you know." The physician didn't know at this time I too was a doctor. How could he have realized what a great step I was proposing to take? Poor man, he could not. He was not a Roman Catholic and could have had no notion of . . . [the] powerful and telling condemnation that takes place in the secrecy of the confessional. . . . He told me that I was hardly the first Roman Catholic woman to consult him on this subject, and he must have been surprised at the extent of the fuss I was making. But I wanted him to understand. I wanted everyone to understand, and I still do. . . .

So I started swallowing my pills on the prescribed date, May 25th, 1962. . . . But, as I swallowed my pills, I was sad. I knew my-

self to be presiding over the burial of . . . an idea to which I had
come to identify myself so closely that in a sense I was burying
myself. Who, I wondered, would lay flowers on my tomb?

In our interview, she further unfolded her anguish: "So I took
the children to church but didn't go to communion. Then Jan
said, 'You didn't go to communion.' I said, 'I told you, when I start
taking the Pill I wouldn't go to communion anymore.' He said,
'All devoted mothers take their children to communion. What
other sins are you committing?' And so I said, 'Look, the sin I'm
committing is seeing that we get some domestic peace. Have we
had any arguments about sex since I started taking the Pill?' He
said, 'No.' I said, 'I told you taking the Pill meant I couldn't go to
communion anymore. It's against the church rules.' He said, 'You
can go. You go to confession, then you go to communion, and
then you start taking the Pill.' But that won't work with the Pill,
you see. If you're using a condom or something like that, you
hold your breath, dash off to church, confess, and get communion, and then you can sigh and fall back again into your sinful
way of living. This is how most Catholics manage. But the Pill
you have to take every day.

"I said, 'That's maybe how a Polish male mind works, but it's
not how this English woman's mind works. I'll take all the children to mass and to Catholic schools, but I can't go.' He got terribly upset."

So Anne went to her confessor, the assistant parish priest near
her new home, announced she was on the Pill, and asked him to
assure her that he would give her the sacraments of confession
and communion.

He said he could not.

To her children Anne expressed her hurt and trouble. Her oldest daughter urged Anne to go with them to school for communion. Demanding of herself a complete forthrightness, Anne
wrote to inform the pastor of school mass, Father James Gaskell,
that she was practicing contraception and that on a date and hour
that she named she would appear to receive communion with her
daughter. As though unrecognized by him, she received the

bread and wine from the same priest who had said he would not give it to her. But she remained unsatisfied with the passiveness of that incident. "I can't as a woman, much less as a doctor, take remedies secretly," she said.

Next February, 1963, her parish was to be honored by a visit from its bishop. Anne wrote for an appointment with him. In their interview, she informed him that she was taking the Pill yet receiving communion regularly. She wanted his assurance that her "receiving" was legitimate. If he needed time to discuss her request with other bishops, while awaiting his verdict she would stop going up to the rail. He declared without hesitation that the matter was plain. She was transgressing an immutable law of God and was guilty of sacrilege. There was nothing further to consider.

She replied just as promptly that that was *his* view, and hers was another. She would voluntarily abstain from seeking communion, she said, if the bishop would personally explain to her children in any terms he thought fit why their mother could not receive. He said he would not do this.

She continued going to church and receiving communion every week, feeling herself no longer bound by the bishop's authority. "I became convinced," she later wrote, "that God wanted me to make this stand, and that it was He who had brought me so near to despair so that I should learn for myself how bullied and wretched Catholic women are."

For a year the Pill brought relative peace to Anne and Jan's redbrick house. When he returned from his voyages, she no longer needed to deflect his advances. Jan's tantrums, which had become violent, now subsided.

The prospect of no new children beyond their brood of seven opened new opportunities. Jan, getting older and on limited earnings, suggested that Anne resume work as a doctor. She found part-time work in the practice of other Wallasey physicians.

"I would see women coming in after the men were paid on Friday. They'd go to the pub and come home to their wives. Then I would dash around treating women after those filthy, criminal, induced abortions. They lived in two-ups and two-downs. The blood loss of those women! The bloodlines on the floor were terrible.

"The women used to say to me, 'How do you manage? You are
a doctor. You go to church. You've got seven children to show
you're a good Catholic. Yet you stopped having babies. You must
be very holy.' I had to say to them, 'It didn't have anything to do
with being holy. It was because I took the Pill.' They'd whisper,
'You take the Pill?' And I'd say, 'Well, it's available now, you know.'
They'd look at me some more: 'But they say it's wrong.' I began to
feel quite stirred up about this."

Then the idea came to her: *Why not open a clinic of my own?*

On the second floor of her hillside house in Wallasey, she
emptied a spare room of furniture, installed a washbasin, and purchased a few medical utensils and supplies. On September 8,
1963, Anne Biezanek accepted the first patient in her private
birth control clinic, the first known anywhere to be opened by a
Catholic physician.

She soon paid another call on her parish priest, trying "to persuade him to give me absolution on all other counts and, if he
could see his way, to suspend judgment on this particular matter.
He refused, though he continued to give me communion whenever I went up to receive. So on I breezed."

A few weeks later the Liverpool *Daily Mail*, under a headline
about a Wallasey woman doctor "defying" her church, printed the
story of Dr. Biezanek's clinic. It was followed by a ninety-second
interview on national television. The headline mortified Dr.
Biezanek. She was aware that in Rome the Second Vatican
Council was meeting and that the world awaited its widely rumored new, liberalized position on birth control. This was a time
for hope, not defiance.

She was soon jarred into learning that change was not about to
arrive. The Sunday morning following the headline and the TV
appearance, Biezanek's parish priest seized the communion plate
from her daughter's hand and said in full voice to the mother,
"You don't get it." Again Anne wrote to her bishop, who this time
was more specific about her offense than he was at their previous
interview: Even worse than "transgressing an immutable law of

God," his previous charge against her, she was guilty of attracting publicity and causing "scandal." This was where she had crossed the line!

Now spurred to militancy, Anne needed to settle her standing with her chosen church once and for all. She wrote to the highest-ranked Catholic in England, John Heenan, archbishop of Westminster, in London. (More than four years earlier, in January 1959, when all the world awaited the church's position on the coming Pill, Archbishop Heenan, a most political and cautious prelate, took a firm position of uncertainty. In a TV discussion of population control, he told a national audience that the church would not object to a birth control pill if it did not transgress divine law.) In her letter, she reviewed her entire adventure, pointing out that she had not been excommunicated, nor had she been hailed into any hearing or trial for her supposed offense. She informed the archbishop that she would appear at his altar at Westminster Cathedral on May 31, 1964, with the intention of receiving communion. She described the style and color of the clothes she would wear, and how she looked. She also informed the press.

When the day arrived, the cathedral was packed and surrounded by an overflow crowd. "There were a lot of prelates outside the church, a lot of television cameras around," she recalled. "I don't know if they were prelates, actually, but they had purple dresses on and were staring at everybody. They never looked at me directly."

Inside, without fuss or even the appearance of special notice, Anne Biezanek received communion. One reporter wrote that the incident "may have been the most publicized and photographed mortal sin ever committed." Later a church spokesman said the officiators had not recognized her.

Dr. Biezanek was satisfied that she had done the right thing. "I don't think the cardinal cared about what I was doing," she said a few days later. "He didn't want to be put on a spot—no fuss. What a loyal Catholic should do, so many think, is just practice contraception and keep quiet about it. But I won't have it that way. I'm a Catholic because I learned that life is something important here before something more important that is to come later. I would still sooner be publicly vilified than to have quietly pretended to be a Catholic. The fact that I believe in standing up for the truth as I see it is something I learned from the Catholic church."

The worldwide news reporting of that morning at Westminster brought Biezanek a flood of letters of support with stamps and postmarks not only from the British Isles, the United States, and Canada, but from Mexico, Pakistan, and the Philippines; from Australia, New Zealand, South Africa, and Rhodesia; from Holland, Germany, Italy, and France; from Cyprus, Japan, Sweden, and Poland, "from countries I have never been to, in languages I do not know, from men and from women, from Roman Catholics and Methodists and Anglicans and Presbyterians, from Jehovah's Witnesses and Quakers, from Buddhists and Moslems, from Humanists and Communists." One letter was addressed to *Dr Anne Biezanek, The Clinic, England.* The post office found her.

While continuing, after her visit to Westminster, to go to mass every Sunday in Wallasey, Dr. Biezanek stopped going to communion. "They said on that occasion that they did not recognize me," she observed, "so I see no point in going again. I am not going to receive communion by stealth."

A British publisher asked Biezanek to explain her challenge in a book. The work that resulted, *All Things New,* published in 1965, attracted a striking set of conflicting critical reactions. In England, Father Alban Byron, a Jesuit, while differing with it in the *Catholic Gazette,* called it "the most extraordinary marriage book I have ever read." In the *Clergy Review,* Canon E. H. Drinkwater, a Catholic theologian, applauded the book for lacking "all those tactful euphemisms and soft-pedaling, those delicate nuances and innuendoes, those discreet circumlocutions, which so often oil the chariot wheels of truth and which those of us who write under constant censorship get so good at." Canon Drinkwater added later in *Search:* "Let nobody imagine for a moment that the author is some kind of nagging eccentric or notoriety-seeker. . . . Here is a book in the same category as Newman's *Apologia.*"

In the United States, however, *Commonweal,* the most liberal of Catholic weeklies, turned down advertisements for the Harper & Row edition. Virtually all other Catholic publications simply ignored the book, except *The Critic,* which asked editorially what had become of responsible publishing at Harper & Row.

Biezanek's book challenged the church not only in its response to the Pill but by brazenly taking on its leading theologians in their seemingly immutable interpretations of the Bible itself:

It is frequently stated by Roman Catholic apologists that the Church has *always* opposed artificial methods of birth control. . . . But what artificial methods have been known (in the sense of being widely known) before the last 50 years or so? . . . What in fact the Church has condemned through the ages is birth prevention by means of abnormal sexuality.

Prior to the introduction of the condom and the dutch cap, there was only one way to ensure that conception did not occur, and that was to pervert, in some way, the very nature of the sexual act. The aim of the particular perversion must needs be to prevent the semen entering the vagina, and this could only be achieved by either the penis never entering the vagina, or else being withdrawn prior to seminal emission.

. . . The argument of the theologians against *coitus interruptus* is based simply on the assertion that such an act frustrates nature. . . . If one inquires further and asks 'whose nature?' they are rather liable to . . . start spelling nature with a capital 'N', and talk about the purpose of the act being primarily reproductive.

The simple truth of the matter is that when two persons engage upon an act of sexual intercourse their primary aim is not children but orgasm. If they conduct the act in such a way as to render orgasm virtually impossible for one partner then they are acting so as to frustrate nature. . . .

The story about Onan in Genesis Chapter 38 [describes] how Onan, in order to avoid conception, spilled his seed upon the ground during intercourse, and was slain by the Lord for the "detestable thing" that he did. These words are very commonly used to prove that birth control is a detestable thing. . . . The only snag is that the men who were responsible for upholding this argument did not, in fact, fully understand it. Not understanding women themselves, they allowed their attention to wander from the needs of women and the nature of women, and found it easier instead to consider nature itself—some impersonal natural force "out there," to which all must needs bow down and do homage. They have made of this nature their god and have made it in their own image and likeness. A god who does not love women.

The year of Biezanek's challenge at Westminster was also the year of the publication of Betty Friedan's vastly influential book, *The Feminine Mystique*. It has become something of a habit to say that publication of that book set off the women's revolution. Ms. Friedan herself has scoffed at that simple cause-and-effect attribution, saying that the imperative elements of the combustion were already there. Her book, however, combined with Simone de Beauvoir's *The Second Sex* and Kate Millett's *Sexual Politics*, set a match to what was ready and waiting to go up spontaneously.

Perhaps something like that also describes the Biezanek challenge. It could be seen as simply a news event and could as easily have disappeared in the daily media stream. (Indeed, Dr. Biezanek soon left the church, was soon left by her husband, and turned to private practice and the rearing of their children.) But the confrontation at the altar of Westminster might also be taken as the ha'penny nail that split the dike that unloosed a deluge across the landscape of the Roman Catholic Church.

The first signs were small. A young English priest declared that the doctor from Wallasey was doing the right thing and that their church was wrong. The Reverend Arnold A. McMahon, a twenty-five-year-old teacher of missionary students at St. Richard's College in western England, dared his assertion in a letter to the editor of the Birmingham *Post*. Feeling compelled to defend Dr. Biezanek, her clinic, and her book, he wrote, "I have come to believe that not only many Catholics use contraceptives—I believe they have the right. Nobody can take this right away from them, for nobody can take away another person's humanity."

When a reaction to his letter erupted, Father McMahon said that "things may happen to me," but if they do, "it will only be fresh sad evidence of how serious the problem is. There is in our church a totalitarianism that is eating away at its heart." Father McMahon was hailed to Rome for consultation and was ordered to a retreat house to contemplate his insubordination.

Within days, McMahon was echoed by twenty-six-year-old Father Joseph Cocker of St. Mary's on the Isle of Wight, who wrote to a Catholic weekly: "No answer has been given to Father McMahon's article, which I thought was splendid. . . . There is not necessarily anything selfish about a woman wishing to love her husband without spending the majority of her youth pregnant." Cocker was promptly rebuked by Monsignor Joseph Mullarky, the vicar general of his diocese.

If there was a date when the dike broke open, perhaps it was March 21, 1963, and the place, appropriately, in the Netherlands. William Bekkers, bishop of 's-Hertogenbosch, speaking on KRO, Holland's Catholic radio station, talked of "birth regulation"—a significant departure from the detested term "birth control"—as "an integral part of the total task entrusted to married people. . . . Man has got to project the number of his offspring against the whole kaleidoscopic background of his married life. There is no particular merit in having a large or small family. . . . The couple, and they alone, decide what the size of their family should be and what span of time there should be between the births of their children. . . . This is a matter for their own consciences, with which nobody must interfere."

What was it that so troubled this bishop that he would venture so far out of line? In an interview with Robert Blair Kaiser of *Time*, Bekkers, avoiding the arcane theorizing of theologians, explained himself simply and boldly: "The people have troubles. They are inquiring. And so I argue this way. If I see people in the church not going to communion because they feel guilty in violating the ban on contraception, and I know they are the kind of people who would otherwise be going to communion, then I say this is a reason for reconsidering the entire question." He concluded with a quiet observation that made ears quiver in Rome: "The context of a dogma can change."

Willem van der Marck, a priest in his mid-thirties, soon published a startling defense of the Pill in a learned Dutch journal, *Tijdschrift voor Theologie*. Its English-translated abstract said: "Fertility control is not only licit and approved by the church, it is a Christian responsibility. As for the means to be employed in fertility control, the rhythm method has until now received exclusive approval, not, indeed, because it is so 'natural' . . . [but] because it does not interfere with the marital act itself. In this respect, the Pill offers no difficulty. The question of its actual goodness or badness must thus be answered [by] the personal attitude and mentality of the married couple."

At Harvard University, Michael Novak, collecting a book of essays anonymously written by thirteen American Catholic couples to "describe the daily concrete realities of sexual life and marriage," which was soon to be published as *The Experience of Marriage*, sent a letter to the fifty-eight-year-old Jesuit weekly *America*. The letter protested: "The present Catholic teaching depends much more heavily on the weight of authority than on the weight of philosophy or reason. . . . At present, many of us find it almost impossible to argue with priests, too many of whom retreat too soon to [Vatican] authority."

Before long, the editors of *America* allied themselves with Catholic physicians who, they said editorially, saw "the compatibility—and even necessity—of some use of contraception in the life of the genuinely Catholic family." It was "obviously too early" to know if such views would prevail within the church, but, the editorial added: "In our judgment, they should. . . . How can couples give themselves to each other and to their children in the unselfish traditions of Christian love when another pregnancy in the family is a constant—and justified—worry?"

Within days after the Bekkers thunderclap in Holland, the rapidly gathering storm centered on the appearance in America of John Rock's book, *The Time Has Come: A Catholic Doctor's Proposals to End the Battle Over Birth Control*. It put forth the arresting proposition that the Pill did not violate the precepts of modern Catholicism because it was a "morally permissible variant of the rhythm method." The Harvard gynecologist pointed out that the Pill merely employed modern versions of natural hormones to extend the period in which a woman was naturally sterile from a portion of a month to several months. Use of the Pill to "regulate" the menstrual cycle—in a sense, to better practice rhythm—was *already* permitted by the church. This logic, Rock urged, was plain for anyone to see: "You can't pass the parental buck to theologians. People must have the guts to take their own responsibility."

That unprecedented reasoning by Rock and other insurgents became a bedevilment to the Catholic hierarchy. Rock advanced an approach to contraception, based on a scientific innovation, that church thinkers had never been required to consider, nor were they notably qualified to do so.

Richard Cardinal Cushing, archbishop of Boston, wrote a review of Rock's book for the Boston *Pilot*, a church journal, that foreshadowed an epidemic of uncertainty. The cardinal first rapped the doctor's knuckles lightly for having failed to abide by a church law that, as Cushing put it, "requires every Catholic who writes on a subject pertaining to faith or morality [to] submit his manuscript to church authority for a so-called 'imprimatur.' " But then the cardinal went on to say, "In this book there is much that is good. . . . [Rock] has clearly demonstrated that the church is not opposed to birth control as such but to the artificial means to control births. . . . Some of his suggestions could contribute to the establishment of domestic peace in our pluralistic society."

But in a *Washington Post* review, Monsignor John C. Knott, director of the Family Life Bureau of the National Catholic Welfare Conference, was more traditional, severe, and blunt: "The cause of honest discussion would be better served if Dr. Rock and all Americans were to face the reality of the Catholic position on contraceptives. It has not changed and will not change."

Rock's thinking on Catholicism and birth control was not to be tossed aside. Its evolution, wrote Loretta McLaughlin, Rock's biographer, was "deeply influenced by the famous French Jesuit anthropologist Pierre Teilhard de Chardin. Rock saw Teilhard de Chardin as 'perhaps . . . the Socrates, the Aquinas of our epoch, and, like them in their day, denied by blind men in high places.' Teilhard de Chardin's writings not only accepted Darwin's concepts of evolution but also tried to converge science and religion. The books had been condemned by the Vatican throughout the author's life. Rock particularly subscribed to Teilhard de Chardin's fundamental thesis, that Mankind, in its continued socialization, was struggling, evolving to a higher plane 'onward and upward toward its goal: Truth.' In that context, unbridled reproduction came to represent in Rock's mind 'a whirlpool of destruction.' "

Gathered in Rome for the Twenty-First Ecumenical Council (which came to be known as the Second Vatican Council, or Vatican II), other European bishops began to join the Dutch skeptics in wondering if the rejection of Rock was not too hasty. A German, Bernard Häring, Pope Paul VI's private theological counselor, said in an interview with Time's Kaiser, "We have a right to help nature." There was no reason, he suggested, why a woman couldn't take the Pill to help regularize her cycle. Whether "Rock's pill" was temporary sterilization (which would be forbidden) or a way to control "rhythm" (which might be permissible) was beyond his competence as a priest, said Häring, but he was aware that scientific advances in the synthetic production of estrogen and progestin were moving ahead rapidly. Häring expressed the unabashed hope that science would soon devise "a Catholic Pill" that would "perfect natural functions," making it possible for a woman to ovulate every month when she wanted to.

As a sign of the times, when Newsweek recognized the convening of the Ecumenical Council with a cover story, the text was accompanied by a photo of Pope Paul VI. But the cover itself displayed a portrait of John Rock.

In September 1964, Look magazine reported "an anguished Catholic debate now coming out into the open . . . in almost every Catholic publication." One example, a letter from a Boston priest appearing in the church magazine Jubilee: "I am not a theologian, but the theologians have to either come up with a better answer, or else I and many other priests will not be able to justify our own consciences in the confessional."

Another: a series of articles in the Arkansas Gazette by the Reverend James Drane, a thirty-seven-year-old professor of philosophy and Romance languages at St. John's Seminary in Little Rock. It said that "we need a new set of rules" about birth control; that celibates should not make those rules because "love cannot be understood from the outside but must be experienced."

Bishop Albert L. Fletcher of the Little Rock diocese promptly relieved Father Drane of his seminary post and forbade him to perform priestly duties in public, saying that the articles had made the author "suspect of heresy"—if validated, the most damnable of offenses for a Catholic.

"I have two sins," Father Drane responded. "First, I said what I said in public forum, and second, I addressed it to the people who need it most, the people trying to love each other in a three-bedroom house." Father Drane addressed a formal appeal to the Vatican, posing a question that possibly no pope had ever been asked by a priest: "Are the ideas I expressed the exclusive questions of the theologians? Or have the people the right to know? I think they have."

Some of Father Drane's other defiant thoughts were soon spread across the nation in *Life* magazine:

> One of the temptations that must always be guarded against in the Church is to make everything about it divine, to make everything sacral, to make everything about it unchangeable, to make everything about it perennial. This amounts to raising to the status of divinity that which is human. . . .
>
> Evidence of the truth of the statement that the Church is human is in its politics. The same type of human political activity that takes place in the state houses takes place in Rome, and no one should be shocked that in a human institution people act like human beings. . . .
>
> Everything possible must be done to push Church authority along, for its members are celibate and do not seem to appreciate the urgency of this issue of birth control. When people are suffering, delay is a form of cruelty.

15

BIRTH CONTROL
VEILED AND UNVEILED

. . . Seeking the bubble reputation even in the cannon's mouth.
—SHAKESPEARE, *AS YOU LIKE IT*

What many remember most vividly of the U.S. presidential election of 1960 was its televised campaign debates, permitting voters the first opportunity to seek out intimate clues to their candidates' personalities. The debates juxtaposed the youth and easy wit of Senator John F. Kennedy against the dark stubble and wary eyes of Vice President Richard M. Nixon. What fewer may recall was that a third man hovered over that race, like a ghost candidate "occupying" an empty chair. The shadowy extra participant was the pope—not necessarily "Good Pope" John XXIII, who reigned that year, but any pope, generically.

Once before a pope had "run" in an American election, as the uninvited ticket mate in 1928 of Alfred E. Smith, the first Roman Catholic to be nominated by a major party. Smith's defeat by Herbert Hoover was probably certain even without papal assistance, but it would be a long time, many Democrats vowed, before they'd nominate a Catholic for president again. Indeed, it was.

The abuse heaped upon John Kennedy at every bend of the campaign trail was that his election would put the pope in ultimate charge of the White House. This was not a simple rerun of 1928—the idea that the primary allegiance of any Catholic was to his church, not his country. In 1960 something new was in the air: the Pill. If the Pope was against it, he'd order Kennedy to be against it. Even non-Catholic critics of the Pill were resentful. Opposing birth control was their American privilege, and they didn't need the pope's help to make their opposition stick.

Kennedy's accusers were by no means all backwoods bigots. Some were also staid Presbyterian and Episcopalian bishops and ministers and members of their lettered flocks. Some were persecution-jittery Jews, and to a surprisingly high degree, university intellectuals. (It has been remarked that anti-Catholicism is the anti-Semitism of liberals.) Kennedy sorely needed to reassure the Protestant majority of his independence but could not afford to do so by alienating the one third of the voters who were of his own faith.

Only a year earlier his predecessor, Dwight D. Eisenhower, had taken a four-square stance opposing government support of birth control, a position that surely pleased the pope, although his reasons were quite different. Said Eisenhower: "I cannot imagine anything more emphatically a subject that is not a proper political or governmental activity or function or responsibility . . . That's not our business." He was soon to have second thoughts.

Kennedy, unable to elude the subject, agreed to a telephone interview with *The New York Times*. It provided the following morning's page-one lead story for papers across the country. The reporter asked directly where he stood on government subsidy of birth control. Kennedy's unhesitating response, establishing his presidential deftness at avoiding a clear yes or no, was that if he were president and legislation came before him, he would decide in accordance with his oath to do "what would be in the interest of the United States. If it became a law of the land, I would uphold it as the law of the land."

On the fuzzier issue of promoting birth control to limit population in foreign countries: "We have to be very careful about how we give advice on this subject. The United States government does not advocate any policy concerning birth control here in the United States. Nor have we ever advocated such a policy in Western Europe. Accordingly, I think it would be the greatest psychological mistake for us to appear to advocate limitation of the black, or brown, or yellow peoples whose population is increasing no faster than in the United States. . . . The question I think we all have to address ourselves to is whether the available resources of the world are increasing as fast as the population. That is the overall question. My belief is that they are, though their management may not be."

Later, in answer to a student's question, Kennedy straddled, but more specifically: "I may be opposed to birth control as a member of my church, but I have no desire to impose my views on others." On television, a viewer challenged him: What if his archbishop confronted him with a straight-out directive? Kennedy answered that one flatly: "I simply would not obey it."

Kennedy did not readily win the skeptics over. One prominent Catholic-born voter, Margaret Sanger, said shortly after his nomination that if Kennedy became president, she would "find another place to live. . . . In my estimation a Roman Catholic is neither Democrat nor Republican, nor American nor Chinese. He is a Roman Catholic." The day after his election, Sanger softened her threat: "Since I made that statement some of my friends who are also very close friends of Senator and Mrs. Kennedy have told me they are both sympathetic and understanding toward the problem of world population. I respect the judgment of these friends. I will wait out the first year of Senator Kennedy's administration and see what happens. I will make my decision then."

By 1964, a year after Kennedy's assassination, President Lyndon Johnson, over strong opposition, put through federal support of birth control for the American poor. A pilot project first involved a few families in Galveston, Texas. Through Johnsonian force and speed, within a decade there was a government-supported birth control clinic in 2,379 of the nation's 3,099 counties. Clinics—located in hospital wards, storefronts, farmhouses, and trailers—offered films and lectures teaching the pros and cons of various contraceptive methods from rhythm to sterilization. More than any other method, women chose the Pill. The clinics supplied the Pills and regular follow-up checks. More than three quarters of "patients," as clinics came to call their clients, had an income of less than $6,300 for a family of four.

In 1970 President Richard Nixon swung into the rhythm of the era by propounding a new "national goal": "adequate family-planning services within the next five years for all those who want them but cannot afford them." That goal reflected a sea change brought about by the Pill but also by a growing permissiveness in many areas of American life, concern over an explosive growth in welfare rolls, and the rise of the women's movement with its new focus on an individual's right to control her own fertility.

By 1973 nearly four million women in every part of the country had received birth control aid at government expense. From the 1964 grant of $8,000 for the Galveston project, the national budget for birth control soared to $190 million in 1973 and more than $400 million in 1975. Those increases, social-welfare advocates argued successfully, would help poor families limit births, thus soon pay off by reducing welfare rolls. The compelling logic fell short of explaining the reality that soon developed. In 1989, the federal government spent more than $21.5 billion to support families of teenage mothers.

Among the converts to the new trend was former President Eisenhower, who said: "Once as President, I thought and said that birth control was not the business of our federal government. The facts changed my mind. . . . Governments must act. . . . Failure would limit the expectations of future generations to abject poverty and suffering and bring down upon us history's condemnation."

Even before federal money had begun trickling into the first

American birth control clinics, foreign-aid funds of rich countries began making their way into poorer ones. Starting with about $2 million in 1960, the world sum multiplied by 1972 to $250 million dollars. And by that time, even America's commitment to global birth control was deep. About half of the world's donations came from the U.S. Agency for International Development (AID).

For all its good intentions, the vast network of U.S. clinics occasionally developed aberrations that appalled its most devoted supporters. In South Carolina, a low-income woman claimed that the only physician available to deliver her fifth baby refused to do so unless she agreed to sterilization after the delivery. In Alabama, a white doctor at a clinic sterilized two black teenage sisters after their mother gave signed permission to do so. Later, the scarcely literate mother said she had not understood what she had signed and filed a five-million-dollar suit against the federal government.

The path of sterilization as a form of birth control had meandered widely. During World War II, to the horror of an outer world that soon learned of it, Adolf Hitler used sterilization as a way of limiting the reproduction of Jews and of "purifying" the Aryan strain. Even before that, Margaret Sanger was one of many won over by the "eugenics" movement in England led by her friends the Drysdales and Havelock Ellis. Those reformers, driven by goals they felt were humanitarian, as against the genocidal intentions of Nazis, believed in improving the human species through encouraging reproduction by persons having "desirable" genetic traits.

By the 1980s, in a dramatic turnaround, sterilization for women and vasectomies for men had become so popular among the relatively educated middle class that they surpassed the Pill as the world's most widely used contraception among married couples.

Hastened by the Pill, the government-backed birth control clinics had soon stumbled over an unexpected and tender political difficulty. Early in 1966, a conference on population and birth control in a predominantly black church in Germantown, Pennsylvania, a part of Philadelphia, listed a question for discussion: "Can we afford birth control as a minority group in the United States?"

The question was doubly surprising because it flew directly into a rapidly rising wind of social and environmental concern: overcrowding. That concern would find expression within a few months through a best-selling book, *The Population Bomb*, by California sociologist Paul Ehrlich.

Of seven panelists in Germantown, all black, two answered no. One of them, a pathologist, demanded: "If we have empty areas in the world and there is an oversupply of food, where is the population problem? This whole thing is the concern of people who don't want too many of a certain race."

The question set the room abuzz. A social-worker panelist, Uvelia Bowen, a board member of Philadelphia Planned Parenthood, worried aloud, "The trouble is that it is either whispered or hurled as an angry accusation. The fear is certainly there, hidden in the thoughts of every Negro man and woman in the United States, whether they express it or not. We are only ten percent of the entire population, and many of us feel our whole strength lies in numbers, in the power of the vote."

Validating Bowen, *The Washington Post* soon quoted Ruby Evans, an official of the United Planning Organization, the capital's official War on Poverty agency, as saying she opposed efforts to reduce the soaring birthrates among the black poor. "It's part of being a woman to get babies," Mrs. Evans told an audience of thirty teenage women, many of them already experienced with premarital pregnancy. (UPO officials later said that despite "the private views of an individual employee," the agency would continue to furnish birth control devices and counseling to the poor.)

The "accusation" went public in December 1967 when the Pittsburgh branch of the National Association for the Advancement of Colored People (NAACP) charged that Planned Parenthood clinics were devoted to keeping the black birthrate as low as possible. The clinics, said the statement, were operated "without moral responsibility to the black race and [had] become an instrument of genocide for the Negro people." Later the statement's author, Dr. Charles E. Greenlee, a physician, allowed that perhaps the term *genocide* was too strong, but he said he would withdraw nothing else.

The strong term caught on, however. "Across America," reported *Ebony* magazine in March 1968,

> black people are raising even deeper queries: Is birth control just a "white man's plot" to "contain" the black population? Is it just another scheme to cut back on welfare aid or still another method of "keeping the black man down"? The questions come mainly from the black ghetto (middle-class Negroes have accepted contraceptive practices well), and they come not only because of concern about "containment" and welfare cutbacks, but also because of a very prevalent idea that birth control actually means "black genocide." That was the key phrase in an anti–birth control resolution passed at last summer's Black Power conference in Newark. Since then, opposition to birth control has grown.

Ebony quoted Pittsburgh's Dr. Greenlee who had, despite his disavowal of *genocide*, grown more militant: "Our birthrate is the only thing we have. If we keep on producing, they're going to have to either kill us or grant us full citizenship." He said that Planned Parenthood "coerced disadvantaged black people to employ birth control by sending workers door-to-door until women feel they are forced to go to a clinic. I don't oppose contraceptives *per se*, but I'm against this 'Pill-pushing' in black neighborhoods where many people are made to feel that they'd better obey 'official suggestions' to visit a birth control clinic or risk losing their monthly welfare checks."

Greenlee's chief adversary, Douglas Stewart, national director of Planned Parenthood's community relations, also a black, acknowledged the widespread emotional current that Greenlee was voicing, as well as a new fear: "Many Negro women have told our workers, 'The Pill comes in two kinds—one for white women and one for us—and the one for us causes sterilization.' This is a very real fear for some women. Perhaps it's because many of them are from the South where black people have heard of instances of unwarranted sterilization by white clinic workers whose attitude seems to be 'Let's get the Negro before he's born.' These women bring their fears and suspicions with them when they move North."

Strange bedfellows were made by birth control politics, too. Greenlee's closest ally in the Pittsburgh campaign against "genocide" was William "Bouie" Haden, head of a black self-help group, United Movement for Progress. Haden was awarded a ten-thousand-dollar grant, the only black leader to receive one, from the Catholic diocese of Pittsburgh to support his neighborhood work.

Another convert to the "genocide" viewpoint was Dick Gregory, the comedian and political militant: "I'm one black cat who's going to have all the kids he wants. White folks can have their birth control. Personally, I've never trusted anything white folks tried to give us with the word 'control' in it. For years they told us where to sit, where to eat and where to live. Now they want to dictate our bedroom habits. First the white man tells me to sit in the back of the bus. Now it looks like he wants me to sleep under the bed." (Gregory and his wife, Lillian, had eight children.)

More than a year after the official baptism of the Pill, birth control as a topic of open discourse remained at the edge of scandal, in some states outright illegal. Yet by September 1961 the new trickle of permissibility permitted *Newsweek* to break a piece of extraordinary news as the lead item on its "Medicine" page:

> Complete information on birth control is now available in stores and stands. Two books at 50 cents each—both written by experts and aimed at people who cannot or will not pay $4 or $5 for the hardcover books that deal explicitly with the subject—are on sale across the U.S. this month. . . . Publishers were a little leery of publishing a popular treatise on birth control. Dr. [Alan F.] Guttmacher's book [*The Complete Book of Birth Control*] was turned down by two houses before Ballantine agreed to bring it out. Both Ballantine and Paperback could still conceivably run into trouble in Connecticut where an 1879 law prohibits the release of advice on contraceptives.

On New York's Lower East Side, Margaret Sanger, a social service nurse, watched a twenty-eight-year-old immigrant mother of three, Sadie Sachs, bleed to death following a five-dollar abortion. "A moving picture rolled before my eyes with photographic clearness," she wrote. "Women writhing in travail to bring forth little babies; the babies themselves naked and hungry, wrapped in newspapers to keep them from the cold. . . . The scenes piled one upon another. I could bear it no longer." (Sophia Smith Collection, Smith College)

By 1916, when she sat for this photo, thirty-year-old Margaret Sanger had organized the Birth Control League of America, launched the radical Birth Control Review, faced jail for doing so, and embarked on a national speaking tour. Foreseeing vividly her life's mission and carefully designing an image to suit it, Sanger engaged a speech coach and "reduced" her age by three years. (Sophia Smith Collection, Smith College)

In 1904, when she was twenty-nine, Katharine Dexter married Stanley R. McCormick, heir of the immense farm-equipment fortune of Cyrus McCormick. Their majestic wedding took place at her mother's Château de Prangins in Geneva. Almost a quarter of a century later, Katharine McCormick lent the château to Margaret Sanger for a reception for the three hundred delegates to the first World Population Conference. (The MIT Museum)

Katharine McCormick at eighty-eight. (The MIT Museum)

"Halfway to the corner . . . shawled, hatless . . . all day long and far into the evening, in ever-increasing numbers they came" to the first birth control clinic in America, opened by Margaret Sanger and her sister Ethel Byrne in 1916 at 46 Amboy Street, Brooklyn. (Sophia Smith Collection, Smith College)

Russell Marker's doctoral thesis was published in the Journal of the American Chemical Society, but he walked out of the University of Maryland in 1925 without his degree. Marker's distressed advisor warned his star student he "would have no future and would end up as a urine analyst." By 1940, without a doctorate, Marker was a professor at Penn State University and an established researcher. (Russell Marker Collection, Pennsylvania State University Archives)

In January 1942, only a month after Pearl Harbor, a nervous U.S. embassy in neutral and often anti-American Mexico urged Russell Marker to forget his yam-seeking mission and go home. When Marker (foreground, at curb) insisted on the pursuit, the embassy helped him hire a botanist who knew no English but who owned a truck. Between Mexico City and Orizaba the truck broke down. When the botanist (beside the truck) abandoned him, Marker doggedly resumed his search alone—by bus. (Russell Marker Collection, Pennsylvania State University Archives)

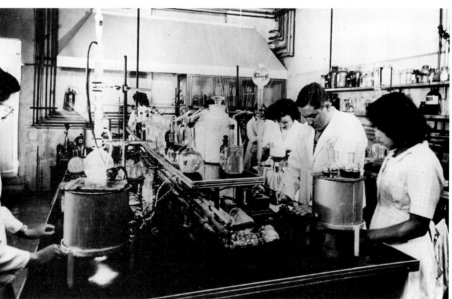

In 1949, at age twenty-six, Carl Djerassi made the wrenching decision to forsake a promising career in the United States to join an obscure, shaky new company in Mexico—a country almost devoid of science research—and to direct a laboratory mostly of young women with high school educations trained as laboratory assistants. His brazen goal: to invent a chemical synthesis of progesterone that was "better" than progesterone. (Chemical Heritage Foundation)

Denied tenure after an outstanding young career at
Harvard, Gregory Goodwin
Pincus, at age thirty-nine,
gambled with a colleague in
1944 on forming the independent Worcester Foundation for
Experimental Biology. By
1967, Pincus had achieved
world renown as developer and
"product champion" of the
Pill. (Worcester Foundation
for Experimental Biology)

A Puerto Rican social worker visiting the home of a mother of ten who had volunteered to go on the Pill during its field trial. Before starting, the volunteer received a physical examination and individual counseling at a clinic. (Suzanne Szasz)

The young wife of a Humacao housepainter waiting with one of her three children for counseling during her participation in the field trials. The young woman's mother, who had six children, also signed up for the field tests and went on the Pill. (Suzanne Szasz)

Dr. John Rock, the eminent gynecologist of Harvard Medical School, "would not dare advance the cause of contraceptive research and remain a Catholic," wrote Margaret Sanger to Katharine McCormick, opposing his participation in the Pill trials. But soon she changed her mind: "Being a good R.C. and as handsome as a god, he can just get away with anything." (Harvard Medical Library)

Early on May 31, 1964, Dr. Anne Biezanek prayed privately at Westminster Cathedral before challenging her bishop to give her communion. After receiving it without fuss or even the appearance of special notice, she was greeted outside by a horde of reporters, who transmitted the news of her confrontation to every corner of the world. One reporter wrote that the incident "may have been the most publicized and photographed mortal sin ever committed." (Praying: Hulton Deutsch Collection Limited; with reporters: © The Press Association Ltd.)

Yet, as *Time* reported in 1961, "Late every night in Connecticut, lights go out in the cities and towns, and citizens by the tens of thousands proceed zestfully to break the law." In fact, as late as 1964, after the Pill's fourth birthday, seventeen states had laws limiting the sale, distribution, or indeed the advertising of birth control devices. Among them still was Connecticut's neighboring state, an irony that stirred John Rock to observe: "It must have amused some of the citizens of the Commonwealth of Massachusetts, with its rigid law against birth control, to discover that the first breakthrough in contraceptive technology in 75 years suffered and survived its labor pains in the environs of Worcester and Boston. Life, indeed, has a way now and then of mocking man's more questionable designs."

Even fervent, powerful Cardinal Cushing of Boston appeared embarrassed by the Massachusetts law: "Catholics do not need the support of civil law to be faithful to their religious convictions, and they do not seek to impose by law their moral views on other members of society."

In the closing days of 1961, New Haven, Connecticut, witnessed a restaging of the forty-five-year-old scene of Margaret Sanger and her illegal storefront clinic in Brooklyn, but with a different cast. Playing the roles of Margaret and her sister Ethel were Dr. C. Lee Buxton, chairman of the Yale Medical School's department of obstetrics and gynecology, and sixty-one-year-old Estelle Griswold, executive director of Connecticut Planned Parenthood. They defied state law by opening four Planned Parenthood clinics, receiving about seventy-five women, all married, who sought advice and treatment.

Buxton and Griswold warned the women that police were expected to arrive and shut the place down. "If they do that," one clear-eyed housewife said, "we'll just sit down here until we get the information we came for." Ten days after the opening, police appeared at a converted mansion on Orange Street where Dr. Buxton and Mrs. Griswold, by prearrangement, awaited them. In the obvious test case, a benign district attorney (poor casting for Anthony Comstock) invited Buxton and Griswold to write their own script for their arrest. Buxton said later, "He asked how we'd like it done. Did we want him to send a paddy wagon with policemen and photographers, or would we prefer to turn ourselves in quietly?" The accused chose the latter and were each charged with being "an accessory to the use of contraceptives."

The state's highest court had already upheld the eighty-year-old prohibition, and it was upheld again when Dr. Buxton and Mrs. Griswold unsurprisingly were found guilty a month after their arrest. They were fined one hundred dollars each. The judge, going along with the open effort to test the law further, overlooked its provision of imprisonment of up to a year. The goal of the defendants this time was a decision by the United States Supreme Court.

More than three years later, in March 1965, the Court finally heard argument on the Connecticut law on which it had twice declined to act in the past.

Justice Potter Stewart demanded of Joseph B. Clark, New Haven's special prosecutor: What is the purpose of the statute barring contraceptive devices?

"To reduce the chances of immorality," Clark replied. "To act as a deterrent to sexual intercourse outside marriage."

"The trouble with that argument," Justice Stewart pointed out, peering down at the New Haven charges, "is that on this record it involves only married women."

Clark, switching his ground, asserted that the anticontraceptive law was a valid exercise of Connecticut's police power. The state had a right, he said, to provide for the "continuity" of its own population by banning birth control devices. Then he added apologetically, "Personally, I'm not too happy with it."

"Well, what *are* you happy with?" Justice Stewart persisted.

Clark said the case was purely one of the state's legislative right: "Does the state have the right to enact laws in this area? I think it does."

Justice Arthur J. Goldberg told Clark that previous decisions had established that a law must be closely tailored to meet the particular abuse at which it is directed. Why did Connecticut need a ban on contraceptives when it already had laws prohibiting adultery and fornication?

The Connecticut law, responded Clark, made it "easier to control the problem."

On June 7, 1965, by a vote of 7–2, the Supreme Court struck down the Connecticut birth control law in what *The New York Times* described on its first page as "a sweeping decision that established a new constitutional 'right of privacy.'" The seven members who declared the statute unconstitutional were divided in their grounds for doing so. Two of them, Stewart and Hugo L. Black, did so on the basis of free speech and press, objecting to striking down laws that were considered "unreasonable," even though the law did not violate a specific provision of the constitution. Even so, Stewart called the 1879 Connecticut vestige of Protestant prudery "an uncommonly silly law." Black found it "every bit as offensive to me as it is to my brethren of the majority." Writing the majority opinion, Justice William O. Douglas made history with the declaration: "We deal with a right of privacy older than the Bill of Rights—older than our political parties, older than our school system. Marriage is a coming together for better or for worse, hopefully enduring and intimate to the degree of being sacred."

It turned out that Cardinal Cushing was not the only Catholic embarrassed by the "uncommonly silly" laws. "Aside from the fact that you can't legislate morality," asserted Monsignor John C. Knott, director of the Family Life Bureau of the National Catholic Welfare Conference, "this [Connecticut] law was a bad one. . . . I am glad it was held unconstitutional." Archbishop Henry J. O'Brien of Hartford called the decision a "valid interpretation of constitutional law," which "in no way involves the morality of the question." Another Catholic notable without regrets was the Jesuit theologian John Courtney Murray of Woodstock College. Even while the law was still standing, he derided it for making a public crime of what the church had regarded as a private sin. It was bad lawmaking, Father Murray had said as early as 1957, because enforcing it would have required a policeman in every bedroom.

One week after the Connecticut decision, the New York state legislature voted to repeal an eighty-four-year-old law banning dissemination of birth control devices and information. New York's new law provided that birth control devices could be sold only by licensed pharmacies but not advertised or displayed, and not sold to persons under age sixteen. (Nine years later, the same legislature killed a move to permit display of contraceptives on pharmacy shelves. The bill's sponsor, Constance E. Cook, watching her bill assaulted by members of both parties, hooted, "I'm surprised they allow people to sell beds in this state.")

Seven years later, on March 23, 1972, the Supreme Court further ruled that a state could not stand in the way of distributing contraceptives to single persons when married couples could legally obtain them. The decision came in knocking down the ninety-three-year-old Massachusetts law. This paved the way for a three-judge federal panel to declare unconstitutional a similar law in Wisconsin, the last remaining state to have one.

Another restrictive practice might have been interpreted as awed respect for birth control, but it too attracted outrage. In New York state, the National Organization for Women (NOW) protested a case of a married woman who was told by a bank that her income would be counted as dependable in a loan application only if she could prove she took the Pill regularly or had undergone a hysterectomy. When the woman offered proof of her husband's vasectomy, she was told that the proof was unsatisfactory. She could become pregnant through another male. In 1975, the federal government issued regulations aimed at eliminating such credit restrictions against women.

In 1976 the New York Telephone Company refused to run an ad paid for by Planned Parenthood announcing the Margaret Sanger Center services of "birth control, pregnancy detection, abortion up to 12 weeks, sterilization, infertility treatment, referral and counseling." The phone company wanted the ad to offer only "information" about the services. Said Alfred F. Moran of Planned Parenthood: "We didn't understand what their mentality was to make this kind of distinction. They can deal [in ads] with sexual therapy and massage parlors, but when it comes to people controlling their own reproduction, their mind goes blank." Three weeks later, the phone company consented to let the ad run.

In 1983 the Supreme Court was still grappling with the debris of the nineteenth-century Comstock Act. By an 8–0 decision the Court declared unconstitutional a federal law that banned unsolicited advertisements for contraceptives through the mail. President Reagan's Justice Department defended the law in the interest of shielding children from exposure to birth control information. Writing for five members of the Court, Justice Thurgood Marshall said that the goal, while legitimate, could not realistically be accomplished by this law. In any event, he wrote, "the Government may not reduce the adult population to reading only what is fit for children. The level of discourse reaching a mailbox simply cannot be limited to that which would be suitable for a sandbox."

The Pill was to clear perhaps the ultimate rung in its long climb to respectability on April 10, 1973, five days before tax-filing day. The Internal Revenue Service declared it tax-deductible as a medicine. While allowing that the ruling might be controversial, an IRS spokesman of firm and uncommonly clear mind decreed: "We can't take moral questions into account."

16

A RUMBLE THAT SHOOK ROME

*Make haste to use your new remedies, before
they lose the power of working miracles.*
—NINETEENTH-CENTURY PHYSICIAN

Besieged by a minuscule pill, the largest religious institution in the world was creaking and straining to hold itself
in one piece.

During the first four decades of the Pill's life, a false-
hood, repeated so often and with such certainty that
scarcely anyone has questioned its truth, has continuously assaulted and hectored Catholics and non-Catholics alike: that the
Catholic church opposes the use of the Pill for contraception and
has banned it.

From the Pill's beginning, wherever true measure has been
taken of the matter, the church has accepted the Pill by substantial majority.

What? demands some astounded questioner. What about the encyclicals, the edicts, the wizened priests, the pope himself?

Perhaps a definition of critical significance needs examining. According to ancient Catholic tradition and teaching, explicitly confirmed in 1965 by the Second Vatican Council, known as Vatican II, the church is not its hierarchy. The church is the flock. *The fact is that the flock has accepted use of the Pill.* If not all or even most have accepted use of it for themselves, the majority has accepted and supported the right of choice of its use by others. If that statement is news to many, this one may surprise still more: There has been steadily rising evidence that the majority of the priesthood agrees with the majority of their flock.

Underneath the remarkable myth that the church "banned" the Pill and that all who use or condone it have "strayed," there lies hidden—deliberately, persistently hidden—this insistent and durable fact: When the Vatican engrossed itself for three years, from 1965 to 1968, in scrutinizing the morality of birth control, a majority of the pope's chosen examiners, the Pontifical Commission on Population, Family and Birth, came down on the side of asserting that birth control was *not* intrinsically evil. As the details that follow will show, only a small protective circle around Pope Paul VI favored "banning" the Pill. Moreover, a majority of the world's Catholics—the flock—defended and held what, according to church law, was higher ground. They acted on what they thought was right.

In the end the tiny medicinal tablet not only divided the flock but split a major faction of priests and bishops away from their brethren, undermining forever the age-old concept that Catholic faith rests upon "obedience."

That singular morsel of history, and the remarkably successful distortion of it, cries to be better known.

For an early glimpse into the myth, in June 1957, about the time the FDA and the church hierarchy itself approved the Pill for menstrual disorders (but not for contraception), fifty young American married couples were surveyed at a large Catholic university in a Southern city. All were members of a family-life discussion group conducted by their priest. All were handed a statement, "In my opinion, contraceptives should be prohibited in marriage." They were asked if they agreed or not. Despite the probable bias of this group in favor of church-approved attitudes, only 35 percent of the young people endorsed the assertion.

Seven years later, in 1964—after government approval of the Pill for contraceptive use—pollster Louis Harris reported that 60 percent of a sampling of U.S. Catholics wanted to see a change in their church's position on birth control. One of those polled, a New Jersey mother of four, voiced the new willingness of many Catholics to take the rules into their own hands: "I don't confess that I take the Pill, because I don't believe it is a sin." A follow-up poll by Harris in March 1967 showed that the American Catholics who wanted the church to lift its ban on artificial contraception had risen from 60 percent to 70 percent. The new poll also registered the opinion that birth control had become the "most pressing" problem facing them as Catholics, and that one in three was using the Pill or a mechanical contraceptive.

Two months later, Dr. Martin J. Meade, vice president and dean of students at Jesuit-run Fordham University, was startled when three undergraduate women brought him a petition signed by 150 others asking for courses in birth control that would be "personal as well as intellectualized," that would be given by theologians, physicians, sociologists, and psychologists. Also they asked for a qualified physician to be available in the university infirmary to answer "any questions" students might have. One of the signers said, "We don't want them to sell the Pill downstairs. We just want information." The *Ram*, Fordham's undergraduate newspaper, supported the petition, its editorial sternly cautioning the administration: "We hope . . . the decision will be one of thought, not culture-bound irrationality. . . . Each year girls are forced to leave Fordham [because of undesired pregnancies], an occurrence that may have a disastrous effect on their future." Dr. Meade called that last point "sheer sensational reporting." He said he knew of only one woman who had become pregnant during the semester, "and she got married."

Four days later the university announced a "full and candid"

sex-education program that would be open to men and women
students as well as faculty and other staff, thus becoming the first
major educational institution in the realm of Catholicism to
launch what one official called a "frank discussion of sex." Its discreet title: "The Permanent Relation Between Mature Man and
Mature Woman."

"Obviously," tiptoed Dr. Meade, "many young people who are
on the verge of making a permanent commitment to living with
another man or woman in marriage should know as much as possible and reflect as much as possible on this relationship." Two
days later, following protests from some Catholics, the president
of Fordham, the Reverend Leo McLaughlin, put out a clarifying
assurance that the new course "is not to be construed as a maverick attempt to topple precepts of the church."

Between 1965 and 1967 the nation was flooded with polls and
surveys showing that Catholic women not only favored birth
control but were using it. Dr. Charles Westoff of Princeton and
Dr. Norman Ryder of the University of Wisconsin found in 1965
that among white Catholic women of all educational groups, two
thirds said they had used contraception other than rhythm. In
1967 *Newsweek* found that 38 percent of Catholics were currently
using the Pill or another artificial device. Under age thirty-five,
60 percent were doing so. Eight out of ten of those under thirty-
five said they would like to see the church approve the Pill. The
magazine quoted one mother of five in Canton, New York: "The
pills are delightful—the best tranquilizers going."

In California a thirty-five-year-old traveling salesman who had
been educated at Catholic schools from first grade through college was one of many troubled churchgoers interviewed by *Time.*
Fred and his wife, Cathy, had three young daughters in parochial
school and a young son still at home. Convinced they could not
afford more children, they had been practicing birth control for
several years. "My parents," said Cathy,

were loyal immigrant Irish-Americans, completely subject to the
pope and to all of his edicts. I am not. As long as the pope is unable to relate his teaching to the needs of all the people, I con-

sider him fallible. I am not convinced that Christ would ever condemn anyone who practiced contraception to save his family
from disaster, or to save his fellow man from the problems of
overpopulation.

As for fears, yes, I'm fearful. I fear the loss of grace, not because
I use birth control but because the church denies me the grace-
giving sacraments. I miss communion most of all, and I cry when
my daughters receive the sacrament. Why do I not receive communion? I suppose it's because I cannot fully tear myself from the
early years of teaching. I don't feel that I am wrong, but where
will I find absolution if my own church says I am wrong?

(Recall Anne Biezanek: "Who, I wondered, would lay flowers
on my tomb?")

In a small, crowded courtroom in Rome with a crucifix on the
wall but no flag, Dr. Alexander Lenard and his publisher,
Salvatore de Carlo, went on trial for writing and marketing a
book, *The Control of Conception and the Limitation of Children*. The criminal law they were charged with violating had been instituted before World War II, at the behest not of the church but of dictator
Benito Mussolini, who added it to the Fascist code when, preparing for war, he was offering prizes to Italian mothers for bearing
the largest number of children. Defense attorneys now appealed
to cultural pride by arguing that it was beneath the dignity of
Italy to punish citizens for publishing knowledge familiar to almost all other educated nations. The trial was postponed twelve
times, and the two defendants were finally acquitted. In Italy, as
elsewhere, doctors were widely writing prescriptions for the Pill,
pretending not to know it was wanted for contraception. In 1965
more than seven million packages of the Pill were sold in Italy.

In Santiago, Chile, medical delegates to the World Planned
Parenthood Assembly were told that at least one and a half million Latin-American women, mostly of the upper and middle
classes, were using the Pill, making it the largest-selling pharmaceutical product in the overwhelmingly Catholic region. One
twenty-seven-year-old Chilean user at the conference told of just

having attended a class reunion of a college run by Catholic nuns: "There were about thirty members of my class there, nearly all married and attractive. Not one was pregnant. Five years ago many of them had a baby on the way, or had just come home from the hospital. It suddenly dawned on me: *They were all on the Pill.*"

A clue to a rising rebellion among the flock in Puerto Rico was the 1960 election, when the government's birth control program was a major issue. The bishops had created a Catholic political party in an effort to defeat the incumbent Popular Democratic party of Governor Luis Muñoz Marin, whom the new party called "Godless, immoral, anti-Christian and against the Ten Commandments." A pastoral letter forbade Catholics from voting for the governor's party because its stand on birth control was "based on the modern heresy that the popular will and not the Divine Law decides what is moral or immoral." The church-supported party gathered only 51,295 votes out of 788,607 cast.

The countryside rebellion in Puerto Rico took devious forms. A monsignor enjoyed telling a writer the story of a priest in a small parish who thundered from his pulpit one Sunday about the evils of a family-planning clinic recently set up in a nearby community. After church, women whispered excitedly to one another about the sermon, and especially about the wonderful new place where they could learn how to prevent babies. Next morning many of them were queued up at the clinic.

A major barrier to the Pill in impoverished countries was its requirement that the user follow the elementary arithmetic discipline of taking the Pill for twenty-five days, and not taking it (or swallowing an inert substitute) for the next five. In affluent, literate countries, the Pill was blocked by an alternative form of birth control in greater favor: abortion. In West Germany, abortions occurred as often as live births—about a million a year. France forbade the importation of birth control materials, but no matter. Abortion was preferred there and was rampant, as it was in Italy and Japan. Dr. Alan Guttmacher called abortion "the most severe pandemic disease in the world today."

While Latin America, predominantly Catholic, accounted for

two million Pill users, that number affected less than 5 percent of fertile women. "Among the masses," reported *Time* in 1967, "baby follows baby with such deadly rapidity that Colombian women crouch on the ground to abort themselves with sharp sticks. In Chile, the victims of bungled abortions occupy 20 percent of the beds in maternity wards, use up 27 percent of the transfusion blood. The situation became so serious that four years ago, with a high death rate among women who left five to ten orphans behind, the Catholic hierarchy tacitly agreed to look the other way while the government backed family planning."

A group of studies in Latin America during the midsixties began to reveal what later studies were to confirm again and again: that the birthrate and the practice—or the absence—of birth control could be predicted by levels of income and of education far better than by the degree of a family's religiosity. The higher the economic and education level, the smaller the family. That was true in Catholic as well as non-Catholic countries.

In the United States, a public policy on birth control designed to appeal to Catholics had been proposed in 1960 by a committee of thirty-four Protestant and Jewish clergymen, chaired by the Right Reverend James A. Pike of San Francisco, Protestant Episcopal bishop for Northern California. The collision on birth control was unnecessary, said the committee, if groups supporting or opposing birth control refrained from trying to impose their position on others.

But that debate, asserted an editorial in the Jesuit journal *America*,

is but one phase of a continuing clash between Protestant pragmatism and Catholic essentialism. Controversy on a matter like contraception so intimately involves the passions that it will hardly be resolved by intellectual discussion alone. . . . Catholics must realize at the outset that Protestants are not going to yield to invocations of the natural law: their minds simply don't work that way. Protestants, on their side, should give up the hope that the Catholic Church will surrender her doctrine on the nature and

ends of marriage. . . . Indeed, if Protestants wish to impress
Catholic moralists with arguments in favor of contraception, they
must appreciate that for the Catholic mind, theory rules practice.
Catholic thinkers can only welcome Protestant attempts to refute
the Catholic position on its proper level of theory. We invite
Protestant thinkers to set forth their own theories. . . .

How does one maintain the institution of marriage as Christian
morality has always understood it? What reason . . . demonstrates
why a man must choose a woman (not another man) for his mate
and cleave to her (and her alone) in a permanent union? The doctrine that procreation is the primary natural end of sex affords an
answer to the question. Until Protestant theory offers an equally
satisfactory answer, Catholic moralists are not likely to take any
Protestant argument very seriously.

Catholic law on the Pill had been set down by Pope Pius XII a
month before his death in 1958. It could be used by doctor's prescription to treat reproductive disorders. But used as a contraceptive, it was morally unacceptable, a form of sterilization. On Palm
Sunday, 1960, the new pope, John XXIII, spoke more benignly
but renewed the mandate. In the Basilica of St. Paul-Outside-the-
Walls before hundreds of children from housing developments
on the western outskirts of Rome, he exhorted parents: "Don't be
afraid of the number of your sons and daughters. On the contrary, ask Divine Providence for them so that you can rear and educate them to their own benefit and to the glory of your
fatherland here on earth and of that one in heaven."

In late June 1964, twenty-seven cardinals gathered in the
deeply carpeted second-floor Vatican palace library of still another new pope, Paul VI. He spoke of change, of pressures upon
the church to change, of the importance of caution in submitting
his two-thousand-year-old church to change. He previewed for
his cardinals a passage he was about to deliver:

The problem, everyone talks of it, is that of birth control, as it is
called—namely, of population increases on one hand, and family
morality on the other. It is an extremely grave problem. It touches

on the mainsprings of human life. . . . The question is being subjected to study, as wide and profound as possible, as grave and honest as it must be on a subject of such importance. . . . But, meanwhile, we say frankly that up to now we do not have sufficient motive to consider out of date, and therefore not binding, the norms given by Pope Pius XII in this regard. Therefore they must be considered valid, at least until we feel obliged in conscience to change them.

Such willingness to display anxiety and tensions within the church was not shared by all in the hierarchy, least of all with some Americans. A popular Sunday-morning television program, *Catholic Hour* on NBC, produced by the National Council of Catholic Men, prepared a four-part series on the church and marriage. The script, written by John Leo, editor of *Commonweal*, reviewed the varying shades of Catholic thought on birth control. In January 1965, hours before airtime, the show was suddenly canceled. The executive head of NCCM, Martin Work, refused to say who had forced the cancellation. A lively rumor blamed New York's Francis Cardinal Spellman. The suppression was defended, however, by the Reverend Francis J. Connell, a retired dean of theology at Catholic University: "While bishops at the Ecumenical Council could discuss the issue completely, the Catholic laity should not necessarily do the same."

In 1964 more than two thousand bishops and higher prelates at the Second Vatican Council in Rome opened debate on a twenty-eight-page draft of "The Church in the Modern World." It was an epochal position paper on the major issues of modern life: war and peace, the distribution of wealth, and, inescapably, birth control. One drafter predicted it would be "fiercely debated," partly because of a startling provision stating that marriage "is not a mere instrument for procreation." The paper also introduced a euphemism for birth control soon to become a favorite in Vatican pronouncements: the "demographic problem"—the explosion of population. Clearly uneasy about rumblings throughout the Catholic world, the drafters urged spouses to be patient in awaiting "practical solutions."

The Ecumenical Council meeting was colored by an extraordinary article, scholarly and footnoted, that had appeared the previous spring in a Belgian periodical, *Ephemerides Theologicae
Lovanienses*, to which, predicted one religion journalist, "future
church historians may well date a profound change in Roman
Catholic thinking on marriage." It was written by an indepen-
dent-minded but highly regarded professor of moral theology at
the University of Lovain, the Reverend Louis Janssens. The article, breaking new ground for a Catholic theology journal, endorsed the Pill as a legitimate means of family planning for
Catholic couples.

Just as with the rhythm method, said Janssens, echoing almost
exactly the reasoning of John Rock, the Pill does not interfere
with the "nature and structure" of intercourse. In the period immediately after childbirth, ovulation is normally suspended,
Janssens pointed out. A woman taking the Pill is simply helping
extend a natural physical process. Thus the Pill is not inherently
evil.

"This is not a maverick talking," commented one U.S. Catholic
theologian, "but a voice from the mainstream." Since 1958
Janssens had been meeting with groups of demographers, sociologists, physicians, and theologians organized by Leo Josef
Suenens, archbishop of Brussels, for conferences on the growing
problem of "population control." Their discussions centered almost exclusively on the theological permissibility of birth control, and particularly the Pill.

The reaction to Janssens, particularly in the United States, was
swift, furious, and predictable. American Catholic thinkers, as a
rule, were far less ready than their colleagues abroad to probe un-
charted territories of theological possibility. Janssens's argument
was "absolutely contrary to the teaching of the church," charged
Catholic University's Father Connell. A leading Jesuit, the
Reverend Edward Duff, wrote with withering certainty that "no
established Catholic theologian is on record as agreeing with"
Janssens. As the Ecumenical Council was soon to learn, however,
Janssens was ventilating a flame of exploratory thought that,
hushed but insistently, had been radiating through the seminaries
and chanceries of Germany, France, and the Netherlands as well

as Belgium. The discrepancy between church teaching on birth control and the actual practice of Catholic couples, said Werner Schöllgen, a theologian of Bonn University, was "the single most important cause for defection from the sacraments among the younger generation of German Catholics."

What had this intrusive Pill done to the peace and intellectual discipline of the hierarchy of Catholicism? How were great thinkers on "natural law" to speak with their accustomed authority when alien issues of biology and chemistry crept in to jam the message?

As signs rose of a sustained controversy, the pope was counseled by his highest-ranking lieutenant, Alfredo Cardinal Ottaviani, secretary of the Holy Office, to stand back in silence. Let the theologians draw their battle lines.

The predicted "fierce debate" at the Ecumenical Council broke out into near-insurrection. The Melchite Patriarch of Antioch, Maximos IV Saigh, spoke in French rather than the Council's official Latin, a defiance in itself, remonstrating that "the immense majority" of Catholics had chosen to go their own way, ignoring the church's teaching on birth control. "Shouldn't the official position of the church be revised in the light of modern science?" he demanded.

"We should affirm," urged Paul-Emile Cardinal Léger of Montreal, speaking the heretofore unspeakable, "that the intimate union of the marriage partners finds its end in love as well as procreation. May this council clearly proclaim the two ends as equally good and holy."

Belgium's Cardinal Suenens added the protest that the church had for too long based its marital teaching exclusively on the biblical injunction to "increase and multiply." Why had it ignored the equally important scriptural instruction, "They shall be two in one flesh"? In an extraordinary challenge, Suenens, tall and lean with graying temples and fiery dark eyes, jabbed his colleagues with the memory of the most embarrassing blunder in the history of the church: the excommunication and life imprisonment of Galileo Galilei in 1633 for teaching the heretical Copernican theory that the earth and other planets orbit the sun. "I urge you, brothers," thundered Suenens, "*let us avoid a new Galileo case.* One is certainly enough in the history of the church!" (Galileo was not to be reinstated by Rome until 1992.)

During his short reign, Pope John XXIII had asked the bishops of the Ecumenical Council to help him bring the church "up to date."

A startling phrase. *Up to date?* Spoken by a *pope?* Hadn't Pope Pius X in 1907 condemned as heresy what he called *modernism,* and denounced the misguided men tainted by it? One who must have blanched at the very words *up to date* was Cardinal Ottaviani, the highest official in the Vatican save for the pope himself: the secretary of the Sacred Congregation of the Holy Office (later called the Congregation for the Doctrine of the Faith). To signify his mission Ottaviani's personal coat of arms, hanging in his splendid waiting room at the Palazzo Sant' Ufficio, was inscribed *Semper Idem,* "Always the same."

And now the traditional church position was defended by Cardinal Ottaviani. His principal argument was that the church had already spoken on birth control and hence was not open to questioning or reexamination.

By the time Vatican II was drawing to a close in 1965, the Catholic church was a remodeled edifice. The mass would thereafter be spoken in the modern local tongue. The rule forbidding meat on Friday was dropped, to the relief of Catholic diners everywhere (and to the dismay of fishermen). Women were excused from covering their heads at Sunday mass, and indeed Sunday mass could be attended on Saturday. Among further liberations, priests and nuns were permitted to wear street "civvies" while off-duty. These upheavals in immutable tradition seemed to confirm that perhaps the new support by some theologians for the Pill was about to become law. In the end, Vatican II adjourned without settling the matter, but it broke historic ground. It approved a statement on marriage that elevated the role of sex as an expression of the love of spouses and acknowledged their right to limit the size of their families by "moral methods" of birth control.

But what methods were moral? The church wrestled with this

issue before, during, and after Vatican II. Pope John had appointed a six-man commission to advise him on the increasingly troublesome question of population and birth control, later described as the first study group in church history created in response to world clamor. But "Good Pope John" died in 1963, before the group had a chance to meet. His successor, Pope Paul VI, had a graceful pathway through the thorny brush. In the Vatican, where the importance of nothing precedes that of precedent, the precedent had been created. Paul VI recharged Pope John's birth control commission and indeed enlarged it. (His announcement in June 1964 said nothing, however, of who the original or added members of the commission were, or where or when they would meet.) The commission's purpose was to advise him—thus the clear implication that he was reserving to himself the judgment of whether there would be a new church position on birth control.

But the fact that the church was to study birth control—and its suggestion of coming change—would itself undermine the ban. Word soon slipped out that priests in Munich, Germany, had been advised that they could offer communion to married couples who used contraceptives "not lightly and habitually, but rather as a regrettable emergency solution." A church spokesman in Munich confirmed that Julius Cardinal Döpfner had expressed "no objection" to that instruction to priests. Perhaps more embarrassing was that Cardinal Döpfner had been identified by a news leak as the vice chairman of the newly expanded commission. Indeed, Döpfner was not the only advocate of change on the reconstituted commission. All the original six members appointed by Good Pope John had been regular participants at Cardinal Suenens's exploratory conferences in Belgium on population control. They would one day become identified as Stanislas de Lestapis, a French motorcycle-riding Jesuit specializing in the sociology of the family; John Marshall, a British physician who had studied the rhythm-combined-with-temperature method; Clement Mertens, a Belgian Jesuit and demographer; Henri de Riedmatten, a Swiss Dominican; Pierre Van Rossum, a Brussels physician; and Jacques Mertens de Wilmars, a professor of sociology at Louvain in Belgium. A striking characteristic of the commission was its worldliness. None of the six was trained primarily in theology.

Soon the new uncertainties began to shape themselves into a new formal theology. At the University of Toronto, the Reverend Gregory Baum, a *peritus* (theological advisor) to Vatican II, advanced the proposition in several published articles that as long as church leaders were divided among themselves, *there could be no church law* about contraception. Therefore, he concluded, lay Catholics were free to follow their own conscience on the matter. He based his argument on the principle of *Lex dubia non obligat:* a doubtful law does not oblige. The use of contraception, said Father Baum in his groundbreaking argument, is a matter of discipline that involves neither the church's infallibility nor divine revelation, and thus is subject to change. Pope Paul's directive that Catholics continue observing the old rules while new rules were being investigated, said Baum, was more in the nature of a request than a command.

Some reactions were eminently worldly. On May 27, 1964, reports reached Wall Street in the early afternoon that the Roman Catholic Church was about to take a new look at its position on pharmaceutical contraceptives. At 1:30 P.M. the Dow Jones ticker carried the rumor. Immediately, volume soared in Johnson & Johnson, G. D. Searle, and Parke-Davis stocks, forcing a temporary halt in trading.

The original six members of the commission actually had met once, in 1963, shortly after the death of Pope John, who had appointed them. Immediately, the commission stepped beyond the prim boundary of overpopulation—or the preferred term "demographics"—and directly into the comparative morality of particular methods of birth control. Pope Paul wanted guidance on the Pill. The commission's problem, however, was that its two physician members could not enlighten the other four as to just how the Pill worked. In 1963, nobody was precisely sure. For one thing, one of their worries had also gripped Gregory Pincus, John Rock, and the field testers in Puerto Rico. What biological impact would the Pill have on women after, say, a decade or more of taking it? Was the church to commit itself to a position based on medical guesswork? Moreover, there were no known answers to the more profound questions of its morality, as they saw them. Does the Pill snuff out a life?

In his twenty-two-page report of that 1963 opening meeting of the six-man commission, its secretary, Henri de Riedmatten, summed up the perplexities. The commission had learned from its physician members that the Pill did not destroy a fetus, since conception had not yet occurred. Looking at it another way, if the Pill blocked the womb's capacity to create a "nest" for a fertilized egg, that made it unacceptable, for it was thus "a veritable abortifacient." By suppressing ovulation it was indeed useful for various therapeutic purposes, among them, ameliorating endometriosis and dysmenorrhea. In doing so, however, the Pill would temporarily make a woman sterile. Then what if a woman, knowing it could do that, took the Pill with a deliberate *intent* of contraceiving? Pius XII had already banned that, calling it "temporary sterilization." But more than one voice in the six-man commission questioned that judgment. The Pill, they pointed out, did not damage the ovaries or choke off any tubes, did not physically alter any organs nor otherwise disable them, and therefore could not be classed as a "mutilating" method of sterilization that Pius XII had condemned.

The group members urged caution until research, then under way, could better explain how the Pill managed to put ovaries into "a state of repose." For these reasons, the report suggested with gingerly respect, it would be "preferable on the part of the Authority [the pope] not to take a definitive position on the morality of using the pill." But if the pope should act swiftly on anything, it should be in taking a stand against governments that were threatening more and more drastic measures to control what they perceived as runaway growth in population. If those moves were not squelched, de Riedmatten's report urged, the church would fail in its role as "the defender of the natural moral law."

That initial meeting had opened many questions but answered few. Paul VI asked the members to meet again, which they did in April 1964.

This time he expanded the group from six members to thirteen. To achieve more theological depth than the first meeting
had produced, the pope's additions included five theologians,
four of them from Rome: the Reverend Josef Fuchs and the
Reverend Marcelino Zalba, both Jesuit priests who taught moral
theology at Gregorian University, the Reverend Jan Visser of
Pontifical Lateran University, and the Reverend Bernard Häring,
often called the pope's personal theologian. The fifth was Canon
Pierre de Locht of Brussels, a personal advisor to Suenens. He also
added two more sociologists: Bernard Colombo of Venice and
Thomas K. Burch, a population scholar at Georgetown University in Washington, D.C.

The expanded commission seemed to show the pope's deter-
mination to admit a wide variety of points of view on the troublesome issue. Father Zalba, for example, had been first to suggest
that nuns in the Congo, then at war, be permitted to take the Pill
to prevent the consequences of rape. The Vatican agreed to it on
the ground that children were "the end" of marriage, but not of
rape. Häring and de Locht were known for having taken more
liberal positions than the others. The seven new members were
instructed to tell as few friends and family as possible that they
were going to Rome and to tell no one the purpose of their travel.

They joined the six original members at the Domus Mariae, an
unheated private residence. According to de Riedmatten's nine-
teen-page report of that meeting, de Locht and Häring argued
that the church's teachings on "natural law" were relatively recent
in history, and that on the subject of birth control the church
should take a longer view. Shouldn't the church go back to the
Bible itself to find the meaning of marriage? Marriage was supposed to help people, not bring them pain. Why should the
church make itself the instrument of bringing difficulty to marriage? One traditionalist asked de Locht and Häring: When
human weakness showed its face, who could be counted upon for
mercy and patience more than pastors? But de Locht rejected
that. Yes, pastors were merciful and patient, but not in response
to human weakness. It was because they were in touch with real
life, because they saw with their own eyes the anguish felt by
people of good will who challenged the validity of the abstract

laws handed down by Rome. He was thus advancing an idea uncommon in Vatican discussion: that real life as it was lived by the
people ought to be a major influence over the shaping of the
ideal.

"But now, Canon de Locht," one member hurled back with a
challenge that electrified the room, "aren't you raising questions
of fundamental theology?"

De Locht's swift reply was even more shocking: "Yes. Why
not?"

It was a sky-opening moment for the members uncertain of
their role, especially the doctors and sociologists. *Yes, why not?*

John Marshall, the London neurologist, later recalled the scene
for Robert Blair Kaiser, whose book, *The Politics of Sex and Religion,*
minutely chronicles the commission, and from which much of
this account of its meetings is drawn: "No one else actually said
this out loud, but I think everyone said to himself at that moment
that it obviously *was* appropriate for this commission to raise
questions of fundamental theology."

Sensing the importance of the moment but wanting to explore
more of its meaning, de Riedmatten ordered a coffee break.

"We were out on this large terrace," Marshall continued, "and I
can recall that a few of us were standing in a small group together,
not really saying very much, and thinking hard about what de
Locht had said, and looking down to the other end of the terrace.
There was Canon de Locht pacing back and forth, saying his
rosary. Whether he was experiencing a crisis of his own at this
moment, I cannot say. But that's what I imagined."

When the meeting resumed, Dr. Van Rossum, the Belgian
physician, apologizing for his lack of expertise in theology, said
he failed to see the difference between making love when a couple knew they couldn't create a baby and taking active steps to
forestall making a baby when they followed their impulses simply
to love each other. One of the bishops promptly invoked his expertise: A couple practicing rhythm was not doing anything "to
frustrate nature," but when they deliberately avoided conception
they did "go against the finality of the act."

Dr. Van Rossum countered boldly. Celibates, he declared—

looking a few directly in the eye—were trying to impose their rules on married people. "You think of sex as something you must avoid in order to be faithful to your vocation. That is all right for you. But our vocation is to love one another."

Words, words, theories, theories. Were words the reality? Were the moral laws, the abstractions, the theories and dogma the reality? Or was what people *did* the reality? Whose law was law? If procreation is the primary purpose of marriage, is love only secondary? If good people make love far more often than they make babies, or want to make babies, or perhaps *should* make babies, then how can procreation be primary? The commission groped for words, words that they could imagine the pope accepting—and that *they* could accept delivering to him. They finally agreed on this: "The group unanimously affirms," de Riedmatten recorded, "that love is at the heart of marriage, and the majority of the members agree that *the love of the husband and wife should not, in any way, be ranked among the secondary ends* of marriage." (Emphasis added.) As unrequested advice to the pope, this was a veritable revolt.

The meeting went on to adopt a dutiful and amicable consensus that rhythm remained "the most desirable means of exercising responsible parenthood." But part of the group balked at condemning the Pill. It was not sterilization, they insisted. Every month a woman's ovaries had a period of natural "rest." Why could she not recreate, under her own control, that *natural* state? That opened the inevitable question, But what is *natural?* Some members argued that famine and disease were perfectly "natural." Nobody on the commission or in the church would oppose science in its remarkable efforts to halt hunger and illness. Indeed, it had done so well that perhaps science was partly responsible for the world population crisis. The commission agreed to agree that "natural law alone cannot provide a good answer to this question." What was needed, their report said, was to look beyond "natural law" to the experience of Christian couples. That would take more time. They decided to meet again later in the year at Louvain.

In the first week of June, the members received telegrams from

Amleto Cicognani, the Vatican secretary of state, hailing them to Rome. It was urgent to discuss immediately the troublesome new thinkers who were causing, the wire said, *"l'inquietude de ses enfants,"* "disturbance to the children of the church." Canon de Locht wrote resentfully in his diary that the disquieted faithful were not children but adults, married couples. If they were disturbed, they had every reason to be.

When the members gathered on June 14, 1964, they found their number had again been expanded, now to fifteen. The two new members were not, like most of the others, of Suenens's ever-questioning crowd, but intimates of the pope himself. They were Tullo Goffi, a priest from Brescia, the pope's hometown, and Ferdinando Lambruschini, a Vatican theologian. Debate continued to center on the refusal of some to put conjugal love in second place behind the production of children. The traditionalists warned: "If love is in first place, then anything goes."

Some days later at the Ecumenical Council, Cardinal Ottaviani, the pope's highest lieutenant, listened with alarm to speech after speech by the "new thinkers" until he felt compelled to rise and restate his reliable theme: *Semper Idem. Always the same.* Nothing, he insisted, had changed! He personalized his certainty: "I am not pleased with the statement of the text that married couples can determine the number of children they are to have. This is unheard of, from previous centuries up to our own times. The priest who speaks to you is the eleventh of twelve children, whose father was a laborer in a bakery—a laborer, not the owner of a bakery, the *laborer.* He never doubted Providence, never thought about limiting his family, even though there were difficulties. . . . Can it be possible that the church has erred for centuries?"

Ottaviani, seventy-three and almost blind, wore his Roman working-class origins the way another priest might flourish in his Lateran University degree. Ottaviani was not only of Rome but of the part of Rome called Trastevere ("Across the Tiber"), whose people trace their ancestry to Caesar's armies. He grew up hearing the postmidnight clip-clop of horses' hooves on cobblestones still heard there today as drivers of the *carrozzelle,* Rome's horse-drawn open cabs, a traditional occupation of the district, return home to their stables. Trasteverinos, proud outsiders who call themselves *Noiantii*, "We others," work hard at preserving their own shadings of accent, self-mocking humor, and earthy phrases. In his relaxed moments, Cardinal Ottaviani eased back into those familiar speech patterns. His simple forthrightness and absence of lordly air played obbligato to his highest-but-one of all church positions. He knew it and used it. A member of the Roman Curia since 1929, he had once called himself the Roman Catholic Church's "old policeman." Church liberals cast him as the incarnation of reaction and obstructionism, but that in no way ruffled his steady and absolute certainty of his grasp of truth and rightness.

In advance of the next commission meeting in Rome, now scheduled for late March 1965, Cardinal Cicognani told Henri de Riedmatten that the pope had suggested another expansion of the membership, this time to include some doctors, perhaps some psychologists. And what about adding some married couples? Could de Riedmatten suggest names?

The March 1965 meeting, with a membership now of fifty-five, included thirty-four laymen and -women and twenty-one clergy, many from the faculties of the great Catholic universities. Two were psychiatrists. Eight were from Belgium, eight from France; from the United States, seven; Germany and Italy, five each; and three from Spain. Eleven came from the Third World.

The commission now included three married couples: Dr. and Mrs. Laurent Potvin of Ottawa, Canada, Dr. and Mrs. Henri Rendu of Paris, and Patrick and Patricia Crowley of Chicago. The Crowleys headed an international organization called the Christian Family Movement. This was another precedent breaker: the first time in history that the church sought informed opinions on marriage from married couples instead of exclusively from the celibate clergy.

The Crowleys' notice of appointment took them by complete surprise. On December 3, with no advance hint, they received a letter from Albert Cardinal Meyer of Chicago that said, "While nothing was indicated about keeping this appointment secret, for the time being, I would suggest that no publicity be given to the appointment until you have further word." The notice thrilled them. Their lives had been wrapped around the church. Patrick, a Chicago lawyer, had gone to Notre Dame. One of their daughters had become a Benedictine nun.

Patty had four children after six pregnancies. Her last delivery, which almost killed her, forced life-saving surgery that barred her from having any more children. "We'd never challenged the church's teaching on birth control," she recalled many years after their appointment. "We believed whatever the pope said." They conferred, despite the advice of secrecy, with Cardinal Meyer and their spiritual advisor, Father Reynold Hillenbrand, and finally with a Jesuit biblical scholar, John McKenzie.

They discussed their discomfort about the secrecy with another commission appointee, André Hellegers, a Johns Hopkins gynecologist. The three worked out a plan that they hoped was not too disobedient. Hellegers helped them design a questionnaire for members of the Christian Family Movement, mainly about rhythm. Hellegers felt that the results would prove not that rhythm failed its users, but that it hadn't really been tried because so few people knew how to practice it. Such an outcome, Hellegers said, "could take some of the pressure off the priests if they knew that their people hadn't practiced rhythm, but only thought they had."

The eight North American members gathered for a preliminary meeting in Baltimore on January 20, 1965. They sent a report to Rome with a few typically American management suggestions. They requested simultaneous translations during their meetings; that all sessions be recorded and transcribed; that all documents be available to every member, and—most American of all—they requested a bigger budget and bigger staff for the commission secretary. They estimated a cost of a million dollars but offered to raise the money if that were a problem. "Hellegers had sources," one of the members, Thomas Burch, the Georgetown University population expert, later said. "He knew the Kennedys. Pat Crowley could have helped too." More than a year later the Vatican was to find additional money.

They also prepared a letter that seemed to scold the pope for insufficiently strong leadership. Despite his previous statements, "confusion among the faithful has not diminished, but has, if anything, increased." With too many conflicting statements on birth control by "experts," the church ran the danger of "following the practice of the faithful, rather than determining it."

Yet could they write that to the pope? Wasn't it the tradition that the church *was* the faithful? Is "following the practice of the faithful" a *danger*? The Crowleys and some other members didn't like the letter, which seemed to pit the church against the faithful and seemed to reinvest in the pope the authority to resolve "confusion" before advice could be offered by this very commission he had appointed for an "exhaustive study." The letter was not sent.

In mid-January the pope added two more members: Bishop Josef Reuss of Mainz, Germany, author of a provocative journal article that sided him with the Belgian dissidents, and as though to counterbalance him, Leo Binz, archbishop of Minneapolis–St. Paul, who had maneuvered the suppression of the television series by the National Council of Catholic Men.

On March 24 the members converged at the outskirts of Rome at the Spanish College on the Via Torre Rossa, chosen because it lent itself to a secret meeting. In a large, gleaming new building not yet occupied by seminarians, there was plenty of room for everyone. Everyone except the five women members. Father de Riedmatten, who greeted the arrivals, said the women were to sleep at the Monte del Gallo, a poor convent down the road. The Crowleys were startled. They thought they had been put on the commission *because* they were married. Now they were treated as celibates.

Well, the meeting was to last only four days. They said nothing.

17

ADVICE WITHOUT CONSENT

When you take your pill
it's like a mine disaster.
I think of all the people
lost inside of you.
—RICHARD BRAUTIGAN,
"THE PILL VERSUS THE SPRINGHILL MINE DISASTER"

That evening all the members gathered in their "conference" room, two classrooms joined by removing a
portable partition, making the room awkwardly long and
narrow. Tables organized in a large U pressed the backs of
the fifty-seven conferees against the walls, with a large
emptiness in the middle. The thought must have been inescapable, at least for those willing to contemplate it, that a new
day had come. The majority around this table, all appointed by
the pope to deliberate and give him advice on a moral issue, were
laymen and laywomen.

Also, some were deeply uncertain about the issue at hand. One
member's English-speaking wife (who was not in Rome for the
commission meetings) was already "on the Pill."

"I objected to the Pill when I first took it," she told me in an interview more than twenty-five years later. "My husband wanted me to take it, so I took it. He didn't want still another child. The question in my mind was, How does it work? And when my husband got asked to sit on the pope's birth control commission, that's what they discussed in Rome at their meetings. I know that one university theologian was alarmed about the moral implications. Now, how did he put it? Would the Pill, by preventing conception instead of destroying the fetus, lead to a fad of sodomy as a lesser of two evils? He was one of the arch-conservatives. The main question was, Does it prevent conception—or does it destroy the fetus after conception occurs? Nobody knew quite how the Pill worked. That's why they had all these doctors on the commission."

To the relief of the Americans, the first meeting was addressed in English by a professor from Notre Dame, John Noonan. A thirty-two-year-old bachelor who spoke in the flat vowels of working-class Boston, Noonan explained he had been writing a history of the church and contraception, which was at that moment awaiting publication by the Harvard University Press. His talk soon established his scholarly independence from church "correctness." Although there was a persistent idea that "the church has always condemned contraception," Noonan said, that was contradicted by classical literature, which told of "potions of sterility" in ancient Rome as well as in Greece. People everywhere, including Catholics, who feared additional children had for centuries practiced coitus interruptus.

Other modern Catholic historians besides Noonan had found similar bones in embarrassing musty closets. Early cultivators of established Catholic thought on women, sex, and marriage included, for example, Saint Jerome, who described woman as "the devil's gateway, a dangerous species, a scorpion's dart." Saint John Damascene called woman a "sicked She-ass, a hideous tapeworm, the advance post of hell." Saint Clement of Alexandria declared it "shameful for her to think about what nature she has." Pope Saint Gregory the Great limited woman's "use" to two essentials: maternity or harlotry. Saint Thomas Aquinas declared that woman "is misbegotten and defective." Married couples, preached Saint Francis de Sales as late as the seventeenth century, should avoid thinking about the act they might have to perform that night.

Gregory in the sixth century also declared that intercourse was sinful even for married couples, not because the act itself is wrong, but because "what is licit is not kept within the bound of moderation." Seven centuries later, Saint Bonaventure liberalized that severe judgment by Gregory. He reappraised marital relations as good—if motivated solely by the desire to have children. From that doctrine arose the church position that interference with conception was wrong because it frustrated the "primary" end of marriage.*

From the thirteenth century to the twentieth, the prohibition of birth control was doctrine, but not written as church law. The unwritten ban ruled until the last day of 1930 when Pope Pius XI issued a definitive statement in the form of an encyclical called *Casti Canubii* (Of Chaste Marriage): "Any use whatsoever of matrimony exercised in such a way that the act is deliberately frustrated in its natural power to generate life is an offense against the law of God and of nature . . . a grave sin." Pressing for perfect consistency of dialectic, Pius XI rejected out of hand any justification for the practice: "Assuredly, no reason, even the most serious, can make congruent with nature and decent what is intrinsically against nature."

Rhythm became known as the church's "approved" method. But those most demanding of doctrinal perfection could not bring themselves to accept that breach of perfection. Father Hugh Calkins, an American member of the missionary order of Servites, known to millions in the 1940s for a weekly Catholic magazine column, asserted that the church "merely tolerates" the method. Speaking for many purists, Calkins deplored the

> vicious Rhythm mentality—a state of mind that won't trust God. . . . It's enough to make God vomit out of His mouth, the creatures who ignore so completely the divine purposes of marriage.

---

*A twentieth-century translation of that view appears in the Catholic Marriage Manual, written by Monsignor George A. Kelly, director of the Family Life Bureau of the New York archdiocese: "The reason why the artificial prevention of births is immoral is written into the very nature of the

. . . These bleeding hearts, especially busybodies-in-law and nosy neighbors, scream protestingly: "Who'll take care of the next baby?" The simple answer is: The same God that takes care of you even when you resist His Will. "But we must give our children security and education." Just because God doesn't give parents and children all today's phony materialistic standards require, doesn't mean He fails them.

Now, to his rapt audience on the pope's commission, the young scholar Noonan allowed that the church was right to make a moral issue of contraception and to support procreation. But then, startling some of his audience, Noonan added that it was right for possibly the wrong reasons: Saint Augustine, for example, had a contempt for sexual pleasure that clearly resulted more from his embattled relationship with his own sexuality than from anything ever uttered by Christ. Augustine's disdain for sex grew into a dogma that connubial pleasure was wrong for everyone, including wedded couples, unless there was "an excusing good." To Augustine, the only excusing good was procreation.

"We hung back," Patty Crowley later said of the lay members in the discussion that followed. "We didn't know what the church wanted. It wasn't until later that we realized that 'the church' didn't know any more than we did." Then it hit her. "*We* were the church."

The next day, the Crowleys distributed their survey of Christian Family Movement couples in the United States and Canada. Most of the polled couples had six children or more, some as many as thirteen. The written testimony of one university-educated father of six riveted the attention of commission members, laity, and celibates alike:

Rhythm destroys the meaning of the sex act: it turns it from a spontaneous expression of spiritual and physical love into a mere bodily sexual relief; it makes me obsessed with sex throughout the

---

sexual organs and the marital act itself. The sex organs were made by God to reproduce the human race. . . . Once you say there's no clear-cut moral law on the subject, then you're saying sexuality has an autonomy all its own."

month; it seriously endangers my chastity; it has a noticeable effect upon my disposition toward my wife and children. . . . I have watched a magnificent spiritual and physical union dissipate and, due to rhythm, turn into a tense and mutually damaging relationship. Rhythm seems to be immoral and deeply unnatural. It seems to me diabolical.

Writing separately, his wife told of her experience with the calendar-and-temperature method:

The psychological problems worsened . . . as we had baby after baby. We eventually had to resort to a three-week abstinence and since then (three years) we have had no pregnancy. I find myself sullen and resentful of my husband when the time for sexual relations finally arrives. I resent his necessarily guarded affection during the month and I find I cannot respond suddenly. I find, also, that my subconscious dreams and unguarded thoughts are inevitably sexual and time-consuming. All this in spite of a great intellectual and emotional companionship and a generally beautiful marriage and home life.

A couple married for ten years with three children, both members thirty-three years old, submitted jointly:

The "natural law," or whatever pseudo-scientific, theological term is used to describe our physical union, has little meaning to us. We cannot fail to love each other unless we overburden this most intimate relationship with musty phrases like "primary purpose," "your bound duty" and "animal affection." As Christians we demand more and expect more from our union than "just gifts from God."
. . . As busy parents raising children, we know few moments of complete harmony and personal communion. Our physical and spiritual union, when it does occur, is just such a moment. It should not be subjected to scientific or metaphysical scrutiny. . . . We do not believe that every time a man and wife feel the need to express their love to each other that it is a "call from God" to raise more kids. Neither is it a resurgence of the base and selfish sex

drive which has yet to be conquered by pure love. We are frail and lonely people holding to the only mutual concern and affection we really know.

Not all responses were suffused with such pain or expressed interest in change. Among other shadings of opinion, one father of fourteen, married for seventeen years, who slightly misunderstood the Crowley questionnaire's search for experience rather than judgment of doctrine, wrote: "Frankly, we are surprised at the thought of a popular referendum on one of the teachings of the church. Do we next vote on the question of abortion, and then maybe divorce? . . . Of course, should the pope, with the guidance of the Holy Spirit, decide that the teaching is in error, then it should be changed. But our vote should not have any effect on such decision."

Another couple with ten children responded: "If one moral law is changed, then why not change the others? Then why should it not be all right to start lying because it makes things easier in business?"

As the birth control commission met and debated, reports of the debates went up to the Ecumenical Council. The council's debates in turn were reported to the pope.

And what of the pope?

Only once did Paul VI permit a public glimpse of the confusion in his mind over birth control and the Pill.

Popes rarely give interviews, but Paul VI gave one to the Milan daily, *Corriere della Sera,* published October 4, 1965. Sitting with his reporter-visitor, Alberto Cavallari, at the end of a long table in his private library, the pope, after a long conversation, slipped into seemingly unguarded rumination about birth control—and about his own awkward discomfort with the subject:

"So many problems. How many problems there are and how many answers we have to give. We want to open up to the world, and every day we have to make decisions that will have consequences for centuries to come. . . . Take birth control, for example. The world asks what we think and we find ourselves trying to give an answer. But what answer? We can't keep silent. And yet to speak is a real problem. . . . The church has never in her history confronted such a problem."

After a pause, he started to say something, then faltered. Suddenly, he blurted out, "This is a strange subject for men of the church to be discussing." Another pause. "Even humanly embarrassing." As though recovering from that surprising revelation, the pope turned impersonal again. "So—the commissions meet, the reports pile up, the studies are published. Oh, they study a lot, you know. But then we will have to make the final decisions. And in deciding, we are all alone. Deciding is not as easy as studying. We have to say something. But what? God will simply have to enlighten us."

The commission, too, appointed to enlighten him, groped toward enlightenment and wished the pope would lead *them* to it. In the midst of this mutual need and quandary, the commission held its one meeting with the pope. It turned out not to be a discussion but a papal audience in which the pope did all the talking. (What one of the five women members recalled most clearly was that twice the pope addressed the gathering as "Dear sons.") He reminded the members of the pressing need for him "to give guidance without ambiguity," that the flock was waiting, but that he understood why the members needed more time. "We are confident," he said, "that you will know how to continue it to the end with courage."

But continue how far, some members worried, when already they had dug further into the twisted briar of medicine and theology than the pope had originally imagined they would go? "At each step forward," Dr. John Marshall later said, "as we got deeper and deeper into the mire, each time de Riedmatten went to see the pope he got the same answer: 'Press on.' This ought to be emphasized. Throughout the whole of the agony of the commission, the pope insisted on going on, going on, 'in complete objectivity and liberty of spirit,' never insisted on our not touching this or that aspect of the question. Pope Paul has not been given all the credit he deserved."

Thus Marshall perceived the Pope Paul of the detached intellect. On the other side, his strong-willed secretary of the Holy

Office, Cardinal Ottaviani, was whispering sedulously into the ear of Paul the guardian of tradition, Paul of the eye trained on history, Paul the stabilizer of the great ship of the church. Paul listened through both ears and seemed forever of two minds. Paul VI was not called the Hamlet among popes for nothing.

Paul had been born Giovanni Battista Montini, to an Italian editor who had met his wife-to-be in 1883 while walking out of St. Peter's in Rome. Young Montini, ordained a priest at age twenty-two in 1920, soon became chaplain to a national student group that opposed Benito Mussolini and his rising Fascist movement. By 1930 he was an insider, developing friendships with cardinals well connected at the Vatican. In 1937 he was appointed an assistant to the Vatican secretary of state, Eugenio Cardinal Pacelli, who two years later was to become Pius XII. In 1954 "Good Pope John" XXIII appointed him archbishop of Milan, where, suddenly sprung from three decades of the special insularity of the Vatican, Montini made himself a familiar sight in the streets. Sometimes inviting indulgent smiles, he descended into mines wearing a hard hat, toured factories, visited Communist districts never approached by his predecessors. If disaffected workers had stopped coming to church, he felt, it was the duty of the church to go to them.

When his patron, John XXIII, died in 1963, the College of Cardinals, on its sixth ballot, elevated Montini to the papacy. Paul VI's penchant for untraditional inclusiveness became something of a papal style. In 1964 he became the first modern pontiff to take to the road for a Holy Land pilgrimage, then to India, where a Bombay newspaper reported, "There had never been anything like it within living memory, and there will never be anything like it for decades to come." That was the prelude to bringing his sermon to the loftiest of mounts, Yankee Stadium, where he celebrated mass in 1965. He appointed the first female doctors of the church, Catherine of Siena and Teresa of Avila, and was also to reach out in 1974 to the worldwide revolution of women, receiving Betty Friedan for a much-publicized private audience. Friedan wrote of it: "He took my hand in both of his, as if he really meant his concern for women. He seemed much more human, somehow, than I had expected, with a warm and caring expression: He wasn't going through perfunctory motions in meeting with me. He seemed strangely intent, curious, interested in this meeting which was going on much longer than anyone had given me reason to expect."

Soon after the commission's meeting with the pope, Father Häring, a commission member, spoke at Notre Dame University in Indiana to a convention of the Crowleys' Christian Family Movement. His text was simple and courageous. Häring told couples that how they express tenderness in marriage is their business and nobody else's. In light of the church's uncertainty on modern methods of birth control, those who had serious reasons for not wishing another child at a given moment of their lives had a right to decide not to have it and that they could use any methods short of sterilization and abortion. The startlingly clear statement by the pope's private theologian appeared on front pages around the world and was widely read as a sure marker of official positions to come.

The fifth session of the pope's commission started on April 13, 1966, two and a half years after it had begun work. The heat was on, and the world was watching, waiting, expecting. Before this, the pope had authorized three- and four-day meetings. Now he gave them two months and a big budget. Previously commission members had to depend on seminarians whispering unreliable translations into their ears. Now they could hire interpreters from the Rome headquarters of the UN Food and Agricultural Organization (swearing them to secrecy), simultaneous translation equipment, and copying machines.

Cardinal Ottaviani nudged the pope to make the commission "safer." In a classic example of the Vatican's subtle rearrangements to accomplish specific ends—a style that knowing priests call *Romanita*—Paul VI expanded the commission by fourteen additional members, for a membership of seventy-one, grown from an original six. All the new members were cardinals and bishops to provide a "new pastoral emphasis to the deliberations." Canon de Locht fretted in his diary, "We already *had* a pastoral emphasis."

The new president of the commission, an unmistakable sign of the shift to "safety," was Cardinal Ottaviani himself. For cosmetic "balance," two of the newly arrived cardinals were named vice presidents. One was a liberal, Cardinal Döpfner of Munich, who had supported his priests in giving communion to couples using contraceptives; the other a conservative, John Heenan, the archbishop of Westminster who "did not recognize" Dr. Anne Biezanek at his altar. Heenan had been promoted to cardinal early in 1965.

Another pair of "balancing" cardinals were Suenens, liberal sponsor of the Belgium seminars on birth control; and Karol Wojtyla of Kraków, Poland. (Wojtyla, usually a reliable conservative, had once committed the slip-up of writing kindly of birth control. Commission conservatives now persuaded him, for the sake of protecting his political future, to forego attending any commission meetings. It was good advice. He never showed his face. A dozen years later he became Pope John Paul II.)

The expanded commission, lay members and cardinals in tense partnership, plunged into questions scarcely approached at the first commission meeting two and a half years earlier: What was the nature of church "law"? Did it "bind consciences," as the traditionalists argued? Does such a law bind under threat of mortal sin? Was the penalty of disobeying such a law a condemnation to hell? Or was the purpose of church law to lay down an ideal against which Catholics could make a personal judgment for themselves?

Once more, according to the established principle Lex dubia non obligat—a doubtful law does not oblige—the pope could not ban all contraception by a mere statement. Canon de Locht seized a central place in the debate by pressing that point: The law is doubtful. Couples as well as theologians knew it, and even bishops and cardinals had articulated their doubts at the Ecumenical Council. Pope Paul himself had publicly expressed his uncertainty. Forcing the abstractions of theology into a modern realm of reality, de Locht declared, "If the church were to say the Pill were acceptable, the Pill wouldn't become acceptable the day the church said so. The Pill would be acceptable in itself. It isn't pronouncements from the magisterium that make things good or bad. . . . You can't legislate morality just by laying down some rules."

Another jolt of reality inflicted upon the commission was a new survey on "the Catholic way" of birth control, rhythm. This second survey presented by the Crowleys was worldwide, covering three thousand couples from eighteen countries, almost all members of the Christian Family Movement.

Rhythm did not come out well. A conclusive majority, 63 percent, said that rhythm harmed their marriages by forcing upon them tension, frustration, sexual strain, loss of spontaneity, arguments, irritability, discouragement, insecurity, and fear of pregnancy. Less than 10 percent believed that rhythm works or felt positively about it. About 25 percent allowed that rhythm works, but they felt negatively about it. About 65 percent agreed with the statement: "Rhythm does not work, and we have a negative reaction to it."

Emboldened, the Crowleys sent copies of their survey to the pope as well as to Cardinal Ottaviani and his two vice presidents. Their covering letter flatly declared, "Rhythm does not promote unity in marriage. . . . Almost all of the couples hope the church's position will change."

After presenting the survey results, Patty Crowley spoke, instructing her largely male audience in certain facts of life most of them had never heard:

No amount of theory by men will convince women that periodic continence is natural. We have heard some men, both married and celibate, argue that rhythm is a way to develop love. But we have heard few women who agree. Over and over, we hear women say that the physical and psychological implications of rhythm are not adequately understood by the male church. . . . The wife who is unsure, who is afraid of another pregnancy, is not a true love-mate and can come to resent her husband, intercourse, in fact, her whole life. . . . Couples want children and will have them generously and love them. They do not need the impetus of legislation to procreate. It is the very instinct of life, love and sexuality.

Crowley's dressing-down stirred Colette Potvin of Ottawa to deepen the education of the gathered celibates. Describing herself as "simply a wife, married seventeen years, with five children," with the further qualifications of three miscarriages and a hysterectomy, she addressed them with phrases that during the infancy of the mid-twentieth-century women's revolution struck their ears like shell bursts: "To understand woman, you need to stop looking at her as a deficient male, an occasion of sin or an incarnation of the demon of sex, but rather as Genesis presents her: a companion to man. Where I come from, we marry primarily to live with a man of our choice. Children are a normal consequence of our love and not the goal."

She asserted that the commission needed to get away from the technicalities of methods and understand what lovemaking meant to a couple—indeed to Colette and her husband, Laurent. "We don't live our lives around a method . . . but [with] marvelous moments when each of us accepts the other, forgives the other and can give the best of ourselves to the other." With a forthrightness seldom heard in the esoteric and theoretical world of cardinals and theologians, Potvin told of the morning after lovemaking with her husband, when she felt more serene, more patient with her children, and filled with love for others. Nothing, she said, contributed more to her family harmony than the spiritual sense arising from "the conjugal orgasm" and its accompanying "rainbow of wonderful tingling sensations."

Her self-revelation was followed by an awed silence until de Riedmatten offered the genuinely humble observation, "This is why we wanted to have couples on our commission."

One severely discomfited listener, a Jesuit theologian from Paris, Father Stanislas de Lestapis, waited till the next day to rebut. He admonished that the commission was starting to "idolize" the idea of the couple: "The couple has become a state of grace, and contraception their sacrament . . . and the result is a sort of intoxication, a practical obliteration of the sense of God, a mystification in the psychological order, a devaluation of procreation."

When the commission had first met, de Lestapis moved among them as an awesome intellectual. Now his stock was going down.

Patrick Crowley scribbled on his yellow legal pad that de Lestapis seemed to be filibustering.

On May 23, 1966, as the commission entered its fourth year of study and debate, scratching its way through levels of tradition, of encyclicals, of "natural law," of the received wisdoms of Augustine and Thomas Aquinas, it ultimately hit bedrock. De Riedmatten gave the floor to two of the most respected scholars of the Jesuits' Biblical Institute in Rome. Father Ernest Vogt, rector of the institute, was an Old Testament scholar; Father Stanislaus Lyonnet was an expert in the New Testament. Each was asked to tell the commission what Scripture had to say about birth control.

Except for God's suggestion to Adam and Eve that they "shall be two in one flesh" and his command to "increase and multiply," the answer of both men to the simple question was: Virtually nothing. The Old Testament had only one reference that could be possibly construed as touching on contraception: that scene again of Onan spilling his seed on the ground. Vogt said that most biblical experts agree the story had nothing to do with marital morality, but with Onan's shirking his duty under Judaic law to carry on the family name. As for the New Testament, Lyonnet listed all the sexual sins mentioned there and said it was impossible to establish that any of those references had anything to do with contraception.

On June 3 de Riedmatten, proving himself the champion of free inquiry as well as an excellent recorder of members' views, now presented to the divided group two new questions:

Does it seem opportune for the church to speak on the birth control question without delay? The response was unanimous: Yes.

Is the church in a state of doubt concerning the received teaching on the intrinsic malice of contraception? Of those voting, thirty-five said yes, five no.

Cardinal Heenan, remaining more the disciple of politics and appearances than of theology, worried aloud that he did not see how the church without deep embarrassment could "downgrade contraception from mortal to venial sin"—from one warranting excommunication to one that could be cleansed by mere confession. And he proposed a way to get rid of the troublesome headache of the Pill by tossing it into someone else's lap: "Science which has brought us the problem . . . must bring us the solution."

Thomas Burch, the lay Georgetown sociologist, felt suddenly fed up. Never having expected to face down a cardinal with sheer anger, he called Heenan's position "hypocrisy." (He detected that even the UN translators were nodding encouragement to him. In fact, when the session ended, one translator sidled over to Burch and quietly thanked him "on behalf of millions of others around the world.") Almost two decades later in an interview with Arthur Jones of the *National Catholic Reporter*, Burch was still indignant:

I think my very strong reaction was, "What the heck's going on here? You guys are saying that you might have been wrong for nigh on thirty to forty years on some details on methods. You've caused millions of people untold agony and unwanted children, brought on stresses and strains and tremendous feelings of guilt. You've told people they were going to hell because they wouldn't stop using condoms, and now you're saying maybe you're wrong. But you want to say it in such a way that you can continue to tell people what to do in bed?" I think at that point the absurdity became very clear. My feeling was that what the church was most concerned with was exerting its own authority. It was not terribly concerned with human beings at some levels.

The next day the cardinals and bishops of the commission met without the lay members. Bishop Thomas Morris of Ireland, a conservative newcomer, listened carefully for nuances and realized to his dismay that out of fifteen bishops recently added to the commission for "safety," *fully nine seemed to have swung over to favor change.* He felt the dark specter of a recantation by the church hovering over the conference table. Morris startled his colleagues by proposing that contraception was too grave a matter for the pope to decide alone, even with the advice of this commission. He called for a "collegial decision"—by secret vote of all the world's bishops. Cardinal Ottaviani, as though abandoning the pope whom his whole being was devoted to guarding, leaped to agree with Morris's God-sent insight. The question, he said, concerned the whole world; therefore all the bishops should decide.

Even Cardinal Heenan balked: "Years ago this would have been a very good idea. Now it's too late. We couldn't keep [the ballot] secret. We would have to give all the bishops all of the texts of the commission and then give them time to study. The process would take at least five years."

Cardinal Shehan of Baltimore, another new addition, dropped a stunning and sobering thought. Such a consultation with the world's bishops might give the eventual decision the force of an infallible statement. "Do we want that?"

Ottaviani called for a secret ballot of the commission bishops. Of the fifteen, only three, including Ottaviani, agreed with Morris on a worldwide poll of bishops. Eleven voted against it.

Ottaviani knew—as did everyone in the room—that this was effectively an 11–3 vote to send the recommendations of the commission's majority directly to the pope. It was a vote to change the church's position on contraception.

These bishops, archbishops, and cardinals had been appointed to restore the wisdom of cool heads, of tradition, the long view. Now, viewing what they had done, they seemed shocked, unsettled by their own vote. Cardinal Lefebvre of Bourges sensed this and spoke to the vanquished three in a tone that was almost pastoral: "I know you are disturbed—as I am—over the annoying consequences of a change. . . . What is important, above all, is to discover the truth. . . . Science has demonstrated that most conjugal acts are infertile, that they must, then, have another meaning and another goal than procreation, and that these acts can contribute to a deepening of love and self-giving. This does not contradict the traditional doctrine. Rather, it deepens it."

Cardinal Suenens, who had hung back, knowing his early seminars in Belgium had tainted him as a pioneer, even a radical, now tried to help the others see that they had not been swept away by some sudden turnabout in moral theology. The change had not come about in just the past few months:

For years, [the theologians] have had to come up with arguments on behalf of a doctrine they were not allowed to contradict. They had an obligation to defend the received doctrine, but my guess is they already had many hesitations about it inside. As soon as the question was opened up a little, a whole group of moralists arrived at the position defended by the majority here. We have heard arguments based on "what the bishops all taught for decades." Well, the bishops did defend the classical position. But it was one imposed upon them by authority. The bishops didn't study the pros and cons. They received directives, they bowed to them, and they tried to explain them to their congregations.

Someone moved to adjourn. As they filed out of the room, Bishop Morris muttered to de Riedmatten that he had condemned many to hell for what this meeting had just condoned. He wanted release from this commission.

Two days later, on June 24, the cardinals caucused separately from the rest of the commission. They wanted their own vote *as cardinals* as to whether all contraception is by nature evil. After forty-five minutes of wrangling over the meaning of the question, they voted. Morris's tentative headcount was confirmed: *Nine of the fifteen bishops voted no, contraception was not intrinsically evil; two said it was; one said it was, but with a reservation; three abstained.*

After more than three years of study, the full commission now decided that they must not simply present a set of conclusions to the pope. Instead they would state for him the pro and con arguments on the four main themes of the study: moral theology, medical evidence, pastoral considerations, and demographic problems.

The drafts produced by the commission's panels—the bishops, the theologians, the medical doctors, and the sociological experts—were combined in mere hours into a text, and it was delivered to the pope by Döpfner and de Riedmatten on June 28. Cardinal Ottaviani followed behind with what became known as the "minority report," even though the commission had agreed that there would be no majority and minority reports.

But Ottaviani felt his minority had a duty to steer the pope away from the commission's conclusions. (Among the supporters of the "minority report," although he avoided attendance at commission meetings, was soon-to-be pope Cardinal Karol Wojtyla of Poland.)

So that left Paul VI in exactly the same void in 1966 that had moved John XXIII to appoint the six original members in 1963. Essentially unguided but better informed, Paul would have to make his own judgment and take responsibility for it.

"Now begins the period of decision," de Riedmatten told reporters. The decision, he added, would not be "sensational." In his heart he knew the pope was turning away from his commission's recommendations.

The world's Catholics waited—and yet they did not. The talk of change had spread so wide and for so long that it drowned the subtle new hints that change might be doomed. Cardinal Döpfner instructed priests of his diocese that couples using contraceptives were not necessarily in a state of sin and should not stay away from communion. At Catholic University of America, Charles Curran, a thirty-three-year-old theology scholar, was teaching seminarians that the church was in a state of doubt. In American Catholicism particularly, freedom of conscience was itself a novel state but becoming more familiar. Throughout America and the Western world, unnumbered priests were telling their people: In this state of doubt, follow your conscience.

The world waited through almost five months of silence from the Vatican. During those months, one of Ottaviani's closest allies, Bishop Carlo Columbo, now personal theologian to the pope, had located a soft spot in Paul's indecision and had successfully urged upon him that a change in moral teaching would be a clear repudiation of his papal predecessors. That appeared to be the argument, exquisitely harmonious with the relentless entreaties of Ottaviani, that overtook all others.

Finally, in November 1966, Paul VI chose a congress of Italian gynecologists and obstetricians as the occasion to say he needed more time to study before making a decision about contraception, and that the ban remained in force. His stated reasoning was a wisp from the grandest tradition of high-functionary billowing of murk and fog: "Because [the recommendations] carry grave implications together with several other weighty questions both in the sphere of doctrine and in the pastoral and social spheres which cannot be isolated or set aside, but which demand a logical consideration in the context of what precisely is under study."

One of England's leading theologians, Charles Davis, was so shocked he immediately resigned from the priesthood. The pope's statement, he said, "illustrated the subordination of truth to the prestige of authority and the sacrifice of persons to the preservation of an out-of-date institution." A month later Davis left the church as well.

The Vatican continued to insist that the old teaching against birth control was not "doubtful law." Its official press officer, Monsignor Fausto Vallainc, with a fine-tuned touch of *Romanita* news management, put out the word that present teaching was still to be considered "certain," and that if the church were to change, the change would be "from one state of certainty to another state of certainty."

The commission report, so long in the making, looked dead.

Across the Tiber a great oblong pool in the Piazza Navona is eternally watered by vigorous streams from the penises of naked cherubs. Just off the piazza stood the Dutch Documentation Center, whose director, Father Leo Alting von Geusau, had served throughout the Ecumenical Council meetings as a crafty leaker of secret documents to the continent's leading news journals. Now he somehow found in his surprised hands a copy of the commission's reports to the pope as well as that of the apparently triumphant minority. Feeling compelled by some inner calling to do something more active than file them in his archive, Father von Geusau knew exactly what to do. "He knew knowledge was power," Robert Blair Kaiser later wrote, "and that publishing these documents would give millions of Catholics power in an area once reserved for higher authority to decide."

Von Geusau approached one of the best of reporters covering

the Vatican, Hénri Fesquet of the French daily *Le Monde*, and offered the documents. To von Geusau's surprise, Fesquet blanched
at the fearful prospect of publishing them. He referred von
Geusau to a *Le Monde* editor. While von Geusau waited for an answer, on December 13 an Irish-born American author and journalist of his acquaintance, Gary MacEoin, turned up in Rome.

MacEoin was taken aback upon learning from von Geusau that
an overwhelming majority of the pope's commission had voted to
change the position on birth control—and had then been crudely
jostled aside. He perceived no moral barrier to leaking the reports
but urged von Geusau that *Le Monde* was the wrong place. "*Le
Monde* won't print the reports in full," he admonished. "It will do a
news story. And then the world will pick up a truncated version
of that story and get things all garbled." Why not offer them, he
suggested, to the *National Catholic Reporter*, the gadfly periodical
published in Kansas City, on the condition that they print the reports in full and provide summaries to the American press? Then
Le Monde could "scoop" Europe with its news story the same day.

The editor of the *Reporter*, Robert G. Hoyt, reached by
MacEoin, leaped at the plan. But all too aware of the backlash it
would invite, he felt that in fairness he needed the consent of his
staff. Gathering them, he asked, "Do we think this is the right
thing to do?" There was not a doubt in the room. It would be their
biggest news story of the decade and, in fact, the essence of what
the *Reporter* existed for: information hidden from the flock by
their own bishops and the Vatican.

On February 2, 1967, MacEoin flew from Rome to New York
carrying the reports. His deal with Hoyt: He would not reveal
how he got the papers and did not want credit. If any income resulted from their publication, MacEoin wanted the money to go
to a young black friend in South Africa who was working his way
through medical school. Hoyt, upon seeing the papers, had no
doubt about their authenticity. He engaged two American
priests, Larry Guillot and Phil Tompkins, to translate them from
Latin and French.

The *National Catholic Reporter* published the complete texts on
Saturday, April 15. On Monday, April 17, *The New York Times* ran

the news at the top of page one, with a full page of text inside. That day, and for several to follow, the story of the commission report and its minority rebuttal remained near the top of the news.

As the documents were going to press in Kansas City, Hoyt mailed a set of copies to *The Tablet,* England's pugnacious Catholic weekly, in London's Great Peter Street, where editor-publisher Tom Burns ordered them set in type immediately. "They just came in the mail without a note or anything to explain," Burns said. "They were postmarked Kansas City, though, so I was able to guess where they had come from. I knew they were authentic." *The Tablet* published them in several installments starting on April 22.

No commission member commented on the revelations. The Vatican neither confirmed nor denied the published texts but signaled that the pope would announce his decision in his own good time.

In October 1967, twenty-five hundred delegates to the Third World Congress of the Roman Catholic Laity listened in St. Peter's Basilica to a stern warning from Pope Paul against creating "two parallel hierarchies of clergy and laymen. Anyone who attempts to act without the hierarchy or against it could be compared to the branch that atrophies because it is no longer connected with the stem that provided it sap."

The next day the congress rebuked the pope so sharply and spontaneously it may have startled itself. When the reading of a panel report on "the family" favored planned parenthood and advocated that the means for achieving it be left to the conscience of the individual, applause swept across the huge gathering.

Aided by a handpicked few of the commission minority, Cardinal Ottaviani took over the actual writing of the pope's decision. The effort stretched over more than six months. Lobbying, both for the commission's majority position and against it, continued during 1967 and for half of 1968.

Early in the morning of July 29, 1968, after claiming for four days that rumors of an imminent statement on birth control were "absolutely false," Vaillanc, the Vatican press officer, called a news conference to release the long-awaited encyclical. It was titled *Humanae Vitae* (Of Human Life). Without touching on the months of study by medical and social-science scholars on the commission, *Humanae Vitae* pivoted almost entirely on the dark threat of sex and distrust of men and women subject to its ever-present temptation. Written by celibate men, it was a document chiefly about men, declaring that artificial birth control would "open up a wide and easy road . . . toward conjugal infidelity and the general lowering of morality. Man, growing used to the employment of contraceptive practices, may finally lose respect for the woman and, no longer caring for her physical and psychological equilibrium, may come to the point of considering her as a mere instrument of selfish enjoyment, and no longer as his respected and beloved companion."

In New York, a night editor of the Associated Press began calling American members of the commission. He woke the Crowleys in Chicago. The late-night ring of the telephone confused the couple. Clearing his head, Patrick Crowley listened to the news of the encyclical, and, asked for a comment, he mumbled simply, "I don't believe it."

Hanging up the phone, he turned to Patty with heartbroken weariness: "Mom, just what in hell did we keep going to Rome for?"

18

"THEY ARE CRUCIFYING THE CHURCH"

*John XXIII said that during the first months of his pontificate
he often woke during the night, thinking himself a cardinal
and worried over a difficult decision to be made, and
he would say to himself: "I'll talk it over with the Pope!"
Then he would remember where he was. "But I'm the Pope!"
he said to himself. After which he would conclude:
"Well, I'll talk it over with Our Lord!"*
—HENRÍ FESQUET, WIT AND WISDOM OF GOOD POPE JOHN

The breaking of the news of *Humanae Vitae* put an indelible mark on the life of Tina Mariello. It had come shortly before she entered the University of Michigan to behold the sexual "revolution." She, her parents, and her younger sister and brother were resting after a hard day at Disneyland in Anaheim, California. Her father turned on the hotel television. When, contrary to world expectation, the evening news announced that the pope was holding firm against all birth control, including the Pill, Tina's father, with slightly reddened face, muttered, "That's ridiculous." He snapped off the set.

"That was the first time," Tina recalls today, "that I ever heard either of my parents question anything about the church, especially anything about the pope. I think that was the first crack in my devotion."

In Washington, where picket lines often go unnoticed in front of the White House, no one could ignore the spectacle at the Catholic University of America. At a rally of more than two thousand in front of McMahon Hall, nuns waved rebellious placards, priests marched in silent indignation, and students, both female and male, sang "Ain't gonna be no bishops no more, no more." Walter Schmitz, dean of theology, declared that the faculty had just voted 400–18 to strike, and they were joined by almost seven thousand students. No classes were held for a week.

The immediate subject of the protest was the dismissal by the board of trustees of Father Charles E. Curran, a thirty-three-year-old theologian, who had recently been unanimously recommended for promotion by both the faculty of sacred theology and the university's full academic senate. The rector of the university, Bishop William J. McDonald, would give no reason for the firing, but everyone knew it was because Curran had put himself on published record in support of contraception. The board was composed of thirty-three cardinals, archbishops, and bishops, as well as eleven laymen.

"This is not the first time a C.U. professor feared by the hierarchy has been fired," said Father Daniel C. McGuire of the department of religious education. "But it is the first time that the entire university has reacted *en masse* to a firing." Physics professor Malcolm Henderson, head of the faculty assembly, added, "That this university should be run like a medieval monastery is dammed nonsense." More than a few traditionalists were surprised to hear Boston's Richard Cardinal Cushing condone the protests: "I would not condemn this man. He must teach all sides. That's scholarship."

Finally Patrick O'Boyle, the university chancellor as well as archbishop of Washington, announced from the steps of Mullen Library, "The board of trustees has voted to abrogate its decision." But he quickly added, "This decision in no way derogates the teachings of the church by the popes and bishops on birth control." Father Schmitz took the microphone to report, to wild applause, that Curran would receive his promotion to the rank of associate professor.

But insurrection among leading Catholics was exploding from all directions. The Los Angeles Association of Laymen accused the pope of "an anti-sexual bias" and said he may have written *Humanae Vitae* "to alleviate the anxieties he has about his own authority." The association said its members "will not leave the church. We will not be thrown out." The Association of German Physicians said, "Paul VI is completely out of touch with reality. Catholic doctors live in the world as it is, and not in an ideal world longed for by the pope."

Some twenty-six hundred U.S. scientists declared in an open letter of mixed accusations that "the appeals for world peace and pity for the poor made by a man whose actions help to promote war and render poverty inevitable do not impress us anymore."

Dr. John Marshall, one of the commission's original six members, wrote to *The Times* of London, which received a thousand letters in a week on the subject. He disclaimed the encyclical's conclusion that birth control opened the door to infidelity and a lowering of morality: "Despite the widespread and long-standing practice of contraception, there is no scientific evidence to support this sociological assertion made in the encyclical. The assertion, moreover, casts a gratuitous slur, which I greatly regret, on the countless responsible married people who practice contraception and whose family life is an example to all."

At Essen, Germany, five thousand participants in the eighty-second German Catholic Conference noisily voted to demand a "revision" of *Humanae Vitae,* while only ninety raised their hands to oppose the resolution.

One could almost feel the church cracking apart throughout Europe. The Catholic bishops of Austria issued a joint statement saying that couples who were limiting the number of their children "for ethical reasons" dictated by their own consciences did not have to confess their practice of birth control as a sin. The bishops of Belgium and West Germany stated similar positions, emphasizing the role of individual conscience in birth control matters. Scandinavian bishops told their priests that couples acting contrary to the pope's ban on birth control need not be considered bad Catholics. Under no circumstances, the bishops added, should a person act against her or his conscience.

The French hierarchy of 120 bishops declared that the use of artificial contraception "is not always guilty." They left it up to each believer to decide whether use would be sinful in his or her case. Clearly addressing the Pill, the French bishops declared, "Science confers an astounding control over creation, and even over man, and new research about love and sexuality has opened up new prospects as to their respective significance."

Cardinal Heenan of London, the ever-reliable straddler, was reminded by David Frost in a television interview that he might one day become pope. Laughing, Heenan replied, "I hope not. But, if so, I'll be very careful about writing encyclicals."

Less jovially, Cardinal Heenan put out a letter to be read in all 240 churches of his London diocese: "The church has compassion on the many for whom this ruling will bring hardship. Those who have become accustomed to using methods which are unlawful may not be able all at once to resist temptation. They must not despair. Above all they must not abstain from the sacraments."

Denis Hurley, archbishop of Durban, South Africa, told the Catholic weekly *Southern Cross*, that the encyclical was "the most painful experience of my life as a bishop. I have never felt so torn in half."

In the Irish Republic, which often calls itself "the most Catholic country in the world"—93 percent are Catholics, and they enjoy claiming an 83 percent church attendance weekly—the defiance of *Humanae Vitae* came not from the bishops but, more shockingly, from elements of an emboldened flock. "There were waves of winds of change blowing through the church, people terribly aware of what was changing elsewhere, waiting for the pope to change his mind about the Pill," recalled Mary Maher, a columnist for the *Irish Times*. During the late sixties and early seventies she had been women's editor of that leading paper while also helping launch the Irish Women's Liberation Movement. Almost a quarter century later, over lunch in a Dublin hotel pub, she relished recalling that unique time: "We thought we only had a small cozy group that could meet and argue theoretics, but suddenly we had a mass movement on our hands. Our biggest demand was contraception. The fact was, of course, that if you had enough money—the middle class simply went across to Britain, got their contraceptives, and came home. Just smuggled them in. But not these poor little working-class families. Little? We're talking families where twelve, thirteen children were not all that uncommon, you know. So we said, This is a civil right, something we must have. Senator Mary Robinson—who is now the president of Ireland—twice introduced the bill to legalize contraception, but it failed. She came very close but not close enough.

"But it was a popular demand. So we had a meeting: What could we do to get headlines on this topic? What we decided to do was take a train on a Saturday morning from Dublin to Belfast—that's in North Ireland, the British part—and load up with contraceptives of every kind, bring them back, and declare them at customs. Our plan was very much modeled on the American civil rights movement. A peaceful protest. It was going to be desperately dignified. At customs we'd hand in our Pills and condoms and caps and say, 'I demand to be arrested.' Before we left Dublin there was a long debate about whether or not we should take only married people. This was Ireland in 1971, and we didn't want to alienate everybody. But then we decided that while we'd keep the respectably married people in the forefront, it would be hypocritical to confine ourselves simply to married people. So we had a few young ones in miniskirts. We had about thirty women, and a couple of men. We hadn't invited the men, but they arrived. Before we left we notified customs that we'd be coming back on the six o'clock train and that we'd be declaring all these things. At Belfast, we phoned the Dublin papers and television stations with a public announcement.

"Well, the plan did not go quite as planned. When we returned to Dublin, I remember seeing one poor customs fellow with his hands shaking. They all got over-the-wall, not knowing what to do. They wouldn't let us declare things. We'd declare louder and demand to be arrested. They insisted on not seeing and not hearing, and trying to pass us through. That was not what we wanted. So the place went crazy. Some women filled their hands with their 'menstrual-regulating' Pills and shouted, 'Look, I'm swallowing my Pill!' People began throwing caps like Frisbees, throwing condoms over the customs counter to the crowd on the other side who had come because they'd heard the announcements. More than five hundred people were there, cheering, supporting us. It was sensational. The BBC was there, international coverage with television cameras. It was a great day, a turning point in many ways. Because after that, despite this exhibition—it was disgraceful, disgraceful, you know, after we had very much hoped to be proper and respectful—everybody was talking, using words they previously never dared to use. Everybody talked about contraception, saying, 'Do you think it's right? Do you think it's wrong? Do you think people should be able to have contraceptives?' I mean, 'contraception' wasn't a word that people had used. It suddenly became something everybody could state an opinion on, argue about, instead of feeling this is a matter so private and personal you could only whisper it to your confessor or your best woman friend. The climate was changed."

A Gallup poll taken only a month after the encyclical found fifty-four out of every one hundred American Catholics opposed the pope's stand, with only twenty-eight favoring it. Eighteen said they had no opinion. More ominous for the future of the church, among Catholics still in their twenties, eight out of ten disagreed with the pope.

Opposition to the Vatican did not form a solid wall, especially in America. L. Brent Bozell, managing editor of the *National Review*, came out for excommunicating those who could not agree with the pope, adding: "Those priests who refuse to accept, and faithfully carry out in their pastoral capacity, Pope Paul VI's encyclical on birth control should leave the church."

The seven American members of the pope's commission, all laypersons, had now seen, close up and firsthand, how decisions were made in Rome. What they saw changed them forever, not by turning them against their faith, but by confirming exactly what the obscure, seemingly rebellious priest of Little Rock, Arkansas, Father Drane, had observed in *Life* magazine the previous year: "The same type of human political activity that takes place in the state houses takes place in Rome, and no one should be shocked that in a human institution people act like human beings."

Two months after *Humanae Vitae*, the seven American members gathered in Washington, D.C., to assess the disaster.

"It had taken us some time to arrive at our new positions," John Noonan later recalled. "We couldn't just walk away because of an encyclical, one we didn't agree with." At a news conference after their meeting, Noonan spoke for all the American members. He expressed their dismay that the pope had not listened to his own commission, that he had consulted only with a dozen or so of the old guard, and that they had prepared the encyclical in secret. Noonan, the scholar, predicted confidently that the church would repudiate *Humanae Vitae* as it had repudiated an 1832 encyclical, *Mirari Vos*, which condemned freedom of conscience.

The "new positions" of the suddenly independent Catholics were not assumed without personal cost. Among snubs and ostracisms they all were soon to endure, the Crowleys were saddened when their longtime spiritual mentor, Father Reynold Hillenbrand, refused to have anything further to do with them.

Within a week after Noonan's press conference, the front page of *The New York Times* reported that half the Catholic diocesan priests in the United States disagreed with Pope Paul's condemnation of birth control. The survey was conducted by the Center for the Study of Man, located on the University of Notre Dame campus. Perhaps the most revealing finding—and most foreboding for the future of the church—was that among older pastors about 95 percent stood with the *Humanae Vitae* ban on contraception, but among younger curates about 95 percent opposed it. Another finding also furrowed brows among archbishops and in the Vatican itself. Fifty-one percent of the priests said that before the pope spoke they had felt that artificial contraception was permissible "under certain circumstances." After the pope spoke, 49 percent held fast to their earlier view. Only 1 percent switched positions, and 1 percent were now undecided. Forty-five percent reported that before the pope's message they had advised married couples that contraception was permissible. The encyclical reduced that figure to 41 percent, shifting 2 percent to undecided. Another 4 percent said that they no longer advised either way, but recited both sides of the issue, leaving the couples to decide.

The backwash of the encyclical spilled into unlikely forums. On the NBC network's *Tonight Show*, Robert Shaw, the fiery English actor and playwright, said he admired Pope Paul's courage in standing against public opinion but added that "the pope, in my opinion, is not only a fool to do it but he has no right to do it. . . . As far as I'm concerned, all that business about infallibility is nonsense." David Frost, who was substituting for Johnny Carson as program host, responded that in the nineteenth century a pope was supposed to have said one afternoon, "I'm not infallible." Then at seven o'clock that evening the pope said, according to Frost, "I'm sorry. I made a mistake. I *am* infallible."

NBC said it received 120 phone calls and fourteen letters protesting the conversation. Charles E. Reilly, executive director of the National Catholic Office for Radio and Television, wrote NBC that Frost and Shaw had "clearly violated basic decency in their unfortunate and cruel exploitation of an issue that many dedicated people are agonizing over."

As the worldwide revolt hammered at the Vatican, on April 2, 1969, Pope Paul finally let out a cry of anguish. The dissidents, he said, were "crucifying the church." He was grieved by "the restless, critical, unruly and destructive rebellion of so many of its followers, including the most dear, priests, teachers, the lay . . . against its institutional existence, against its canon law, its tradition, its internal cohesion, against its authority. . . ."

The torment lapped at the conscience of James P. Shannon, the auxiliary bishop of St. Paul–Minneapolis, and it turned out to be fateful for him. Shannon had served as president of the College of St. Thomas in St. Paul before he became a bishop in 1965, just in time to join the last session in Rome of the Ecumenical Council. Upon his return Shannon was assigned by Archbishop Leo Binz to "translate" for the priests and flock of the

Twin Cities the reforms voted by Vatican II. He welcomed it as "the most exciting intellectual experience I ever hope to have in my adult life."

Shannon relished spreading the word that the council had settled the birth control question "in favor of the couple." Then came the shock of *Humanae Vitae*. He went to Binz to confide his resistance to the encyclical. Binz instructed, "Once the pope has spoken, it's over. You're a bishop and you gave your word." Returning to his parish, without giving his assistant priests a reason Shannon removed his name from the list of regular confessors.

A few days later a young woman came to the parish house. The incident, Shannon said, changed his life, not because it was unique but because he knew it was being played out in confessionals and rectories across the earth. He later recounted it to Robert Blair Kaiser. The woman, from another parish, said she and her husband had been "good Catholics" all their lives and had practiced rhythm. They soon had an unplanned baby, then another. "I'd love to have ten. I'm the happiest mother you have ever seen," she told the bishop. Her husband had "an ordinary job, and he'll probably never have a better one." After the second baby came, the woman was told by her husband, "If this is the future for us, I can't afford this. We can barely make it now."

"Something went wrong with our relationship," she said to the bishop. "He became distant with me. And I became distant with him. We're both affectionate, loving people, and we learned that that leads to babies.

"Yesterday was my husband's birthday. We don't have much money. We don't ever go out to dinner. But I really splurged yesterday. I fed the babies late so that by the time he got home they'd be in bed and I'd have everything nice for him. I had candles on the table, I had a bottle of wine chilling in the refrigerator. He came home and was so pleased. I was doing the finishing touches at the stove and he came up behind me and put his arms around me and he hugged me." The woman paused, then blurted out, "I ran into the bedroom and slammed the door and cried myself to sleep. He ate all alone last night."

Through tears she gazed at Shannon and said, "My marriage. I don't know what's going to happen to my marriage. This is the loveliest, sweetest, gentlest man. He can't go on like this. I can't go on like this. I can't leave him. And he can't leave me. I don't know what to do. I don't know where to turn."

Shannon, at first speechless, recited his post–Vatican II speech, and how there were new rules now, which had the endorsement of the pope. She could do what she felt she had to do, and in good conscience. He recalled seeing the clouds leaving her face.

She asked, "Would the pope agree with what you're telling me?"

Shannon said, "Madam, if he were here now, I have to believe he would. He's a pastor. He's no less a pastor than I am. He would have to say that the merciful Lord knows your situation and is with you."

The woman appeared released. Shannon saw her to her car, went back into the house, took off his cassock, put on his suit, and drove straight to Archbishop Binz, who had been one of the last two members appointed to the papal commission, and who, as noted, had maneuvered the suppression of the television series by the National Council of Catholic Men.

Shannon said to Binz, "For the first time, I told a devout Catholic that it would be permissible for her to practice a safe contraception on a regular basis."

Binz said he didn't know what to say. Shannon went home, thought things over, then sat down and wrote a letter to Pope Paul. He said he could not "in conscience give internal assent, hence much less external assent," to the papal teaching that "each and every marriage act must remain open to the transmission of life." He said he could not believe God binds people to such rigid standards. He added: "I must now reluctantly admit that I am ashamed of the kind of advice I have given some of these good people, ashamed because it has been bad theology, bad psychology, and because it has not been an honest reflection of my own inner conviction."

Then he went to see Binz again, who was puzzled at Shannon's wanting to tell the pope. Binz pointed out that his letter would be on Shannon's record in the Vatican. This would not be a strong recommendation for his future in the church. Shannon ruefully agreed.

The next summer he resigned the priesthood, in which he had spent twenty-three years.

In 1976, eight years after the encyclical, Father Andrew Greeley's National Opinion Research Center at the University of Chicago published a survey of U.S. Catholics with the arresting conclusion that *Humanae Vitae*, besides directly causing drastic declines in religious devotion, was costing the church almost a billion dollars a year in income.

"It is rare for a social researcher to be able to explain a phenomenon so simply. But that is what happened," flatly asserted Father Greeley. "We don't speculate that the cause of the Catholic decline was the birth control decision, nor do we simply assert it. We prove it with the kind of certainty one rarely attains in historical analysis."

According to Greeley's figures, Catholics who had welcomed the earlier reforms of Vatican II tended to become more active in the church than those who opposed them. Only after the birth control encyclical did church attendance begin to plummet. Compared against a 1963 survey by Greeley's group, his 1974 findings showed that weekly attenders at mass had dropped from 71 to 50 percent. Those going to monthly confession sank from 38 to 17 percent. In 1963, 70 percent of those polled held that the pope drew his authority in a direct line from Jesus. In 1974, only 42 percent had that opinion.

Greeley concluded that the birth control encyclical was "a shattering blow" to the loyalty of U.S. Catholics, and he predicted that future scholars would regard it as "one of the worst mistakes in the history of Catholic Christianity."

BOOK IV

19

"Pill Kills!"

I give you bitter pills in sugar coating.
The pills are harmless;
the poison is in the sugar.
—Stanislaw Lec, *Unkempt Thoughts*

Rumors of serious side effects of the Pill began to spread early on—scattered, uncertain reports first in newspapers and, before long, in solid professional medical journals. Here and there among the ten million American women using the Pill, fifty million worldwide, there erupted disturbing accounts of nausea, water retention, weight gain, and breast tenderness, then far more serious matters: deep-vein thrombosis (blood clots that can kill if they lodge in the lung), heart disease and heart attacks, elevated blood pressure, strokes, gallbladder disease, tumors of the liver, even depression.

By the fall of 1961, the Searle Company had collected a file of 132 reports of thrombosis and embolism in Pill users, including eleven deaths. Were these cases that were normally expectable in the general population and merely coincident with Pill use—or were they *caused* by the Pill? When afflictions are relatively rare and scattered, how is one to separate cause from coincidence? There was talk that the government might withdraw its approval of the Pill for general contraceptive use.

Within a year, on August 6, 1962, Norway halted sale of the Pill on the basis of an article in the *British Medical Journal* reporting that four Pill users had developed thrombosis, although no ill effects from the drug had yet been recorded in Norway.

Soon the Soviets decided to ban the Pill. Under pressure to explain, Dr. Boris Petrovsky, Minister of Health, said in 1967, "I study all the literature about the oral contraceptive question, and I note there is not a single country in which oral contraceptives are regarded as absolutely safe." After a three-year comparison against the Pill by the Soviet Academy of Medicine, in 1969 Petrovsky approved the IUD as "the lesser of two evils."

In the U.S., the FDA stood firm in its approval of the Pill, not persuaded that the small number of blood-clot occurrences outweighed the widespread benefits and satisfaction reported by Pill users. But, under pressure from a new kind of lobby—women's health activists led by a magazine journalist, Barbara Seaman—the FDA ordered manufacturers to insert in their Pill packages a complete listing of known side effects, not a routine procedure at that time.

Meanwhile, the Pill was building its lobby of defenders, medical professionals among them. They argued that the evidences of the Pill's dangers were statistically small. One argument was that the projected number of deaths from the most serious known risk, clotting disease in the lungs, was 1.3 in 100,000 women. A considerably higher number of women would die from the complications of pregnancy itself.

Evidence also mounted that the Pill was bringing certain desirable side effects. One study, published in the *Journal of the American Medical Association* in 1962, had produced early evidence that the Pill provided important defenses against reproductive cancers. Dr. Robert A. Wilson reported a study of mammary and genital cancer in 304 women, ages ranging from forty to seventy, who were treated with estrogens for periods up to twenty-seven years. According to norms, eighteen cases of cancer, either mammary or genital, were to be expected in the group, the author estimated. Instead, no cases occurred.

In medical matters, uncertainty breeds uncertainty, uneasiness stirs uneasiness—and, in America, both generate lawsuits. In the first few years of the Pill, more than one hundred court claims were filed against its manufacturers.

The uneasiness was unexpectedly quickened by a new element that had nothing—yet, in a sense, everything—to do with the controversy. In 1961, almost simultaneous with the FDA authorization of the Pill, another wondrous medicament burst forth in Europe to emancipate women. It was a new sedative which, among its blessings, magically erased the most annoying side effects of pregnancy: nausea and vomiting. Women and their doctors throughout the continent gratefully leaped to embrace the new drug. Its name was thalidomide.

Within a year after the Pill was licensed, unexpected side effects of thalidomide began to come unmasked to the horror of the world. Some ten thousand babies, five thousand in Germany alone, emerged from their aghast mothers with handless arms, armless hands, legless, footless. In the United States, a cautious FDA had delayed approving thalidomide, despite its early popularity in Europe and elsewhere, and now, peering back at its narrow escape, heaved a deep breath of relief.

While the vast tragedy of thalidomide spared American babies, it left an American public highly sensitized to side effects of new medicines.

In a 1969 book, *The Doctors' Case Against the Pill,* the Pill's leading critic, Barbara Seaman, gave voice to a frustration that many women had felt over centuries—not being taken seriously by their doctors, having their complaints brushed aside as "women's nervousness," being soothed with sweet syrup of "trust me." The medical industry's ready assurances that discomforts and diseases linked to the Pill were not real—"trust me"—suddenly boiled over into a long-brewing wrath. Seaman attacked the common ways in which the Pill was prescribed as a violation of the principle of "informed consent."

Indeed a 1970 Gallup poll revealed that two thirds of women taking the Pill reported never having been warned by their physicians of any risks.

Seaman criticized the Puerto Rico field trials of the Pill as insufficient in both duration and numbers of women, claiming that the 1960 decision to approve Enovid was based on studies "of only 132 women who had taken the pill continuously for a year or longer."

Widely published records show that Searle had filed its application to license the Pill as a contraceptive on October 29, 1959, reporting the experience of 897 women, representing 801.6 woman-years and 10,427 cycles. Indeed tens of thousands of additional women had already been using precisely the same drug for regularizing their periods and for other gynecological purposes for more than two years, with FDA approval. Yet the FDA treaded cautiously, almost timidly, in handling the application to use the same Pill as a contraceptive. It invoked a rule that permitted it to postpone the decision on any application for six months. Indeed, with the Pill, that extension ploy was used twice.

Dr. Pasquale DeFelice, the FDA reviewing officer in charge of the Searle application, reconstructed the episode several years later in an interview:

> When a new drug application came in for the birth control pills, it was—needless to say—revolutionary for that indication. It was a whole new bag of beans. Everything else up to that time was a drug to treat a diseased condition. Here, suddenly, was a pill to be used to treat a healthy person and for long-term use. We really went overboard. Even though the Pill had been through more elaborate testing than any drug in the FDA's history, there was a lot of opposition. Everyone was afraid of the Pill. No other pill ever was put through anything like the tests for *the* Pill. . . . The FDA had to come out of the licensing absolutely clean. We got all the studies, every page of them, from the projects in Puerto Rico

and Haiti. Some of the women had been using the pill for as long as five years. . . . But we were in no hurry to put the FDA stamp of approval on it. . . .

As much as I was impressed by Rock and the study data, we got [I.C.] Winter [of Searle] to promise to do lab tests on five hundred cases detailing any blood-clotting mechanisms. . . . Since the Pill created a pseudo-pregnancy I thought we'd better double-check.

Moreover, the Pill that continued to be assailed after its May 1960 FDA approval was not the same Pill that first came under attack. Its makers were continually discovering that lower active dosages worked as well, and more safely, than higher ones, so dosages kept reducing. Searle's first approval for Enovid as an oral contraceptive was for a ten-milligram dosage.

The lowered doses were approved soon after they passed field tests of their own for effectiveness.

Seaman's book spurred Senator Gaylord Nelson of Wisconsin to call a Senate hearing in 1970 specifically about the Pill. It produced few new facts but led to an explosion of terrifying headlines: PILL MAY CAUSE STROKES, EXPERT TELLS SENATORS (*Philadelphia Inquirer*), PILL AND CANCER—WHAT MEDIC SAYS (*San Francisco Chronicle*), SENATE PANEL TOLD THE PILL CAN KILL (*New York Post*). One tabloid was most succinct: PILL KILLS! Judie Brown, president of the American Life Lobby, joined in with a slogan: "Contraception means better killing through chemistry."

The 1969 annual convention of the American College of Obstetricians and Gynecologists acknowledged the new women's health awareness by staging a "Great Debate" about the merits of the Pill. It included a speech by Dr. Hugh J. Davis of Johns Hopkins University that was particularly combative: "Everyone has the right to select his [*sic*] own form of suicide. . . . It is perfectly true that we all have to die of something, but there is no need to be in a hurry about it. When the apologists for the Pill told me that the risk of death from taking the Pill is really insignificant because it is no greater than the risk of dying in an auto crash, far from finding this reassuring, I rushed out, bought an extra set of safety belts, doubled my disability insurance, bought a St. Christopher's medal to be on the safe side, and took my wife off the Pill."

An ironic embarrassment was yet to befall this particular doctor's case against the Pill. In the 1970 Nelson hearings he was spotlighted as the opening keynote witness, where he declared: "Never in history have so many individuals taken such potent drugs with so little information available as to actual and potential hazards. The synthetic chemicals in the Pill are quite unnatural with respect to their manufacture and with respect to their behavior once they are introduced into the human body. In using these agents, we are in fact embarked on a massive endocrinologic experiment with millions of healthy women."

Dr. Davis's chief qualification for his appearance, besides his known opposition to the Pill, was his expertise on the subject of IUD's. In fact, an IUD he participated in inventing, called the Dalkon Shield, was about to appear on the market. The invention soon turned out to be the misfortune of his life and, worse still, a calamity for many women. The shield's manufacturer, A. H. Robins Company, was forced in 1974 to disclose in letters to 120,000 doctors that septic spontaneous abortions in midpregnancy had been suffered by thirty-six users of the Dalkon Shield, resulting in four deaths. The tragedy for women turned into a harvest for personal-injury lawyers, who ran full-page ads in newspapers across the country seeking users of the Dalkon Shield. After a decade of scouring for victims, the lawyers produced a staggering total of 195,000 successful injury claims. A federal court ordered the Robins Company to set up a trust fund of $2.47 *billion*. Lawyers for some of the women challenged the proposed trust fund as inadequate. The company abandoned the victims by going bankrupt.*

Three years after the Dalkon Shield disaster, and apparently

*Another financial fiasco victimized Great Britain's 450 thalidomide children and their parents. Most reached the age of eleven without receiving a penny of compensation because the case was crawling through the courts. Meanwhile, newspapers could report nothing about their plight because British law forbade reporting on a pending court case for a decade. The thalidomide children were almost forgotten. Outraged at the "secret scandal," Harold Evans, editor of *The Sunday Times* of London, deliberately broke the law (and sacrificed the advertising of Distillers, makers of inspired by it, a Minnesota woman, Esther R. Kociemba, sued G. D. Searle Company, which made the Copper-7 IUD. Eleven years earlier, the suit said, Mrs. Kociemba's doctor had inserted a Copper-7 IUD and twelve days later found a pelvic infection that left her unable to have children. She was awarded $8.75 million to be paid by Searle.

Clearly, these legal side effects of medical side effects did not increase the eagerness of manufacturers to put new IUD's on the market. Moreover, their willingness to risk huge expenditures for researching *any* new contraceptives appeared forever dampened. One reproductive scientist soon observed, "The U.S. is the only country other than Iran in which the birth control clock has been set backward."

In 1986, the largest study ever conducted on the subject—more than nine thousand women ranging in age from twenty to fifty-four—concluded that use of the Pill does not increase a woman's risk of breast cancer even if she has a family history of the disease. The study, directed by Dr. Richard W. Sattin of the Centers for Disease Control, was published in *The New England Journal of Medicine*. What the study did not answer, said Dr. Sattin, was the effect of long-term use of the Pill when taken by adolescents.

The consequences of the Pill scare were mixed. While the Pill remained the American "contraceptive of choice" during the 1970s, sales dropped by 20 percent. Not so across the sea, however. In England the Family Planning Association, which had monitored use of the Pill since 1961, reported in 1977 that there was absolutely no statistical evidence to show that women were rejecting the Pill. More and more women were now using it because "they see it as the only 100 percent reliable method of contracep-

---

thalidomide) by running a campaign that included a large picture of armless Philippa, age ten, under a headline, OUR THALIDOMIDE CHILDREN: A CAUSE FOR NATIONAL SHAME. Evans's action forced a legal crisis: balancing the gag law against the public's interest in freedom of the press. After a long and circuitous legal battle, the action of Evans and *The Sunday Times* resulted in both a substantial financial settlement for thalidomide victims and historic reform of the gag law. The battle is told in detail in Harold Evans's 1983 book, *Good Times, Bad Times.*

tion. In 1969, there were just over a million users. In 1976, there were more than three million."

In the United States, confidence in the safety of the Pill rose again in the early eighties. The FDA reported that approximately 10.7 million American women were now on the Pill, easily the most popular method of nonsurgical contraception.

Another major consequence of the Pill scare was that women's health became a permanent and central focus of the rising women's liberation movement, and Barbara Seaman won recognition as its principal pioneer. Not all doctors, needless to say, have since learned to listen, explain, and take the complaints of their women patients seriously. But an ever-expanding number of women have become conscious of evaluating their doctors and of picking and choosing carefully among them.

In 1990 the FDA, through its official monthly magazine, *FDA Consumer*, summarized thirty years of monitoring the Pill into a quasi-official inventory of "what every woman should know" about it. (The author of the research summary, Sharon Snider of the FDA staff, identified herself as a fifteen-year user of the Pill.) At the risk of some repetition, this report is extensively quoted here as a three-decade summary of the Pill's known benefits and risks.

Over the years, more studies have been done on the Pill to look for serious side effects than have been done on any other medicine in history. While fears about blood clots, heart attack, and stroke have largely been laid to rest by safer, low-dose birth control Pills, current research suggests that healthy, non-smoking women have little if any greater risk of these serious health problems than do women who do not use the Pill.

Questions about the Pill's association with cancer, however, remain. Some widely reported recent studies support the hypothesis that in certain groups of women the risk of breast cancer increases with oral contraceptive use. A larger number of studies, however, found no significant increased risk. Nor is it definitely known yet whether or not the Pill causes cervical cancer in some groups of women. So far, a cause-and-effect relationship has not been established. . . . The advisory committee concluded the Pill hadn't been in use long enough to draw valid conclusions about its carcinogenic effects. For example, it would be at least 10 years before the risk of uterine cancer could be accurately assessed.

The advisory committee's finding on the possible relationship between the Pill and blood clotting and cancer were supported by a World Health Organization scientific group, which independently had reached the same conclusions. . . . But the Pill has been found to help prevent two major types of cancer—cancer of the ovaries and cancer of the endometrium (lining of the uterus).

An FDA advisory committee, reviewing the relative risks and benefits of today's birth control Pills, recommended that an earlier upper age limit of 40 for use of the Pill be lifted for healthy, non-smoking women, thereby making this popular, effective means of contraception available until menopause. . . .

In 1982, a new version of the Pill, called the "biphasic" Pill, was introduced. Two years later, three "triphasic" Pills were introduced. These "multiphasic" oral contraceptives are low-dose Pills in which the ratio of progestin to estrogen changes during the 21 days the Pill is taken.

In 1986, use of high-dose estrogen Pills had been drastically reduced—to 3.4 percent of the oral contraceptive market.

Nevertheless, some 400,000 women were still using high-dose estrogen Pills. Most were between 30 and 39 years of age, the age group most at risk of serious side effects.

In 1988, at FDA's urging, the three drug companies still manufacturing high-dose estrogen oral contraceptives voluntarily withdrew from the market all remaining products containing over 50 mcg estrogen. . . .

Today's oral contraceptives are considerably safer than the Pill of the '60s because they contain less estrogen and progestin. Over the years, the amount of estrogen has been reduced to one-third or less of that in the first birth control Pills, and the progestin has been decreased to one-tenth or less. . . . The present benefits are associated with oral contraceptives containing 50 mcg of estro-gen. As yet, no scientific data are available on the effects of those containing 30 to 35 mcg or less.

Today's Pill is still not safe for all women. The risk of serious illness and death increases significantly for certain groups:

• Women who smoke—particularly those over 35 who are heavy smokers (more than 14 cigarettes a day)—have a signif-icantly increased risk of heart attack and stroke. This risk in-creases with age. Women who use oral contraceptives are strongly advised not to smoke.

• Women who are obese or have underlying health prob-lems, such as diabetes, high blood pressure, or high choles-terol, also have a significantly increased risk of serious side effects from using the Pill.

• Women who have a personal or family history of blood clots, heart attack, stroke, liver disease, or cancer of the breast or sex organs should not use oral contraceptives.

Women who become pregnant while on the Pill should imme-diately discontinue taking it because of a risk of birth defects in the child.

Uncertainties remain about whether the Pill causes breast or cervical cancer in some groups of women. Despite many studies over the years, there is still insufficient evidence to definitely rule out these possibilities. . . .

One of the major problems of the studies to date . . . is that all the data reflect the effects of the higher-dose Pills (those contain-ing more than 50 mcg of estrogen). No studies have been done on the recent low-dose Pills, and none are under way. The cancer-Pill issue is very complicated and therefore difficult to study, and the research is expensive. . . .

Though some safety questions remain unsettled, for most healthy women the Pill provides a safe, effective means of birth control with some possible beneficial health effects.

In the late 1960s, thirteen major drug companies—nine of them American—were investing in fervent competition to dis-cover new and better approaches to birth control. By the early 1980s, only four large companies in the world remained in the research-and-development race, only one of them American, Ortho Pharmaceuticals. A survey in 1988 showed that improved birth control had totally disappeared from a list of the thirty-five top-priority topics of research. The causes of this plummeting interest appeared traceable to the earliest years of the Pill.

Dr. Gabriel Bialy, head of contraceptive research for the National Institute of Child Health and Human Development, has called the position of anti-Pill feminists "naïve": "Every drug has certain side effects. People should have expected this, but they didn't." Granting that the first Pill to come on the market "represented an overdosage," Bialy said, "The companies didn't understand the effects of the hormones, and needed a product that would be unquestionably effective. So they guessed too high. . . . But the next thing you know it was a male conspiracy, and the Pill was bringing death to women. Generally when people pick their cause, objectivity is not the prime requirement."

"This is not a very encouraging atmosphere," said Roderick L. Mackenzie, chairman of Gynopharma Inc., a small contraceptive manufacturer in New Jersey. "After the baby has been thrown out with the bath water, then they say: Hey, why is everybody out of business?" Peter F. Carpenter, president of Alza Corporation's development branch, added: "My sense of what's happening is that things are getting polarized, a time when everybody but the fiercest bows out of the argument. I find it a very ominous development because in a democracy we need to find a common ground. Contraception has become politicized. We as a society do not deal well with controversy, and leave it to those on either end to fight it out. . . . As a consequence here we sit in the most advanced society in the world, and a woman here has fewer contraceptive choices than any woman in the developed world."

"It is entirely possible," concluded Dr. Allan C. Barnes, vice president of the Rockefeller Foundation, "that if the ideal contraceptive were developed today, it would never be introduced in the United States."

By the time the *FDA Consumer* published its 1990 summary report, it was clear that all sides of the Pill debate had a piece of the truth. Millions of users were finding it satisfying, liberating, and apparently safe. Substantial numbers were finding uncomfortable side effects that outweighed its benefits. To a small number it brought serious afflictions, of which all users needed to be clearly warned.

The long journey once undertaken by Margaret Sanger toward a "perfect contraceptive" has not yet led there. As a later chapter will suggest, perhaps it never will. It may be that the most the human race may hope for is an ever-so-slightly declining risk from the potions we invent, while never quite achieving the holy grail of pregnancy-free sex.

20

Parents Revisited

*T*he world is always worth saving. It has been done many times,
and it will have to be done many times more.
But it isn't "sacrifice" to help save it. It is self-expression.
It is an interesting game, saving this old world,
a live people just can't
keep out of it.

—An entry in Margaret Sanger's scrapbook

Margaret Sanger was spared *Humanae Vitae* and *The Doctors' Case Against the Pill*. On September 6, 1966, she died in wealth and in torment. Death came in Tucson, Arizona, where she had built a home in the early thirties after a recurrence of the tuberculosis that had plagued her early adulthood and killed her mother.

The chief source of her wealth was her husband, J. Noah H. Slee, who had died in 1943. He had become a warm stepfather to her sons, providing the money for one to go to Yale, the other to Princeton, and for both to go to medical school.

The source of her torment is not as easily pinpointed. Following a heart attack in 1953, a physician had put Sanger on

Demerol to ease her pains. It was a new drug with a capacity to addict that was not yet understood. Sanger became enthralled by it and, despite coaxing and conniving by her two medical-doctor sons to keep the drug out of her reach, she remained addicted for the rest of her days. Said Margaret Lampe, daughter of the older son, Stuart, who for the first twenty years of her life had lived next door to Sanger in Tucson: "She'd go from doctor to doctor to get it. She would go to San Francisco to visit so she could get Demerol. Daddy wouldn't give it to her, or he would water it down."

Lampe was in many ways her grandmother's favorite companion and confidante. In Sanger's last, lonely months of illness she was taken to a nursing home and one day said to the young woman, "It's very hard work to die. It's hard to let go and come to peace." She didn't dwell long on such talk. One day she turned to advising her granddaughter "about not being afraid of men, about enjoying sex, something my own mother would never have talked about."

"Earlier," Lampe recalled, "when she had been recovering from a heart attack, we were in Hawaii together. We got her a little record player, but turning on the machine was just more than she could do. So I came every morning to turn it on. I was twelve or thirteen years old and had just gone out with my very first date— to an afternoon movie, with the boy's parents chaperoning. Mimi asked me, 'Did he touch your breast? It's perfectly fine if he wants to do that.' I was shocked, just speechless. I mean, even to think about holding hands, much less kissing or any of the other things, was more than—. Then she said, 'I just want you to know that sex is such a normal, wonderful part of life. Just enjoy it and don't be afraid of it.' After that, I wondered many times as an adult if she had been told as a child that sex was terrible. Or if she had heard her father having intercourse with her bedridden mother and heard her crying, 'Stay away from me,' or whatever, because her mother was pregnant every single year during those last years of her life. I can see her remembering her mother as so frail and such a wonderful woman, and I wondered as a child, Why did her father have to have sex with her? Why did she have to have so many children? Because her mother lost seven, born dead. Eighteen pregnancies. Mimi found her mother's weakness to be absolutely abominable. Weakness in females was something she couldn't bear. No man was going to run her life—ever.

"At the age of eighty-two, she was still interested in men. I remember sitting at her dining-room table one evening after she had hired a male secretary. He was probably forty-five years old, and he was at dinner with the family. After dinner he left, and I remember my father glaring glumly at my grandmother down at the other end of the table and saying, 'Mother, when are you going to give these things up?' She sat up in her chair and, in this lovely voice, said 'Stuart, hopefully never.'

"When my grandmother died, my Uncle Grant was devastated. He was in his late fifties, four years younger than my father, and he seemed to go to pieces, just cried and cried. What seemed to crack open at last was that he had *always* been grief-stricken, tormented, about his mother. He was unbelievably proud of her accomplishments, of who she had become. But when you get below that, both boys were so wounded. Grant, the younger one, was his mother's favorite, the more sensitive of the two, but she was not there for him. She wasn't at home when he needed her, and then she left the country to stay out of jail. Neither of the boys went to school or even learned to read and write until my father was eleven years old. Grant always felt she was more interested in others than in him. Later, he would tell of the turmoil growing up, of being shuffled from one aunt to another, of Mother being in jail, of embarrassment at newspaper articles, being taunted by kids at school. He grew up upset about her affairs, the numbers of men in her life—always this overhanging feeling of shame.

"You know, neither Grant nor Daddy would ever say that they were part Jewish. Bill Sanger, their father, was born Orthodox Jewish. He had grown up in a kosher house. But neither Daddy nor Grant would ever speak of it, would not admit it for anything in this world.

"Daddy, her older son, was not tormented as Grant was by Mimi's death. He handled the past by being unable to remember almost anything that happened in his childhood. He just blocked it all. He would tell two stories—only two, no more. One is of both boys coming home from prep school to see their father for Christmas break. Bill Sanger meets them at Grand Central Station and takes them to his apartment on the Lower East Side. He had no money, no food, no fuel. It was freezing. They went up to the roof of the apartment and cut down a flagpole to burn in the fireplace. And the other story my father would tell was of Mimi writing to him one summer at Cape Cod to say she was coming up to Truro. So he walked from Truro to Provincetown to meet her and waited five hours. She never came, and he turned around and walked back. He was a sandy redhead, and next day his legs were so burned, first on the front and then on the back, that he couldn't walk. Those were the deep disappointments of his childhood. He learned that her words meant nothing. Her dedication to the movement was all-consuming.

"Yet I think these two surviving children of Margaret Sanger also had a philosophical view of their mother. They understood that in order for a movement to begin, only a blind-sided radical can stir up the truth. Then, in time, more moderate people come along. They also understood that it's impossible to place on one individual all the roles of being a radical and a good mother and a good wife and a good friend. It cannot be done. They understood that philosophically. But then, taking it to the next step down, they never accepted . . ."

After Margaret Sanger's death, Stuart and Grant opened their mother's safe and found a letter she had written to their father. It was not to be mailed until after her death, but Bill Sanger was already dead. The letter asked his forgiveness for leaving him and hurting him. The sons destroyed the letter. "They didn't want the world to know still more," says Lampe, whose father told her of it. "I think they did it because the pain for those two sons was too much."

It was as though a particular gene of Margaret Sanger, who had learned to pull down the blinds as her train passed through the not-to-be-remembered shames of Corning, was passed to her sons.

Jailed and driven from the country for printing and mailing

birth control information, the Mother of the Pill would have been surprised that only five years after her death the United States issued a postage stamp honoring family planning. *The New York Times* noted that the issue "may be considered as a salute to the late Margaret Sanger . . . although it is not likely that the Postal Service will acknowledge any such connection."

Little more than a year after the loss of her celebrated collaborator, on December 28, 1967, Katharine Dexter McCormick died in Boston at the age of ninety-two in a kind of obscurity. In contrast to Sanger's death, which was front-page news, the passing of the woman who envisioned and personally financed one of the colossal scientific and social breakthroughs of the century was not recorded by an obituary in *The New York Times, The Boston Globe,* or the *Los Angeles Times.*

On a Saturday in early 1966, Anne Merrill, the young scientist at the Worcester Foundation, went in to the lab to clean up some odds and ends. Gregory Pincus dropped by her lab bench. Faintly embarrassed and asking her not to mention it, Pincus requested that Merrill take a sample of his blood for a white-cell count, explaining that some minor disorder had been requiring that he take an occasional transfusion, and he wanted to know if he needed another.

"I never saw such funny blood in my life," she recalled. "In fact, a little later, when I took a count just before he went into the hospital his last time, I remember calling them to say, 'You better take a good look at this, because I can't find *any* white cells at all.' "

On March 28, 1966, Pincus wrote to his friend Dwight J. Ingle, flagrantly camouflaging his worrisome knowledge: "Oscar Hechter tells me that you have been informed that I am a seriously ill man. This is entirely untrue, and I wish you would do your utmost to scotch this rumor. At the moment I am healthier than I have been in many years. As nearly as I can make out, a minor setback which occurred about a year ago has been the source of the unfounded rumor."

Shortly thereafter, it became known that he had myeloid meta-plasia, a rare disease of the white blood cells. "He assured his friends," Ingle later wrote, "that modern therapeutic measures would allow him to live into retirement years. He continued to work and travel."

On August 22, 1967, less than a year after Sanger's passing and four months before McCormick's, Gregory Pincus died in a Boston hospital at the age of sixty-four.

Twenty-five years later Mary Ellen Johnson, the technician who assisted Anne Merrill, recalled: "When we closed down Pincus's lab after he died, we had, in little boxes like this, bottles of pills, the actual Searle Company pills that we had used in Puerto Rico way back when, and we had to dispose of them, throw them away. I picked one up and said to Anne, 'I'm going to keep one of these, just as a souvenir'—never dreaming that any-one would ever want something like that. I brought them home and stuck them in the medicine cabinet. For years I just kept mov-ing them around with me. A little while ago, the Smithsonian Institution came around looking for the original Pill for an ex-hibit, so I donated it."

"When Pincus started," observed Dr. Howard J. Ringold, a re-search chemist at the Worcester Foundation before joining Syntex, "if you had asked ninety-two people out of a hundred whether an oral contraceptive was feasible and acceptable, the answer would have been 'No.' Pincus had the vision to see that the idea would work, and he had the drive to see it was carried out. Of course after it was all over, there were people who said, 'I showed that ovulation-inhibition was possible in the rat,' and so on. But saying the accomplishment is possible is not the same as having the foresight and the imagination to actually do it."

Shortly after his friend's death, Oscar Hechter, who had coau-thored with Pincus a classic work on the synthesis of adrenal hor-mones, wrote:

Pincus for me represents the prototype of a *new* scientist, whose life and achievements merit critical examination and analysis. On a planet rapidly being irreversibly transformed by science and technology in ways not clearly foreseen, we desperately need information about the mechanisms by which individual scientists change the world. Pincus and his life merit a critical case history, because if new Pincuses arise in the future, they will have a powerful impact upon the world. . . . To oversimplify, some scientists become great by making important contributions to knowledge—discovery in the laboratory—and others become great as organizers and by making important applications of knowledge. Gregory Pincus, a scientist-statesman, was one of the latter.

Soon after its approval by the FDA, the Pill was the subject of a lecture attended by Katharine McCormick. She wrote Margaret Sanger about it. Her letter touched the essence of the conflict that had hounded the professional life of her friend Pincus. In academia the persistent demand is for "basic" or "nonmission" scientific research—work motivated solely to add to the world's fund of new knowledge. In contrast, "applied" research, often holding a position of lower esteem within the academic world, directs itself at a specific solution to a specific human problem. The traditional defense of the higher status of "basic" research has been that the world's great technological advances—miracle drugs, space satellites, the computer chip—would not have happened without the basic research that preceded them. So the two pursuits disdainfully glower at each other. McCormick wrote Sanger that science has "been working for years on *basic research* of the human reproductive system—all for sound knowledge and learning doubtless, but for practical aid *nothing*—and the abortions merrily went on before their very eyes! Personally I doubt if *they*, at Harvard, would ever have found an oral contraceptive!"

In 1967, somewhat defensively, the National Science Foundation launched a study hoping to determine, once and for all, to what extent basic, or "nonmission," research did actually contribute to life-changing technological innovations. The NSF, of course, had been committed to promoting basic research. So no one was terribly surprised when its study found that of the "key events" leading to practical inventions, fully 70 percent "resulted" from research that had no predetermined mission. Only 10 percent of the "basic" discoveries that later had practical results, the NSF found, were made in the laboratories of profit-making industries.

A later study, in 1973, traced ten inventions, including the heart pacemaker, hybrid corn, and the VCR, as well as the Pill, which led the NSF to refine its conclusions: Science research did not lead to practical invention inevitably or quickly, and certainly not easily. Its new conclusion seemed to point directly at the role filled by Gregory Pincus. NSF found that in nine out of the ten cases studied, an additional critical element was present: a "technical entrepreneur" who pursued the innovation in the face of opposition or indifference by fellow scientists, or despite predictions by businessmen that the product would fall flat in the marketplace. The "strongest conclusion" of the study was the importance of the special kind of scientific entrepreneur whom it called the "product champion."

"The role played by Gregory Pincus in the development of the Pill," James Reed, the Rutgers birth control historian, has observed, "fits perfectly the model outlined in the later NSF studies. He drew G. D. Searle and Company into the project while other drug houses were refusing to consider marketing a contraceptive. . . . He traveled all over the world spreading the good news of his invention while many scientists and physicians were still skeptical."

"Scientific research," wrote Walter Spieth of the Georgetown Clinical Research Institute, "is only good or bad, not basic or applied. Good science consists of original thinking about careful observations in interesting situations, and frequently a practical question is the stimulus for it. The false dichotomy of 'basic' and 'applied' too often reinforces a preciousness or snobbery in the graduate student that narrows his vision for years after leaving the academy. Some never recover."

A passage from the book by Dr. Anne Biezanek may be an apt eulogy to Gregory Pincus, as well as to Sanger and McCormick:

All honor is due to those . . . who many years ago saw birth control as preeminently a woman's problem, and devoted so much time and thought and energy towards perfecting a technique that would remain easily handled, comfortable, reliable and at all times within the control of the woman herself. Of all the great works that men have undertaken for women, [the Pill] must surely rank amongst the noblest. The work was instigated at a time when financial reward was very slight and the risks considerable. It was done in the teeth of social disapproval and in the teeth of great apathy on the part of the medical profession at large. The first men who devised those things at such risk to themselves must one day be named and honored by all women as "the instigators of the revolution."

Although the Pill enormously fed and fattened the G. D. Searle Company, its arrangement with Pincus called for not a penny of royalty to him or the Worcester Foundation. After his death, despite urgings for more generosity by friends of the Pincus family, Searle paid only three hundred dollars per month in benefits to Pincus's widow. In 1974, seven years after his death, the Searle Company "honored" Pincus with a four-hundred-thousand-dollar endowment for a chair in his specialty, reproductive studies, but not at the Worcester Foundation as he surely would have preferred. Searle gave the money to Harvard, which had turned down Pincus for tenure.

After he was betrayed—as he perceived it—by his principal partner at the infant Syntex Corporation, Russell Marker, the genius alchemist of synthetic progesterone, determined in 1949 never to practice science again. And he never did.

For the next twenty years, word flitted around the tight world of chemistry that Marker, the eccentric, had utterly disappeared. As those years slipped by, and as the number of those who recalled him dwindled, vague guesses became speculations, which became rumors about whatever happened to Russell Marker. One lively story was that Marker was dead. A writer later reported that Marker said to someone, "I didn't care what they were saying. I had to make a living and had other interests." He said he would not reveal what those interests were, but then offered a thin explanation: "Some things were useful to the [Mexican] natives." Perhaps that comment is what led to one fanciful whisper heard around Penn State that Marker had turned to gunrunning for the rebel descendants of Pancho Villa.

Forty-five years after quitting his science career, at the age of ninety, Marker broke his silence in an interview with me. In it he explained repeatedly, as though making an apology, that after leaving science he "had to make a living" to support his family. What struck me about his revelation was not the need for apology but its ironic connection with birth control, which had been of no professional concern to him.

Through someone he had met in Mexico City, Marker was hired for an undercover postwar role arranged jointly by the American, Mexican, and Swiss governments to accomplish a secret diplomatic interest in Austria.

"Right after the war," Marker reluctantly explained, "we had a lot of soldiers that stayed behind to occupy the defeated countries, one of which was Austria. The girls there went out with Negro soldiers and some of them became pregnant. But the Austrians didn't want to keep those black babies. They just wanted to get rid of them. So the United States government, affiliated with the Swiss, made an arrangement with Mexico. An orphanage in Veracruz took those babies in. I averaged four trips a year between Austria and Mexico. The work lasted four or five years. Afterwards, my wife and I visited Vienna several times. At the present time, you don't see many colored people in Austria."

So, I suggested, it was a matter of trying to find a safe, healthy place for these kids?

Missing the intent of my question, again he inserted the apologia: "It was more or less for making money for my family. I don't know what has ever happened to those children. I dropped out of that when I got on to the silver market."

Getting "on to the silver market" is another chapter in Marker's flight from science—yet a roundabout path to remaking contact with it.

In about 1949, a few steps from his hotel residence in Mexico City, Marker befriended the owner of a leather shop, an erudite

Russian who had emigrated via Paris. One day the man led Marker to a back corner of his shop and displayed two sets of rococo English silver. Marker was taken with them, but their price was far out of his reach. The owner explained that only a few such pieces existed, and most were in museums. Marker asked if his friend knew any Mexicans who did fine work in silver, and the shopkeeper named a few. Marker looked them up, showed them photographs from books of some of the world's finest rococo silver pieces, and put them to work making exact reproductions. Before long he expanded to commissioning the reproduction of Mayan plaques. Marker, who had given every appearance of business naïveté in his partnership at Syntex, from which he departed virtually penniless, was soon conducting a robust commerce in art reproductions.

Robust indeed. In the mid-1980s, Marker renewed contact with Penn State—and with science—for the first time in almost forty years, although he had been living quietly, virtually secretly, less than a half mile from its campus. To the Penn State College of Science, whose employment he had long ago abruptly quit in midsemester to pursue his yam root in Mexico, he now just as abruptly presented a gift, including a quarter-million dollars to endow six lectures every year by invited scientists of extraordinary accomplishment. (He specified that the first Russell Marker Lecturer was to be Carl Djerassi.) Also he endowed the Russell Marker Chair in Chemistry.

As though making a defiant display of his success to the world of science—the world he had abandoned, but which he seemed to feel had abandoned him—he next heaped, as though in a potlatch, a large money gift on the University of Maryland, which had refused him his doctorate. The money paid for two annual Russell Marker Lectureships in Chemistry. Maryland reciprocated the thoughtfulness with an honorary degree. At the age of eighty-five, after three chemistry historians had named him "the father of the Mexican steroid industry," he was, at last, Dr. Marker.

Landing his final jab, Marker told an interviewer: "The reason I established these lectures here in science was because in the entire time I spent at the University of Maryland I never came across what was considered a top-rate scientist."

On December 4, 1984, in his simple farmhouse retreat on a secluded hillside in Temple, New Hampshire, John Rock died at the age of ninety-four. With self-effacement but accuracy, he once told a visitor there, "I am not the father of oral contraceptives. If anything, I am the stepfather."

Gregory Pincus's chief research operative, M. C. Chang, died on June 5, 1991, at the age of eighty-two. He was buried near his lab, in Mt. View Cemetery, Shrewsbury, not far from the grave of Pincus. His widow saw Chang in a different light than Rock had viewed himself. Upon her instruction, her husband's stone was engraved: M. C. CHANG, THE FATHER OF THE BIRTH CONTROL PILL.

Carl Djerassi remains hugely alive as he watches his fatherhood of the Pill being reappraised. After a career of denying paternity, Djerassi appears to have relented by popular demand. Or, at least, his publisher has done so. On the cover of his 1992 memoir, *The Pill, Pygmy Chimps and Degas' Horse*, appears a subtitle, "The Remarkable Autobiography of the Award-Winning Scientist Who Synthesized the Pill." The back-cover blurb, holding him more specifically responsible, calls the author "Father of the birth control pill. . . ." then promises that the book will tell "how Djerassi, at the ripe old age of 28, within a twelve-month period, leading a small team in an obscure laboratory in Mexico City, used a locally grown yam to synthesize the first cortisone and *then* the Pill." [Emphasis added.]

Close, but not so.

And Frank B. Colton, who synthesized the same drug as Djerassi's at about the same time for the G. D. Searle Company, is scarcely acknowledged anywhere—as father, stepfather, or anything else.

HOW DO YOU
MAKE A BOMB IMPLODE?

*T*he moon is nothing
But a circumambulating aphrodisiac
Divinely subsidized to provoke the world
Into a rising birth-rate
—CHRISTOPHER FRY, *THE LADY'S NOT FOR BURNING* (ACT 1), 1949

In June 1992 the World Health Organization disclosed one of the most captivating statistics in all statistical history: Sexual intercourse occurs more than 100 million times daily. (The WHO omitted detailing how the survey was conducted.) The result is nearly one million conceptions with every turn of the earth. To all concerned with controlling what they view as runaway population, surely the report billowed with futility and gloom.

The "simple truth of the matter," as Dr. Anne Biezanek had written, is that two persons engaging in sexual intercourse primarily seek "not children but orgasm."

An equally simple truth is that when two persons engage in an act of birth control, their primary aim, unless their personal lives are driven by some inordinately blinding social duty, is not to limit the growth of world population but to keep themselves free of child.

Margaret Sanger and her allies had found early on that they needed the banner of population control to raise birth control and the Pill to the status of "decency." But they also found that every new decade made population control more genuinely imperative.

As the Pill approaches its fortieth birthday, its career in that regard seems to demand an evaluation: Has the Pill—or any modern form of artificial contraception—fulfilled its large-scale social mission? Has it dampened the world explosion of births? If so, how much? If not, why not?

They are confusing questions, for the answers are now appearing to turn back on themselves. In the third decade of the Pill, an answer was emerging quite visibly: While a boon to the private lives of its users, the Pill as a population controller appeared something of a flop, especially in the poorest, fastest-growing countries.

An almost unchallenged conclusion arose from post–World War II population statistics and grew to high fashion after the 1968 publication of Paul Ehrlich's *The Population Bomb*. Ehrlich predicted with more alarm than prescience that well before the turn of the century, billions—not merely millions, but *billions*—would die from widespread famine, raging epidemics and pandemics, or nuclear war that would be made inevitable by a desperate worldwide struggle to control scarce resources. The solution to the population problem was to lift Third World nations out of poverty.

At the first United Nations International Conference on Population and Development in 1974 in Bucharest, when world population had reached four billion, delegates from Western nations pressed hard for birth control clinics and for targets of growth limitation. The Third World contingent balked. That was a cheap fix, they said, to cover over the colossal inequities between the rich and poor nations. They raised the slogan "The best contraceptive is development." European and North American demographic history had consistently shown that population growth slowed in close tandem with economic development—and its partners, literacy and technical education.

Carl Djerassi came at the problem from another direction. In a 1970 article in *Science*, called "Birth Control After 1984," he found himself using a computer metaphor to explain why the population problem appeared so unyielding. Birth control, he said, needed to be separated into two parts: hardware and software: "*Hardware*, for me, meant all the actual methods used by people— steroid contraceptives [the Pill], abortion, condoms, sterilization. . . . *Software* covered the more difficult political, religious, legal, economic, and sociocultural issues that individuals and, ultimately, governments must resolve before birth control hardware is employed." As with computers, before the hardware could be effective, the right software needed to be installed.

By the 1990s, long after demographic experts gave up on the devices of contraception alone, a turnaround in the numbers had begun and many began taking a second look. The *combination* of ready access to contraceptives *and* strong political leadership— the hardware and the software together—seemed responsible for the first widespread substantial change:

• In Bangladesh an average of seven births per woman in 1975 had dropped to about five in the nineties. In this poorest of all the world's countries, three of every ten women had begun using contraception. That achievement is underscored by a software comparison with Pakistan, just next door and far less poor. In two decades since the two countries split, the government of Pakistan for religious reasons (Pakistan follows a stricter form of Islam) failed to encourage family planning. Only one in ten Pakistani women were using birth control, and birthrates remained high.

• In India the average woman in 1980 was mother of 5.3 children. By 1991 the average was down to 3.9. In this vast country where birth control so recently seemed hopeless, almost half of all couples were using contraception.

• In Indonesia the number of children per woman dropped from 4.6 in 1980 to 3 in 1992.

• In Thailand the number of children per mother dropped from 4.6 in 1975 to 2.3 in 1987.

• In Colombia the number dropped from 4.7 in 1975 to 2.8 in 1990.

• In the Philippines 22 percent of couples defy the dominant Catholic church by using contraception. Compared with 5 children per woman in 1980, the 1992 average was 4.1.

According to a 1990 Population Council estimate, today's world has 400 million fewer people than it would have had without family-planning programs. The World Health Organization estimated in 1992 that the world fertility rate—the average number of children per woman—in developing countries declined from 6.1 in 1970 to 3.9 in 1990. To point out how rapidly some countries have changed, the WHO said that while the United States took fifty-eight years to drop its fertility rate from 6.5 to 3.5, Indonesia accomplished the same drop in only twenty-seven years, Colombia in fifteen years, Thailand in eight years, and China in seven years.

Repeat: *Some* countries have rapidly changed. Across the developing world, declines in fertility rate have been steeply uneven. Most successful has been East Asia, where some 70 percent of couples are believed to be using one birth control method or another. Between 1965 and 1970, birth control users leaped from 18 million to 217 million. Africa has been least successful, however; in the same period, use of contraceptives there increased from 5 percent to only 14 percent.

In the late sixties, one in ten Third World women were using some form of contraception. By 1993 fully half of Third World women of fertile age were "on the Pill" or using some other means. Compared to having six children two decades earlier, Third World women were averaging four.

Those and similar successes in other countries led Bryant Robey of Johns Hopkins University, editor of *American Demographics*, to advance in the early nineties a revised slogan: "Contraceptives are the best contraceptive."

Where can it be hoped that these shifting numbers are leading us? From where have they led us?

Ten thousand years into antiquity, an estimated one million people roamed the earth. Two thousand years ago, Christ was born to a world that had propagated to 275 million. In about 1830, the planet achieved its first human head count of a billion. By 1930, that billion had become two; by 1975, four billion. By the turn of the century, predicts the United Nations, the world's people will redouble again to eight billion. At the going exponential rate of growth, by the year 2050, when half of today's humans may be alive to see it, world population is projected to reach a level of crowding that strains the contemporary imagination: 21 billion.

While the momentum may slow, in the short run it cannot be stopped. That is because the proliferation of today's new babies will soon grow into a vastly expanded generation of parents. Even if fertility-age women of the Third World started today to replace only themselves, total population of those countries would still increase by more than 50 percent before growth scaled the mountain and settled down to zero.

More than one fourth of the world population lives in one country, the People's Republic of China, which has a land area only slightly larger than that of the United States. Of those Chinese, 80 percent live in rural areas. Hence, of all the world's humans, one out of five is a Chinese peasant.

In 1982, after the population of that nation surpassed a billion, the Chinese National People's Congress adopted a new constitution mandating that family planning was a duty. The government translated that duty into an order that a Chinese couple was to have only one child. But even that would not halt the growth. At the time of the order, two thirds of all Chinese were under twenty-four and had yet to marry and produce the lone permissible child. So by the year 2000, even if the policy succeeds, inexorably the population will be more than 1.2 billion—an *addition* about equal to the entire population of the United States in less than twenty years.

China's audacious plan is to maintain the one-child policy for a hundred years, which should eventually halve the population. After two generations, if all goes right, no Chinese youngster will grow up with a brother or sister. The children of these children would be bereft not only of siblings, but of first cousins, aunts, uncles—families of just two parents and four aging grandparents.

To the Western mind, this is unimaginable. Only an autocratic, tyrannical central government could think up such a satanic plot.

Or is there another way to see it?: Only a society that set population reduction as a *serious* goal could think it up. Let the flabby, soft-headed, soft-hearted Western "democracies" prattle their soothing ideological phrases about population control until their voices drown in the inevitable flood of their own people.

Is China's villainous way worse than its inexorable alternative—a head count roaring toward 21 billion?

China's bold shift—in software—probably leaps beyond anything Djerassi imagined when he introduced his metaphor. Granted, that country's burden is atypical, extreme. But it may, in more ways than meet the eye, illustrate the rigors that any country may be forcing upon itself in pledging, truly *intending*, to control the size of its population. A closer look at the cost that country decided it could not avoid undertaking may be instructive.

"In the beginning our policy was difficult to put into practice," said a woman in charge of enforcing the one-child policy of Changzhou, a "model town." The official, a Dr. Chen, interviewed for *Nova* for a 1984 broadcast on PBS, explained that workers must have their factories' permission to get married. To get such permission, they must be twenty-four or older, receive instruction in family planning, and pass a written test to earn their one-child certificate card. The card is the official license to become pregnant. If lost, it cannot be replaced. Without it, the mother loses entitlement to prenatal visits with a doctor, and the baby cannot be registered at the police station. In return for the parents' one-child commitment, a child will be entitled to important benefits throughout life: thirty dollars a year from the state, a free education, priority for a job when he or she grows up.

"I've put you in the quota for this year," Dr. Chen tells a

woman. "If you don't get pregnant, see me, and I'll put you in next year's quota."

"Some people," Dr. Chen explained, "wanted a second child and tried to do it secretly. There was a pregnant woman in Wazan factory. We persuaded her to have an abortion. We took her to the hospital. That night she changed her mind and escaped. She ran off to Shanghai. The Shanghai people helped us find her, and we brought her back to the hospital for the abortion."

A factory official elaborated: "If the pregnancy is their fault, then we don't pay them bonus. If it is not their fault, we pay bonus." A doctor added: "If a woman with a coil gets pregnant, then we say it was not her fault. However, for a woman on pills it is her fault. She'll lose bonus and 'model worker' status." Which says something about official faith in relative modes of hardware.

Each team of sixteen women in a workplace has its birth-planning worker. She is constantly alert for anyone who might be pregnant or even thinking about having a child. Said one: "We watch for women who start to eat less or who get morning sickness. If a woman isn't as active as she usually is, that's a sign of pregnancy. It's very difficult to escape the attention of us family-planning workers."

Doesn't that make for resentment?

"No, the workers are grateful for my concern. Our relationship is good."

Outside the factory, Madame Chen had a second network, usually of older women, on the watch for younger ones who might be thinking of having a second child. One of these "Granny Police" is known as Granny Wu, who said: "You must find out what's really going on in your families. Especially where there is a husband or wife home on their annual visit. This contraceptive is specially for such people. It's called the 'Visiting Spouse Pill.' The first day the wife takes two. Then one a day—every day. You must work hard, not only in your own streets but with anyone you meet.

"Mrs. Wang Shou Ying was not taking her pills. Granny Yang told me and I went to do my work. First I asked her if she was taking her pills. She said that she was. I asked her to get her packet of pills and show me, but she couldn't. So I gave her some pills. I told her to start taking them after her next period. I told the husband that as they went to bed to give her a glass of water and tell her to eat her pill."

It is rare to witness a government fashioning severe remedies that won't show results for thirty, fifty, a hundred years down the road. Yet China, having firmly decided that the alternative is unacceptable, has indoctrinated the Dr. Chens in their villages clearly and firmly. "Our work would be easier," Dr. Chen said, "if we could allow two children. But we think of our country's future. We have to keep our population under 1.2 billion in the year 2000."

The clear message is matched by ruthless enforcement. In Hunan Province, a classified government report, confirmed by local authorities, was reported on the front page of *The New York Times* in 1993. It concerned a young woman, Li Qiuliang. Two months after her baby was born it lay buried in a tiny grave. Li had been certified by her local officials to give birth in 1992. But on December 30, her pregnancy was only seven months advanced. So the officials took her to a rustic rural first-aid station and ordered the doctor to induce labor.

Li's doctor protested. Her family pleaded. But family-planning workers insisted that the deadline had to be met. After nine hours of forcing the labor, the baby died. Li, at twenty-three, was left physically damaged and emotionally crushed.

Soon after declaration of the one-child policy, authorities sometimes descended upon women pregnant without permission and dragged them to abortion clinics. H. Yuan Tien of Ohio State University studied fertility and contraception in four locales of Sichuan Province and reported in 1980 that in Chengdu, the capital of Sichuan, 58 percent of all pregnancies were terminated by abortion. The rate in Beijing was reported to be similar. Some women fled to relatives' villages, returning only after the child was born. The focus then shifted to compulsory sterilization after a first child. Posses swooped down on villages once or twice a year, rounding up women who already had a child and bringing them to a nearby clinic where they were fitted with IUD's or sterilized.

By 1990, world protests led Tan Sushun, director of the Sichuan College for Family Planning Professions, to declare, "The government absolutely doesn't allow forced abortions. When a woman is pregnant outside the plan, we give her education and tell her not to give birth. But if she really wants to, she can." She acknowledged that some village officials still do resort to force, against the rules.

A special difficulty of the one-child policy is that it collides directly with an ancient imperative of Chinese peasant culture: bearing sons. Not to extend the male line is to dishonor one's ancestors. Chinese peasants rely on sons as old-age insurance. Who else will support them in their late years? Daughters become members of their husbands' families, and thus traditionally help support their in-laws. Parents often do not want to use up their one-child ration on a girl.

Biology dictates that for every hundred female births there should be about 105 or 106 male births. But in 1989, for every one hundred reported girl births, there were 113.8 births of boys. That ratio implies that about 8 percent—900,000—of newborn girls appear to have vanished from the statistics. Somehow the parents get away without reporting the birth, then try again. Another explanation is the growing use of ultrasound in Chinese hospitals. Peasants find out from the doctor, usually with a small bribe, whether a fetus is male or female. If it is female, they abort and start all over. A third, less frequent contribution to the imbalance, according to Zeng Yi, a leading demographer, is simply infanticide. The parents instruct the midwife to keep a bucket of water beside her. If a girl emerges, she drowns the baby immediately and reports a stillbirth.

An extreme case of that practice involved an illiterate peasant named Liu Chunshan of Shandong Province. He already had a four-year-old daughter when his wife found herself pregnant in violation of the one-child policy. A soothsayer told Liu that the new baby would be a son. Unable to resist that improvement in his fortunes, Liu threw his daughter down a well. "Without a son," he explained later, "the generations cannot be passed on and we remain childless." Caught and convicted, he was sentenced to fifteen years in prison. It turned out the soothsayer was wrong. Liu's wife gave birth to a second daughter.

Other societies, perhaps all, will surely be too timid—or too humane—to follow China's way. What the Chinese seem to be daring the world to contemplate is that countries following other, "more humane" pathways will find themselves fearfully crowded.

22

For Women a Not-a-Pill
and for Men a Not-Yet Pill

There are no perfect contraceptives.
Let's not make the perfect the enemy of the good.
—Sharon Camp, Population Crisis Committee

And yet, despite its creeping pace, new development continues, in the United States as well as elsewhere.

A recent transformation of the Sanger-Pincus Pill may launch it into a whole new life. The original Pill was born after Carl Djerassi and Frank Colton separately concocted a progestin that would work when swallowed and Pincus adopted the new compound as a contraceptive—indeed, a contraceptive *pill*. The recent invention uses the very same progestin found in the Pill, but its gift is that it is *not* a pill.

The new not-a-Pill, called Norplant, is made of six soft matchstick-sized capsules that are inserted under the skin of a woman's upper arm. It is done during a fifteen-minute procedure in a doctor's office with the aid of a local anesthetic.

Norplant's prospect of changing birth control forever is contained in its power of slow release. The implant remains effective not for merely a single day but for fully five years. Its failure rate is less than 1 percent. Norplant has been called "teenager-proof" because it exempts users from having to worry about remembering to take the Pill. (Like the Pill, however, it has a notable drawback: It provides an excuse not to use a condom for prevention of AIDS and other sexually transmitted diseases.)

Norplant was developed by the nonprofit Population Council, which began clinical trials in 1975, reaching half a million women volunteers in forty-six countries, ranging from Sweden to Bangladesh. Its first government approval was given by Finland in 1983. In 1991 Norplant won FDA approval for use in the United States. Thereupon the Population Council licensed Wyeth-Ayerst Laboratories of Philadelphia to manufacture it. In its first year, more than 200,000 American women began using Norplant, compared to almost 11 million on the Pill.

"Norplant is the most effective reversible contraceptive on the market," wrote Deborah Franklin in the July 1992 issue of *Health* magazine. "It's still too soon to have detected rare or long-term side effects, but there is no evidence of cardiovascular, respiratory, or central-nervous-system problems in the short run. Also, the hormone in Norplant—a form of progestin—has been used in higher doses in oral contraceptives for more than 20 years, so there's good reason to think that surprise side effects will be few."

About 60 percent of Norplant users report menstrual changes, especially bleeding outside the normal cycle, in the first six to nine months after insertion. Then the cycle stabilizes. In another 10 percent of users, periods may cease entirely. Rarer side effects include persistent headaches and acne, also reported by some users of the Pill. Like the Pill, Norplant is commonly linked with weight gain or, less often, weight loss. Most weight changes are minimal—less than five pounds.

Women are advised against Norplant if they hope to get pregnant within a few years. While Norplant is removable with immediate return of fertility, its removal is more difficult than insertion and may require two additional visits to the doctor. It is also deemed inadvisable for those told to avoid the Pill, who have been breast-feeding for less than six weeks, or who suspect they may be pregnant. Like the Pill, Norplant is precluded for women who have had breast cancer, liver damage, blood clots, or undiagnosed abnormal vaginal bleeding, and for those taking certain antiseizure medications.

Norplant's effectiveness received a back-of-the-hand compliment in 1992 from Governor Pete Wilson of California when he proposed implanting low-income women at little or no cost. By that year thirteen state legislatures had considered bills to compel women on welfare to take it. Some judges and officeholders went further, urging Norplant for women convicted of child abuse or drug abuse. In one case, Judge Howard Broadman of Tulare County, California, Superior Court told a mother of five that she wouldn't have to go to prison for a child-abuse conviction if she agreed to use Norplant. That coercive sentence was appealed, but when the defendant violated her probation by testing positive for cocaine, she was sent to prison, making the appeal moot. The American Medical Association, the American Civil Liberties Union, and reproductive-rights groups have since opposed involuntary "Norping" of low-income women.

Norplant is covered by Medicaid in all fifty states. Planned Parenthood clinics were surprised to discover that, although only 12 percent of its clients are eligible for Medicaid, fully 69 percent of its Norplant recipients did so with payment by Medicaid. The implant's eager acceptance by women of low income startled clinic officials. In Planned Parenthood clinics in Baltimore and New York, 90 to 100 percent of implanted women were on Medicaid. Only 10 percent nationwide paid out of pocket, and only 1 percent were covered by private insurance.

"Because Norplant is covered by Medicaid," Jane M. Johnson of Planned Parenthood pointed out ironically, "poor women now have better access than working-poor women and some middle-class women, who are expected to pay the full cost up front."

Norplant's cost is relatively high. In 1994 the kit was priced at $365, with many private doctors charging an additional $500 or more for insertion. Its high price troubled American health experts, since Wyeth-Ayerst sold the kit in other countries, including several in Europe, for less than $100.

At the unexpected news of Norplant's popularity among the poor, Wyeth-Ayerst spokesperson Audrey Ashby hastened to announce that 61 percent of Norplant sales were to doctors in private practice. "We don't want it to be looked at as a product only for a certain segment of the population."

In a forbidding brick fortress in inner Baltimore, the Paquin School houses an experiment like no other. All of its three hundred students are pregnant teens or recent mothers, many of them aged thirteen or fourteen. Early in 1993, they became the first students in America to be offered Norplant at school.

Consuelo Laws, a Paquin senior and at nineteen a mother of two, said after "wearing" Norplant for a year, "Without it I'd probably have more children. I want to complete my education."

In an echo of the Pill's early days, black Baltimore minister Melvin Tuggle fears that Norplant is a tool for genocide: "One third of us are in jail and another third is killing us, and now they're taking away the babies. If the community, the churches and our white brothers don't stand up for us, there won't be any of us left."

But Gracie Dawkins, a hopeful school counselor says, "Our dream is to prevent them from getting pregnant again until they're at least twenty-one."

Norplant's strong appeal for the young was shown in a survey of 21,276 women who received the implant through Planned Parenthood. A vast majority, 89 percent, were under thirty, and 22 percent were nineteen or younger.

Another long-term contraceptive that is shorter-term and less expensive than Norplant, called Depo-Provera, had been on the market for almost thirty years as a cancer drug when an FDA advisory committee recommended in June 1992 that it be approved for contraception. Since its introduction in the 1960s, more than ten million women outside the United States had used it. The synthetic hormone, medroxprogesterone acetate, developed by the Upjohn Company of Kalamazoo, Michigan, received its approval as a contraceptive in October 1992.

Depo-Provera is injected into a muscle in a doctor's office. The injection remains effective for three months.

Because of concerns about a possible link with cancer, its approval was consistently opposed by the National Women's Health Network and by Dr. Judith Weisz, a reproductive biologist of Penn State's University Hospital at Hershey, Pennsylvania, who chaired a three-person FDA panel to investigate Depo-Provera's side effects. After the more expensive Norplant was approved, Weisz, repeating that no one knew just how Depo-Provera worked, warned that it was "going to be a poor person's Norplant." The FDA refused approval of Depo-Provera three times. Sharon Camp of the Population Crisis Committee responded that at least ten thousand U.S. women had been using it, and that she herself had taken it for thirteen years. She told the FDA: "Let me put this very bluntly. You, the FDA, have now screwed up three times on this issue. Don't do it again."

As a cancer drug Depo-Provera had been on the market for twelve dollars a dose. As soon as it was cleared by the FDA as an injectable contraceptive, Upjohn multiplied its price almost by three, to $34 a dose, or an annual cost of $136, roughly the same as women had become accustomed to paying for the Pill.

In mid-1993, the World Health Organization endorsed two new once-a-month injections, Cyclofem and Mesigyna, which in trials had shown fewer side effects than the longer-lasting Depo-Provera.

A few months earlier, in 1992, scientists at India's National Institute of Immunology in New Delhi announced another birth control shot for women that would stretch the three months of protection offered by Depo-Provera to a full year. The major novelty of their injection is that it is the first contraceptive to use a woman's own immune system for rejecting a fertilized egg before it sticks to the wall of the uterus. Many women have since favored that self-produced protection over steroids or hormones that are chemically created outside the woman's body. Dr.

Gursaran Prasad Talwar, a leading immunologist who developed the vaccine, claims that it protects women "without any side effects."

Dr. Rosemarie Thau of the Population Council, welcoming the vaccine, said that its precise action wasn't known but speculated that it acts before the egg is fertilized. If that is proved true, it might allay the concerns of antiabortion forces who oppose contraceptives that act after fertilization.

The Population Council has closely studied this vaccine approach in clinical trials in Australia as well as India and was gearing up in 1994 to start a larger trial in Sweden. A major question to be answered is how reliably vaccinated women make antibodies sufficient to expel a fertilized egg.

Another approach in its early stages is introduction of an agent that would go after a woman's *un*fertilized eggs. Bonnie Dunbar of Baylor College of Medicine is working on one to stimulate antibodies that would attach to the protein coating of an egg. These clustered antibodies, in turn, would form a physical barrier, like a coat of armor, against sperm.

As innovations keep coming around the corner in a slow but steady parade, they raise an aura of constant hope that the next discovery will be the long-sought panacea. But the panacea is not to be found in "hardware," cautions Dr. Jacqueline Darroch Forrest, chief of research of the Alan Guttmacher Institute. "We are fooling ourselves if we think a technological fix is going to get us out of our dilemma of high rates of pregnancy and abortion."

Giving voice to what has emerged as a complaint of many women, Pepper Schwartz, a sexual-behavior sociologist at the University of Washington, has protested, "Since the Pill, men tend to assume that *women* will take care of protection."

The perception is laden with complexity and contradiction. Dilys Cossey, chair of the U.K. Family Planning Association, expressed a view, held by many, that the Pill at last enabled a woman to take and keep control of her own protection against pregnancy: "Women on the Pill could choose when to have sex, whether to have sex. They could choose when to have children or whether not to have children."

One contrary view, heard in both Europe and America, was expressed to me by an unmarried woman in Germany: "If a man knows I'm on the Pill, I lose my strongest excuse for saying no." More widely and strongly felt is a resentment that the woman must be the partner who risks swallowing exotic substances that have uncertain long-term consequences.

More subtle, less definable, but profoundly troubling to many is the question of *responsibility*. Why should the woman have to be alone in carrying the responsibility? So what if a man doesn't *like* putting on a condom, or loses a bit of sensation, or whatever? Neither do women like diaphragms, or the risks of the Pill or of forgetting to take the Pill.

Comes the inevitable answer: Yes, but isn't the woman better off having the *control* than not having it?

And to that comes the deeply felt, albeit unspecific, reply: But why must she have *all* the responsibility?

It is an argument that may go on forever, to be resolved only in a way that satisfies particular partners.

But it leaves a glaring question dangling in the charged air: *Why isn't there a pill for men?*

The answer is perfectly obvious. At least, many have readily declared it obvious: There is no pill for men because science is run by men.

It's not that simple, says the highest-ranking scientist in the field, the chief of contraceptive development at the National Institutes of Health, Nancy Alexander: "Turning off the continuous production of 200 million sperm per day presents tougher problems than stopping the development of one egg a month."

The longtime health columnist of *The New York Times*, Jane E. Brody, expanded that point in a 1983 article, "Why a Lag in Male-Oriented Birth Control":

> . . . Research has not yet produced effective, easy-to-use contraceptives for men. . . . Women can choose from among eleven contraceptives.
>
> Many feminists believe male chauvinism led to this inequality. Many experts agree it has been a major factor though not the only one. Women themselves bear some responsibility. Margaret

Sanger and other leaders of the American birth control movement in the early 1900's wanted to give women control of their own fertility; they ignored male contraception. Through the years, advocates of birth control have continued emphasizing female contraception. Moreover, men perform the bulk of contraceptive research and some people believe that male researchers feel more comfortable tampering with a woman's physiology than with their own.

Brody then emphasizes the same point made by Nancy Alexander of NIH, although her numbers differ:

Behind the sexual politics and proclivities of the medical profession, however, basic biology remains the greatest barrier to developing male contraceptives. A woman produces only one fertile egg a month. A man produces perhaps 30 million sperm a day, any one of which can impregnate an egg. In plain terms, it is simpler to disrupt production of one egg than millions of sperm, particularly since a drug that reduces sperm production 90 percent might still leave a man fertile. It may also be safer for the fetus. A woman is born with a lifetime supply of eggs. But a man's sperm form daily and can be genetically damaged during that formative process by, say, sperm-blocking chemicals.

The organ that produces sperm also makes the male sex hormone testosterone, which is responsible for libido, potency and secondary sex characteristics, such as a deep voice. Medical experts have found that shutting down sperm production also shuts down testosterone, resulting in impotence and loss of libido. (Experimental contraceptives have their own side effects, including shrinking of the testicles, breast enlargement and increases in blood cholesterol.) By contrast, ovarian function need not be wholly suppressed to prevent ovulation. And since a woman's libido is not a function of ovarian hormones, but of other sex hormones produced in the adrenal glands, side effects are more easily avoided.

The process of conception can be disrupted at various points: the release of the egg, the migration of sperm to the egg, fertilization, the implantation of the fertilized egg in the uterine wall. Male fertility can be interrupted only at the site of sperm production, maturation or release, and even these sites can prove unworkable. Since it takes sperm about three months to develop to the point of release, a drug that suppresses the production process would not become fully effective for three to six months. A return to fertility would be similarly delayed. . . .

[The National Institute of Child Health and Human Development] is spending about twice as much on developing contraceptives for women as for men. The Population Council has allocated its money in a similar fashion.

A detailed portrait-at-work of that rarity in science, a pill-for-men researcher, has been drawn by Nancy Marie Brown, for *Research/Penn State* magazine. Dr. Joseph Clemon Hall, a former Penn State biochemist now at North Carolina State University, wrote his Ph.D. dissertation, as a National Institutes of Health Graduate Research Fellow at Kent State from 1980 to 1985, on glycoproteins associated with sperm. He began studying the properties of the sperm's plasma membrane the next year while a postdoctoral fellow at Penn State, and he received funding from the National Science Foundation for 1987 to 1990. In 1991 he won the NSF's five-year Presidential Young Investigator award.

"When I was young," he said, "I was a member of a gang in New York City and it was doubtful that I would even finish high school. One day a counselor, who happened to be Jewish, pulled me to the side and said, 'You're making B's in classes without even studying. Just imagine what you could do if you did study.' And she proceeded to tell me how her people had succeeded in the days of the 1920s and '30s when they experienced a lot of racism and prejudice." Hall is African-American. "I remember the things she told me. Essentially, it was, 'Never make excuses for failure, because you become a failure. Learn from failures. Never allow your "competitors" to become more qualified than you. Seek out what made them qualified. And lastly, never stop learning. If you stop learning, you're going to fail simply by being stagnant.' I remember those things, and that is what drove me into science."

Brown's profile of Hall also merits extensive quotation:

"We want to create a blind sperm," Joe Hall begins, "a sperm that cannot recognize an egg. That's a very difficult task—poking the eyes out of a sperm."

Hall is working on a male Pill: a fast, safe, and completely reversible way to make men equal partners in family planning. His is one of four labs in the country with a similar mission, and the only one focusing not on the testis, where sperm are made, but on the epididymis. . . .

"This organ is a single tubule more than 50 meters long—in the rat. . . . A major problem with using this organ as a target site for a male Pill is that no one has yet identified a cellular function that can be blocked or offset in an organ-specific manner."

The enormous changes in sperm en route can't come from within the sperm cell itself, Hall explains, since that cell is "biosynthetically dormant" when it leaves the testis. "That is, sperm do not make any new biomolecules—proteins, lipids, or sugars—after they leave the testis. Nature has nicely shut down the biosynthetic processes of the sperm to prevent any mutations, or mistakes, in the DNA of the cell.

". . . If you look at the history of birth-control research very closely . . . you see essentially two factors that come into play. First, the science is primarily a male-dominated endeavor, and as a result males tend to focus on females—"

"That is," interjected the interviewer, a woman, "mucking about with women's bodies."

"Sure. And that focus is quite evident. The other aspect," continued Hall, "is a moral dilemma. When you are involved in a research endeavor that 'prevents' conception—and you have congressmen who are actively involved in funding research—there's a lot of pressure by fundamentalist groups to block such research. . . .

"In fact . . . it wasn't my male colleagues who got me interested in this problem. . . . It was my wife. Basically, having had five kids, we, as many couples do, had the dilemma of: Who is going to get fixed? . . . Is she going to get, or am I going to get fixed? . . . We had to deal with that problem for a good three or four months before deciding that it would be better in her interests for her to get fixed, because of other family problems. For example, her family had a history of ovarian cancer and cervical cancer. But until she found that out, we were really battling. It was basically, 'I'm not going to get fixed, you get fixed!' 'Well, we can't have any more kids!' And she finally said, 'You're a biochemist, why don't you try to come up with something?' That was my third year of graduate school, and that started me down this path of researching male contraception."

The same year Hall won the Presidential Young Investigator Award, he found a sperm "eye." . . . Hall identified it as one of the glycoproteins on the sperm's surface that allows it to recognize and bind to the egg. It might be just the molecule to be plucked out by a male Pill. . . .

"If I had a laboratory in an industrial setting," Hall muses, "there would be fifty technicians and scientists assigned to me to help study the secretory pathway of these cells and to compare them with other cell types, such as the gut. . . . For things that are luxury items, like birth-control pills, industry doesn't want to spend millions of dollars on biology."

Hall's student, Dianne Jacobetz, an honors student in biochemistry, spent all of last year purifying one enzyme produced by one type of cell in a rat's epididymis, and describing in precise, molecular detail how the enzyme functions. Carol Peterson, another undergraduate in biochemistry, is spending this year purifying and describing a second enzyme. John Kochins, a graduate student in reproductive physiology, has taken on a third enzyme. At this rate, Hall estimates it could take his laboratory 10 to 15 years before the biology of these enzymes involved in the fertilization process is completely clear. In the rat.

To work on male contraception these days, Hall admits, you have to accept the fact that "you're going to be spending a large portion of your life in the scientific wilderness."

Is the quest for a male pill ultimately bound to disappoint—just as the Pill did not meet the needs (esthetically or medically) of many women?

To look closely at one possibility, a pill for men may be hidden in imidazoles, drugs used to relieve women of fungal and vaginal infections. These drugs, tried on lab animals, have been found to enter male sex fluids, where they kill sperm. Another may lie in waiting in a drug commonly prescribed for lowering blood pressure, nifedipine (sold as Procardia and Procardia XL). Researchers at North Shore University Hospital, Long Island, have discovered that nifedipine disables the membrane of the sperm, preventing it from binding to an egg and fertilizing it. The effect is reversible. Florence Hazeltine of NIH says nifedipine "may be a possible male contraceptive [which] we need more than you possibly could imagine." But Harvey Florman of the Worcester Foundation for Experimental Biology is wary: "Don't forget that nifedipine lowers blood pressure and slows the heart rate, two effects that may be unacceptable . . . for healthy people."

Still another research interest is in a native Chinese herbal remedy called *Tripterigium wilfordii*. Men who continuously take extracts of this herb become infertile. Although scientists haven't run any U.S. clinical trials yet, they suspect the infertility wears off soon after the man stops taking the tablets.

These items of news may bring hope for the woman who yearns to "share responsibility" for contraception. But a large question must inevitably confront her: Can she trust that *his* pill worked—totally? Did it disempower or kill *all* the millions of sperm that a single ejaculation sent swimming ardently toward her egg? Could an especially hardy one have made it through? If somehow one arrived partly disabled, might it conceive a child that is unwhole? Inescapably the pregnancy remains hers. Can Nature be outwitted by having her "share" that responsibility—with anybody?

Yet the male pill, if and when it comes, will strikingly enlarge the palette of contraceptive choices.

23

THE MORNING AFTER

*E*ach of you is constituted in your mother's womb for forty days as a semen,
then it becomes a bloodlike clot for an equal period,
then a lump of flesh for another equal period,
then the angel is sent, and he breathes
the soul into it.
—THE PROPHET MUHAMMED

or nine years it had dangled in the mind of M. C. Chang like an untied shoelace. After he and Gregory Pincus had begun drudging their way through the two-hundred-plus Pill compounds to choose one for the Puerto Rico tests, Chang made a mental note of a piece of unfinished business he wanted to pursue as soon as he was free. Finally, in 1959, he decided that this was the time.

Certain compounds, he had once observed, caused a rabbit to expel an egg from the fallopian tube *after* it was fertilized but before the egg implanted itself in the wall of the uterus. Chang had tested hundreds of agents and found ten that reliably turned that trick to his satisfaction. Would they work in humans too? If so, there would be a pill for approaching birth control through a new door. Instead of preventing conception by keeping the egg unavailable to the sperm, this new pill could prevent conception *after* the sperm penetrated the egg.

Prevent conception? If the sperm indeed penetrates the egg, hasn't conception taken place? Or does conception happen upon *implantation* of the fertilized egg in the soft uterine tissue?

Perhaps that was a valid question for philosophy or religion, but Chang was a biologist. If the goal was to prevent a human pregnancy, this new pill might do it. Chang explained the idea to Pincus, who became interested.

While they did not understand how these new discoveries worked, Pincus explained what they thought they did understand: "Our theory is that the compound causes [either] muscular contraction of the uterus or secretion in the fallopian tubes which speeds the egg along."

By 1966, after their Pill was a worldwide success, the two men had completed four years of testing the new substances on animals. They isolated three compounds that consistently prevented pregnancies.

They began calling their discovery the "New Pill." It promised dramatic advantages. Unlike the "old" Pill, which required fastidious taking for twenty consecutive days each cycle, the New Pill would offer the astonishing convenience of needing to be taken only once, a day or two after coitus. Indeed, early tests showed that taking their New Pill the "morning after" was not the only way to bring about its remarkable advantage. When swallowed as a regimen once a week, it was a reliable contraceptive but far less troublesome than the old Pill. In addition, the New Pill offered a powerful safety benefit: It did its work locally and discretely in the reproductive organs without affecting the whole hormonal system. "It is always better," stated Chang, "to reach a specific target than the whole system."

They showed the New Pill to John Rock. Rock's frigid response made it clear he shared none of their desire to pursue it further. Catholic thinking held that conception takes place when the sperm and egg *meet*, not when the conceptus is implanted. "Therefore," concluded Rock, "I feel this pill is an abortifacient."

The New Pill faded from the work of the Worcester Foundation. Perhaps that disappearance was caused by the troubled difference between the old friends, Pincus and Rock. Or perhaps any attempt that could be accused of causing abortion was simply too dangerous under the glare of Massachusetts law. In any case, a recent inquiry at the Worcester Foundation produced no recollection or record of what happened to the New Pill.

Research biologists plumb and probe crannies of life so infinitesimal and with such absorption that they often know little about the wee crevices roamed by their fellow researchers. As they grope ever so privately, sometimes they bump into one another in the dark.

In 1970 a French biochemist with a medical degree in endocrinology was spending his working days probing his private cranny of science. He knew as much as anyone in the world about something called the progesterone receptor. Hormones, the messengers of regulation, function only if they seek out and find the corresponding molecules to which their messages are addressed, called receptors. Flowing around the body, hormones relentlessly pursue these receptors as bees hunt flowers. Less poetically but more precisely, they are keys seeking locks, and each key is useless until it finds the lock—the receptor—that it fits. When a key succeeds in entering a lock, no other key can get in.

If a receptor can be fooled into accepting as its key something other than its true counterpart hormone, too bad for the hormone. Having no place to go, it is useless and disappears from the body. The fake key, because it bars entry of the true one, is called an *antihormone.*

French biochemist Étienne-Émile Baulieu, laboring in a laboratory of the Roussel-Uclaf pharmaceutical company in Romainville, France, was looking for an antihormone for progesterone. Inspired by a brief visit to Worcester and a working contact with Gregory Pincus, and fueled by a $750,000 population-control grant from the Ford Foundation, Baulieu wanted an antihormone that would defraud its way into all those progesterone receptors, leaving genuine progesterone with no place to go. Without progesterone, the uterus could extend no hospitality to a fertilized egg. Incapable of knowing that it was—or was not—pregnant, the uterus would drive the stranger, the real progesterone, away. Thus a pregnancy would be impossible.

Near Baulieu's workbench, another Roussel chemist, Georges Teutsch, was running experiment after experiment in search of a stand-in, an antihormone, for cortisone. In 1980 Teutsch came up with what appeared his most successful find. Baulieu immediately recognized that Teutsch's formulation—labeled Roussel-Uclaf 38486, later to be abbreviated to RU 486—was the exact receptor plug that for ten years he had been seeking!

His top employers at Roussel, almost all Catholics who disapproved of work on sex hormones, recoiled. Corticosteroids, *oui*; progesterone, *non*.

Baulieu was scheduled soon to be inducted into France's loftiest scientific body, the Academy of Sciences. Finessing his bosses' opposition, Baulieu chose that august occasion—on April 19, 1982—to announce the creation of what in every corner of the world has come to be known as the "abortion pill." Of course, Baulieu called it no such thing. His preferred terms were an "unpregnancy" pill or a "contragestive."

After years of laboratory testing and of inside arm-wrestling at Roussel, Baulieu tried the drug on eleven human volunteers seeking abortions, most of them nurses or medical students ranging in age from eighteen to twenty-four. Two days after receiving the drug, each woman was injected with prostaglandin, a fatty acid that forced uterine contractions and speeded the ejection of the embryo. Nine of the women expelled their embryos without trouble. The other two required surgical abortions, one of them needing a blood transfusion. The two failures, Baulieu believes today, resulted from insufficient knowledge of proper dosage. Adjustments in the dosage soon gave RU 486 a high degree of reliability.

In France, RU 486 has been legal since 1988 but is closely monitored and strictly regulated. It may be given only within seven weeks of the first day of a woman's last menstrual period (although by 1994 there was talk of extending that to ten weeks). A woman wanting it in France is required to visit one of the country's 850 licensed clinics or hospitals four times over a three-week period. After a gynecological exam to make sure the pregnancy is in an early stage, the woman is instructed to embark upon a one-week period of "reflection." Returning to the doctor, the pregnant woman is then required to sign a formal request for an abortion. In addition to that government document, she must also sign a Roussel form indicating that she understands that if she does not follow the abortion through to completion, a malformed fetus might result. (As yet, among the small number of babies born to women who took RU 486 but did not complete the dosage, no defects have been found.)

Upon signing, the woman is given three RU 486 tablets, each about the size of an aspirin and containing 200 milligrams of the drug. About two times in every hundred, the RU 486 alone may induce a virtually immediate abortion. Except for those unusual occurrences, she must return within forty-eight hours so that a doctor or nurse may witness her taking prostaglandin. The timing is critical. The woman is encouraged to remain for four hours even if the contractions and expulsion occur earlier. Most women experience what seems like a heavy period that begins within an hour or two and lasts around ten days. Eight to ten days after the prostaglandin, the woman must visit for a last exam to make sure that the expulsion is complete.

Every pill is registered with the government and must be accounted for.

Doctor fees for an RU 486 abortion run to about $250, compared to about $300 for a surgical abortion in France, both 80 percent reimbursable by the French national medical plan.

In 1991 the French eliminated the biggest disadvantage of using RU 486 when it authorized testing of several oral forms of prostaglandin, which would make the entire procedure oral. They chose Cytotec, created by Searle to treat stomach ulcers. The earlier method for prostaglandin, by injection, had caused uterine contents to be expelled within four hours in 47 percent of cases. Cytotec reached that four-hour goal in 61 percent. Total effectiveness rose slightly from an impressive 95.5 percent to 96.9 percent. Cytotec reduced instances of nausea and diarrhea, common side effects of the injections. It even benefited the French national health-care system. A shot of prostaglandin cost twenty-two dollars. Cytotec cost only seventy-two cents.

Cytotec's success with RU 486 drew from its maker, Searle, a reenactment of its timid double-talk in introducing the Pill, while gleefully accepting its profits. "Searle has never willingly made [Cytotec] available for use in abortion," a spokesperson wrote in a letter to *The Wall Street Journal* early in 1993. "It is not Searle's intention or desire to become embroiled in the abortion issue."

The United Kingdom approved RU 486 for pregnancy termination in 1991. For women of other countries yearning for the right to make the RU 486 choice, the first crack of opportunity opened in February 1994. London's Marie Stopes Health Clinic began providing RU 486 for nonresidents, including Americans. In addition to a five-hundred-dollar clinic fee, the cost of airfare, hotel, and meals clearly keeps the London connection well beyond the reach of most nonresidents.

Many women previously denied access to RU 486 may have harbored illusions about its ease. A month after the Stopes clinic began offering the service, Nina Darnton of *The New York Times* interviewed some of the first Americans traveling to London for an RU 486 abortion.

She described Aimee "propped up in her hospital bed wearing an oversized T-shirt that made her look younger than her 17 years. She grimaced. She had never known pain this bad, she said. Her long black hair was moist with perspiration and her face looked pale and tired. Taking the RU-486 pill for an abortion was worse than she had expected."

Minutes after expelling the tiny sac that contained the one-eighth-inch-long fetus, Aimee murmured to Darnton, "I felt like I was dying. I hurt so much. I had contractions coming so fast, and I was sick to my stomach and dry heaving. I couldn't stop trembling and I felt so hot."

Aimee's distress was not unusual among the first American users. "All," Darnton reported, "agreed the experience was more difficult than they had imagined."

Catherine, a twenty-eight-year-old editor from Texas, testified: "They told me there might be some pain, but I guess I still sort of expected a magic pill that would just take it all away. I was surprised by how much it hurt."

Afterwards, however, Darnton observed, the women uttered words of relief and said they were glad they had chosen the RU 486 route.

Frances Perrow, a clinic spokeswoman, said that the Marie Stopes staff is "a bit mystified" at why American women are willing to spend airfare and hotel expenses as well as the clinic fee— all totaling more than fifteen hundred dollars—for the RU 486 option. "After a surgical procedure," Perrow allows, "there may be some cramps. But with this, there can be six hours of severe cramps," similar to but worse than the strongest menstrual pain, "plus diarrhea and vomiting, plus coming all the way to the U.K. on a gray March morning, you might well ask, Is it worth it? These women think yes."

One woman who voted yes, Theresa, a twenty-four-year-old waitress from Boston, readily spent a small inheritance left by her mother to pay for the sudden trip to London. "I live below the poverty line myself," she said. "It was worth it to me because I couldn't bear the thought of a doctor going in and tearing something out of my body. It's such a violent intrusive act. This felt more natural. My body did all the work."

Aimee's trip was especially expensive because the seventeen-year-old flew from Philadelphia with her mother. "I didn't want to just zip in, be put to sleep and zip out in two hours with it all done. In a way, that would have been too easy. This was a big painful decision for me. I would have felt irresponsible if it had just been over with like that. I wanted to remember this all my life. I never want to do it again."

Demands that the series of clinic visits for RU 486 be simplified come mainly from abortion-rights advocates in America, where elimination of any barrier is, for some, an issue of ideology and "rights." Even in advance of the U.S. legalization of RU 486, some Americans talk of "simplifying" the visits down to just two and permitting its administration by nurse-practitioners—or permitting sale of RU 486 and prostaglandin as prescription drugs so a woman can take them at home with no clinic visit, no doctor.

Baulieu, the product champion, strenuously objects. He warns against "the cousin syndrome"—a woman getting the prescription and passing it on to a cousin or friend who, unexamined, might have an undetected tubal pregnancy, potentially lethal and not abortable by the drug. Baulieu does support a two-visit plan, in which the woman at the time of examination takes the initial pills, takes the second set on her own two days later, then returns to the doctor to make sure they have done their work completely. That method relies on a faith, questioned by some, that every woman will surely follow through on her commitment and if she falls into a rare instance of excessive bleeding or another complication, she will get herself to a hospital.

In a conversation with science writer Thomas Bass, Baulieu argued that "women are mistaken if they think they can take [RU 486] under any circumstances without supervision of doctors or nurses. 'Women are free at last!' they say. But this isn't true. . . . Five out of a thousand women have extrauterine pregnancies. RU 486 will not abort these pregnancies, which, if left untreated, are fatal. To demedicalize abortion by removing doctors from the process—it's insane! This is not a miracle pill. It's merely an instance of medical progress. . . . It's insulting to women to say that abortion will now be as easy as taking aspirins. It is always difficult, psychologically and physically, sometimes tragic."

At the Bicêtre Hospital, a former madhouse and prison outside the southern gates of Paris, Étienne-Émile Baulieu's long office table, from which he directs a staff of sixty, is strewn with puppets and masks amidst heaps of papers. Paintings line his walls, and his office bookshelves house the complete works of Louis Pasteur, a set of the *National Geographic*, and thirty-five bound volumes of his own publications.

His office window overlooks a courtyard to the hospital's family-planning clinic, where most of the passersby are women on their way to RU 486 abortions. Gazing down at them, Baulieu contemplates the life of the Pill and of its possible successor, *his* pill. He speaks in mixed phrases of awe and sadness, calling the Pill "the most important medical discovery of the twentieth century."

In immaculate English garnished with a zesty French sauciness, he explains: "Among many thousands of drugs, it alone is called 'the Pill.' [But] the Pill has *failed* in the sense that something that is so effective and generally so well tolerated is used by such a small percentage of women. Out of one billion women in the world of childbearing age, fifty million, or a mere five percent, take the Pill. People grow tired of taking it month after month. . . . The Pill demonstrated the ability of science to transform our lives. It's the forerunner to other chemicals, good or bad, that will alter brain function and behavior.

"Because the Pill, condom, diaphragm, and every other means of contraception have failed, RU 486 itself, *unfortunately*, will *not* be a failure. . . . Since all the remaining methods of contraception are imperfect, there will always be a need for RU 486."

Moving ebulliently through a lofty and glamorous world, Baulieu lives, like Carl Djerassi, with the style of an artist, although one friend says he is more like P. T. Barnum. In his sixties, Baulieu sails and skis. His circle of close friends has included Indira Gandhi and Sophia Loren. When in Washington, he stays at the home of the French ambassador, and like presidents, he has been a guest of Walter Annenberg in Palm Springs.

At sixteen, Baulieu joined the French underground to resist the Nazi occupiers of France. Early in the war, his mother, a lawyer and active suffragette, changed the family name from Blum, identifiably Jewish, to Baulieu. His father, who died when Étienne-Émile was only three and a half, had been a prominent doctor in Strasbourg, the first physician in Europe to treat diabetics with insulin. Young Baulieu's anti-Hitler commitment led him into membership in the Communist party. Offended by the Soviet takeover of Hungary, Baulieu quit the party, but that piece of his past dogged him. He was later barred from studying at Columbia University by the U.S. government's denial of a visa. The denial stuck until the Kennedy administration's State Department relaxed the ban.

"After Kennedy was elected president," he recalls, "I got a visa right away and went [to] New York, which is where I met Pincus.

I'd actually shaken his hands once before, when he visited Paris. But I was put off by his royal air. It was like shaking hands with God. I was initially disappointed that Pincus, although a very good scientist, was doing more politics than science. I didn't realize at the time that he was making a revolution.

"When he invited me to give a talk at his [Worcester] laboratory, my reason for going wasn't the Pill, which I knew little about. I went because he was a famous man who might be able to launch my career. It's thanks to Pincus that I became part of a network of people involved in fertility control. He flew me to Puerto Rico. . . . He turned out to be a very persuasive man, but I never promised Pincus I was going to work on the Pill. It was not reproductive biology that interested me, but the mechanism of hormone activity. I wanted to look at sex steroids from the biochemical viewpoint."

Pincus arranged for Baulieu to get a seat on a World Health Organization committee—"a clever maneuver to have me meet demographers, gynecologists, social scientists, and other people working on the Pill. Pincus thought this would enlarge my vision and make me more inclined to contribute to fertility control. And he was right. . . . I'd been seduced by Pincus's way of thinking."

Baulieu soon attained world renown for identifying the uterine receptor for progesterone. Discovery of that receptor invited the next step in fertility control: an antiprogesterone or antihormone that would prevent the uterus from embedding a fertilized egg.

Pincus arranged in 1965 for Baulieu to give a presentation at the Ford Foundation on estrogen receptors in the rat and its possible implications for fertility and population control. "The Ford people," as Baulieu tells it, "were stupefied. . . . I told them I'd work on sex hormones and their receptors, but I wasn't promising to invent a new pill. I told them to be confident that sooner or later something would come out of basic sciences. They were nice and clever enough to give me money, enough money to last for ten years."

Baulieu was given a seat on the government committee studying whether France should legalize abortion, which it did in 1966. Opposing the influence of Roussel-Uclaf, he began a campaign in France and around the world for legal approval of RU 486. Six years after he announced the discovery and tests on more than seventeen thousand women, the French government announced that RU 486 would be made available for public use.

Furious reaction broke out in the press, especially an outpouring of letters from enraged Roman Catholic doctors. The church sponsored a mass-protest parade through the streets of Paris. A month later, a battered Roussel-Uclaf pulled the drug from the market, saying it was not in business to engage in "moral debate."

That very month thousands of physicians were gathering in a medical congress in Rio de Janeiro. A majority of the congress dispatched a petition to the French government demanding reversal of Roussel's decision. In less than forty-eight hours, health minister Claude Evin issued a government edict staked on new legal ground that was remarkable for a Catholic country: When the government first approved marketing the drug, he declared, "RU 486 became the moral property of women." He ordered Roussel, one third of which is owned by the French government, to resume distribution.

As critics lay alert to pounce on it, the drug stumbled into its first close scrape. In 1991 one woman among its first seventy thousand users, a thirty-one-year-old voracious smoker attempting to abort her thirteenth pregnancy, died of heart failure after her post–RU 486 prostaglandin injection. Thereupon the treatment was banned for heavy smokers and for women age thirty-five and older. The Health Ministry further recommended use of a reduced-dosage oral prostaglandin, leading to the French adoption of Cytotec.

Over the next three and a half years, the new medication successfully ended pregnancies for one hundred thousand French women. Of those seeking abortions early enough, about 85 percent chose RU 486 over the alternative of surgery. When asked later, almost all expressed satisfaction with their choice. In another survey of women who have had both types of abortions, 80 percent said they preferred RU 486.

Sweden and the United Kingdom soon licensed RU 486, and India began testing it. China was experimenting with a close chemical copy. Canada, watching those successes closely, waited to see what the United States would do.

In the United States the shrill and unyielding conflict that has centered around abortion clinics spread to Baulieu's pills. Pamela Maraldo, president of the Planned Parenthood Federation of America, vowed: "We will not allow anti-choice zealots to deny RU486 to American women." The Reverend Keith Tucci of Operation Rescue National shouted back: "When they invent new ways to kill children, we will invent new ways to save them."

Anyone who doubts the creative potential of the opposition to RU 486 would do well to study the inventiveness of one man in defending American shores against it. The Reverend Ken Dupin would seem out of place in a who's who of antiabortion zealots. For one thing, in his small church planted against a Blue Ridge mountainside in Roanoke, Virginia, he advocates education in family planning and believes the Pill has been a good thing.

Ken Dupin's Sunday sermons rarely attracted more than 150 attenders and the religion editor of the local paper had never heard of him until RU 486 gave Dupin a high moment of fame. Late in 1988 he learned about RU 486 during a sleepless night watching CNN. "As crude as it may sound," he says, the subject of abortion, which had not been much on his mind, hit him that moment upon realizing that his daughter "was old enough to be a candidate."

Saying things like that, Dupin searches his listener through almond-shaped eyes that gleam with innocence. His medium height and build, sandy hair, almost invisible wheat-hued brows, and easy, bland manner all conspire to make him unnoticeable.

Dupin is revolted by "the smorgasbord of options" the world offers his young daughter. When he gets worked up he searches hard for the right phrase and often just misses: "I have moments of absolute tyranny when intellectually I am convinced in my mind I am right. I don't know how to explain this. But there is almost no limit to what I will seek until it prevails."

In one of his moments of absolute tyranny, Dupin foresaw that the abortion drug, just approved in France, would soon appear "in every household in every nation." That prospect, for him, was "the ultimate disdain. This was our last opportunity to have any semblance of control over the issue."

Although he was inexperienced at it, challenging enemies across international boundaries did not especially intimidate Dupin. Living not far from the nation's capital, he had "seen plenty of diplomatic license plates on the interstate." He simply picked up the phone and called the French embassy, asked questions about the company that made RU 486, and got answers. Then he began thinking of strategy. He knew he didn't want to be one of those protesters on CNN who, after their parade, go "to the park to drink wine and eat cheese—that's pure Parisian." He wanted Roussel-Uclaf to feel his pressure in a "grab their throat" style of politics.

Dupin assembled a council of some of the best minds he knew in Roanoke—a sociologist, a stockbroker, a physician, a management consultant—about a dozen in all. They organized homework assignments, and Dupin volunteered for most of them. At the Roanoke Public Library he pored through pages of "that big green book," the *Reader's Guide to Periodical Literature,* and other directories. He studied, for reasons he does not fully explain, the texts of NATO treaties and Greenpeace contracts. His council soon realized that Roussel-Uclaf was not the right power center to be aiming at. The proper target was its parent company, Hoechst A.G., a multinational, multi-industry behemoth based in Frankfurt, Germany. Dupin promptly called a Washington friend who worked for a congressman. The friend enlisted the Congressional Research Service to find out what contracts Hoechst had with the U.S. government. One of Dupin's council had an acquaintance in Frankfurt, who helped fill out profiles of Hoechst's top executives and even board members.

The council, enjoying its role, invented a name for itself: the Robins Carbide Reynolds Fund—drawing from the names of three of America's most hugely sued corporations. Its high-spirited humor cooled when one of the companies threatened to take the group to court, so it condensed its name to the less culpable RCR Fund. Then it began to act like a fund—not the foundation kind, but like a mutual fund. Through one of Dupin's council, they enlisted a Wall Street brokerage to get online and find out the names of mutual funds and investors holding sizable chunks of Hoechst stock.

Meeting mostly at Lily's, a bistro at the Roanoke Marriott, the RCR "board" drew up a declaration of demands upon Hoechst. Knowing that one couldn't be too careful, they had lawyer friends comb through it, pro bono. "There's a fine line," Dupin pointed out, "between extortion and lobbying." Carol Pogash of the *San Francisco Examiner*, who was later shown the document, has described it:

> In late December 1988, the official-looking declaration arrived at Hoechst headquarters in an immaculate white folder with Cabernet-colored rococo script spelling out the impressive-sounding RCR Fund. A drawing of the world appeared in the right-hand corner, indicating RCR's alleged global reach. The document didn't mince words. It attacked the [RU 486] pill as a "human pesticide" and said it was "unconscionable" for a German company to market an abortion pill in the Third World, where it would . . . be used for race-related population control. RCR said its knowledge of the Hoechst empire was vast. In a particularly chilling clause, the declaration claimed its members knew the location of Hoechst's chief executive officer Wolfgang Hilgers' "vacation home in the Alps."

Like an ultimatum from the UN Security Council, the declaration named the steps that global-reach RCR planned to take if Hoechst committed the aggression of offering RU 486 outside of France. It would unleash a boycott of products made by Hoechst subsidiaries in the United States. (As one example, Celanese, maker of synthetic fibers, was owned by Hoechst.) It would engage missionaries to search out Third World women injured by RU 486 to join in worldwide liability lawsuits. It would publish newspaper ads informing Hoechst employees of public-interest law firms that stood ready to represent them in charges against Hoechst of sexual harassment, discrimination, and personal injury. It would furnish Hoechst customers the names of competing suppliers "who better demonstrate moral conscience." It would provide U.S. environmental agencies with samples of pollution collected at Hoechst's American factories.

RCR served up a list of major world banks and brokerage houses they had "advised" about the stresses that would be faced by Hoechst. It demanded a face-to-face meeting with Hoechst, preferably in London or Geneva, but specifically not in France or Germany. As Dupin later explained, "We weren't so sure we wouldn't step off the airplane and be arrested for extortion." In late December, they instructed Hoechst precisely to reply before January 6, 1989, 11:59 P.M., Eastern Standard Time.

Hoechst's response, from its senior vice president for communications, D. Von Winterfeldt, was prompt (three days before the deadline), formal, and astonishing: "Needless to say that we are dismayed at the tone of your declaration. . . . We want to contribute toward a more understanding society with a greater meeting of minds. As such we see no reason why a meeting with you should not be arranged."

While capitulating to this bizarre leg pull, Hoechst yanked back. Von Winterfeldt requested to see RCR's member list, its advisory council, and a rundown of its previous activities, financial statements, and publications.

That called for some emergency action by the high command at Lily's in Roanoke. The council suddenly changed its name from RCR Fund to the grander RCR Alliance and immediately sought allies. The first to join was the Moral Majority. Other well-known antiabortion groups, however, declined. The National Right to Life Committee suggested it might find itself subject to conspiracy charges under the RICO Act, the antiracketeering law used to rein in the Mafia and Hell's Angels. (Indeed some years later the Supreme Court ruled that the RICO Act could be used against groups that forcibly block entrance to abortion clinics.)

"One wonders what Hoechst's board of directors would have thought," Carol Pogash observed, "if it had known RCR's 'world headquarters' consisted of a fax machine installed in the floral-print bedroom of Ken Dupin's daughter. Many nights, as she slept in her canopy bed, whining messages from Hoechst spilled out of the fax. . . . Sometimes the sheer gall of what his group was doing overwhelmed him." The feeling led Dupin to philosophize about power, which he said "exists independent of you and me. It essentially comes down to who reaches up and grabs it."

The meeting with Hoechst never took place. After a few more weeks of electronic haggling, the negotiation concluded, as astoundingly as it started. A telex to Roanoke from Hoechst announced: "It is not our intention to market or distribute RU-486 outside France."

In Paris, Catherine Euvrard, head of communications for Roussel-Uclaf and the only woman on its board of directors, gave Carol Pogash "what must rank as one of the most startlingly frank interviews ever given by a public-relations executive." Euvrard called her colleagues "male chauvinists" who "never, never, never think about women. . . . They never pronounce the name of women. Abortion is ninety-nine percent the problem of women alone. It's not their problem—they're men so they don't care." Her fellow directors, she said, are "afraid of controversy, afraid to fight. For them, it is always the problems of politics, money and corporate image. . . . Men speak and men decide. And women suffer."

A year after Dupin's colossal international triumph, the mild-mannered minister left Roanoke, as unforewarned as his housebreak into history, to disappear into a job shepherding a Midwestern congregation that meets in the gym of a junior high school.

Baulieu was outraged that his blessing for the women of the world could be blocked by the timidity and self-interest of one corporation—indeed the one that controlled his discovery. Embarking on nothing short of a personal crusade every bit as determined as Dupin's, Baulieu regularly commuted to the United States, making speeches, talking on TV, visiting scientists, seeking a base of indignation among the influential.

The influential joined up. Peg Yorkin of Los Angeles poured $10 million of her reputed $100 million fortune into the Feminist

Majority Foundation, declaring that the RU 486 "genie" is "out of the bottle." To get it to American women, she said, "we are prepared to do whatever we have to do." Yorkin threatened a counterboycott by pro-choicers of Hoechst's vast medicine chest of products if RU 486 was not released.

In the summer of 1989, seven months after Dupin challenged corporate Europe, America's organized women decided to do some challenging of their own. A delegation headed by Eleanor Smeal, former president of the National Organization for Women (NOW), flew to Paris to take a firsthand reading of the Roussel-Uclaf people. The company hosted them politely, with soothing observations that in the long run "no one could stop scientific progress." In July 1990 Smeal returned to France with eight hundred pounds of baggage: 115,000 petitions to combat the Roussel-Hoechst notion that the American public opposed the right to abortion. Also she brought a resolution from the American Medical Association calling for a test of RU 486 in the United States. This time the hosting was more than polite. The company president, Edvart Sakiz, met with them in the wood-carved executive dining room where white-gloved waiters served them eggs Benedict and mineral water. An all-star team of American women's leaders, physicians, and scientists flanked Smeal, including none other than Carl Djerassi.

From Paris the determined group proceeded to the summit: Hoechst headquarters in Frankfurt. The Germans too made a show of civility, like bankers turning down a mortgage.

Back in America, with Dupin gone from the scene, the National Right to Life Committee rose above its fears of the RICO Act and lined up for counterattack. It unleashed a massive postcard campaign that touched a still-inflamed nerve in the history of German drugs by likening the "death pill" to thalidomide. But something else touched Hoechst's sensitivities more poignantly. The postcards linked Hoechst to its corporate ancestor, I. G. Farben, the developer and maker of the gases used in Hitler's mass-extermination camps. The Right to Life group also demanded a meeting and got it. It was led by the noted opponent of the women's movement, Phyllis Schlafly. "Don't misunderstand the American psyche," the delegation warned. If RU 486 crossed American shores, opponents of abortion "would create a firestorm."

Some pro-choice activists called for a novel way to neutralize the "firestorm." If Hoechst would submit RU 486 for approval in the U.S., they would promote not a boycott but a buying spree of products of Hoechst subsidiaries. They called the reverse effort a "girlcott."

With no postured courtesies or diplomatic evasions, the Hoechst officials put the activists on flatfooted notice: As long as abortion remained a subject of high controversy they would not permit RU 486 in the United States. Whether they would "permit" it was academic: The Bush administration had already extended its antiabortion position to include banning importation of the drug.

In 1992 a decision to court-test that ban was made by a small New York–based group, Abortion Rights Mobilization, led by Lawrence Lader, a birth control activist and Margaret Sanger biographer. It selected a pregnant twenty-nine-year-old California social worker, Leona Benten, as its volunteer challenger. The Lader group flew her to England, helped her obtain a dose of RU 486, then routed her back to the U.S. through New York's Kennedy airport, where customs officials, alerted by the open publicity of the test, seized the tablets. (Ironically, in the kitchen refrigerator of Lader's Fifth Avenue apartment, today he keeps a plastic jar two thirds full of white powder. It is part of a bootleg batch of RU 486 the Abortion Rights Mobilization commissioned a chemist to concoct. The Bush ban did not bar possession, but importation.) The ensuing legal battle over Benten's violation, as hoped, moved speedily up to the Supreme Court. But the Court refused to order the government to return the RU 486 to Benten. She hastened home to California for a surgical abortion.

Two days after his inauguration in January 1993, President Bill Clinton ordered authorization of "testing, licensing and manufacturing" of RU 486. In April, with the new political atmosphere, Roussel-Uclaf, at the urging of the FDA, announced it would license the drug to the Population Council, a nonprofit organization based in New York, which would run clinical tests and later find a company to manufacture and market it in the United States. Promptly the Oregon and New Hampshire legislatures volunteered their states as test sites. Although testing a new drug generally requires seven to ten years, some predicted that U.S. approval of RU 486 might come in as little as two, owing not only to the new administration's support but to the drug's clinical track record in France. The forthcoming testing, meanwhile, would permit at least two thousand American women to use the new drug.

The Population Council quickly raised four million dollars for the trials, also organizing a round-table meeting with women's health organizations to design a broad socioeconomic mix of races and ages among participants. Their American tests would skip directly to the new, all-pill version of RU 486, bypassing the injected prostaglandin.

In his personal effort to solve the problem of finding a manufacturer, Dr. Baulieu has taken steps to establish a nonprofit company that would both manufacture and distribute RU 486, not only in America but worldwide. Since the single drug would be its entire source of sustenance, the threat of a boycott would be made moot.

Like the Pill, the new RU 486 works both sides of the pregnancy street. While freeing women of unwanted pregnancies, RU 486 may also help eliminate infertility. It has shown early promise as an effective treatment for endometriosis and fibroid tumors, painful and leading causes of female infertility, often compelling hysterectomy. Studies also indicate that RU 486 may fight off breast cancer and Cushing's syndrome, a metabolic disorder that can be life-threatening, as well as possibly reducing the need for Cesareans. Further, there are tangible hints that the antihormone may help treat brain tumors known as meningiomas.

After the inauguration of President Clinton, the National Institutes of Health solicited grant requests for RU 486 research, a sharp turnaround from previous years when RU 486 was held off like a poison.

Study of the whole class of antihormones may lead to weapons against cancer. The Food and Drug Administration has approved clinical tests at the Breast Center and Cancer Institute at Long Beach [California] Memorial Medical Center of forty seriously ill women whose cancer has spread beyond the breast. In breast cancers that contain a progesterone receptor, RU 486 appears to go into the cancer cells and turns off the mechanism that causes them to divide, says Dr. John Link, the center director. As a further benefit, RU 486 produces milder side effects than chemotherapy, but they may include nausea, vomiting, fatigue, and rashes.

The National Academy of Sciences, noting that RU 486 has proved its safety through use by scores of thousands of women in Europe, recommended that it be approved quickly for use in the United States. An editorial in *The New England Journal of Medicine* branded the movement to block RU 486 in the United States a "disgrace."

An astute observer of politics once remarked that advocates of causes most often play their politics like the game of bowling—hurling straight at the target—instead of like billiards—getting around barriers by angling off the sides. In that light, advocates of RU 486 might do well to ponder the early story of the Pill. In the late 1950s, while its use as a contraceptive was still in full debate, the Pill successfully charged into the marketplace as a treatment for gynecological disorders. Once it got through the gate, it was impossible to banish. That story might suggest that the most direct route to American acceptance of RU 486 may not be as an abortifacient, with all the accompanying political alarms and barriers, but as a cure for illness.

In 1992 *The New England Journal of Medicine* reported a study in Scotland that showed that a single dose of RU 486 taken within seventy-two hours of unprotected sex works as a *postcoital contraceptive*—not dislodging the fertilized egg but preventing its implantation. By the definition of many, that is not an abortion.

But word is now out and spreading widely that a simpler morning-after pill has been on the market, available in any drugstore in America and most other countries, for twenty years. It has never been openly announced as such and has remained the "secret" of a growing number of college women. Many gynecologists claim not to know about it.

Two tablets of certain brands of the standard birth control Pill—of the older, high-dosage variety—taken as soon as possible within seventy-two hours of intercourse, plus two more pills half a day later, bring on a menstrual period. Among those who know about and have used this underground method, Ovral has been the favored brand. The low-dose Pill (Lo/Ovral, Nordette, Levlen, Triphasil, and Tri-Levlen) works too, with slightly different dose regimens. Because its hormone content is lower, the dose is doubled—four tablets after intercourse, followed by four more twelve hours later. Timing is critical. If initiated more than seventy-two hours after intercourse, the method simply won't work. When done right, the method is about 70 to 80 percent effective. That is less effective than every other birth control method, but it is usually called upon when the user has no other acceptable choice.

No manufacturer has applied for FDA approval to use the Pill as a "morning-after." Why should they throw themselves voluntarily into the bloody arena of abortion accusations? They don't have to do a thing. Once a drug has been approved for one purpose, doctors may legally prescribe it for others—as for years they prescribed the Pill, approved for menstrual disorders, for sub rosa contraception.

The Pill has been used the "morning after" for years in England, France, Germany, Canada, Sweden, and Hungary, among other countries. It is believed to be the leading reason the Netherlands' abortion rate has been one fifth of that in America. In the United States many college health services have long made it available to students "in trouble," and hospital emergency rooms for a decade have routinely prescribed it for rape victims. Pharmacists discreetly avoid asking questions.

Women who have been advised to avoid the Pill for health reasons should not use this method, warns Robert Hatcher, director of Emory University's Family Planning Program and lead author of *Contraceptive Technology*. Studies of women without Pill risk factors have turned up no serious complications. But intense side effects, including nausea or vomiting, breast tenderness, and bloating, are common.

If and when RU 486 is approved in new countries, including the United States, it is likely to become the preferred morning-after pill. Compared to high doses of the Pill, the aftereffects of RU 486 (without the prostaglandin shot) are considerably less unpleasant, and its failure rate is virtually nonexistent.

And what is the woman to do the morning after if she can't get RU 486 and prefers not to fetch a prescription for Ovral? If she's willing to widen her margin of risk yet a bit more, according to newer research, she might beget herself to the nearest fruit market.

A scientist at the University of Sussex in England recently published a finding that the tropical fruit papaya is both an effective contraceptive and powerful enough to abort pregnancy. Researcher Tharmalingam Senthilmohan said the pink-fleshed, mangolike fruit has long been used in India and Sri Lanka as a traditional remedy for contraception and abortion, with no reports of side effects.

Papaya contains a very strong enzyme called papain, which interacts with progesterone, said Senthilmohan. If one whole fruit a day is eaten by a pregnant woman, she will probably abort within a week, perhaps only two or three days. Abortion could be virtually assured, the Sussex scholar found, by eating unripe fruit that contains higher papain concentration.

Malcolm Topping, a senior lecturer at Sussex in chemical and medical sciences, sees no reason why papaya could not become the basis for a new form of contraception for the West. "It clearly works," he has said. "Indeed the question for us is why hasn't it been taken up before."

24

THE AGE OF BIOINTERVENTION

W e have passed through the gateway of a coming half century of new ethical conflicts and moral anxieties for which nothing in our past has prepared us. We have entered the Age of BioIntervention.

The invention of the Pill transported all of us, in a most personal sense, into a new epoch of seeming mastery over our bodies and ourselves. It was the first product of science to alter, for our pleasure and convenience, the way the human reproductive system functions. From the story of what we have done, perhaps we may better understand what we are about to be able to do.

Already we know how to bring forth a baby from a mother

who provides the womb while another has furnished the egg. We rapidly progress toward dispensing with fathers, seeming to need only their sperm deposited in plastic dishes. Now that we can ascertain the sex of a baby before its birth, how far ahead lies the advance *choice* of its sex? Will humans successfully resist that gift of science? And how soon will our new ability to splice and reshape our genes—in a sense, the reinvention of ourselves as a species—offer us exotic choices and social, religious, and ethical trials that we cannot yet comprehend?

If birth is the gateway to life, the Age of BioIntervention involves us with the exit as well.

Death is no longer final in the way it used to be. It was once a truism that everybody dies of heart failure. But heart failure no longer means we die. When a weak and weary heart simply cannot go on, we now open the remains of some poor departed—one whose heart beat sturdily while its unfortunate owner died of something else—and, of all things, we *transplant* the good heart of the dead into the needy cavity of the all-but-dead. Thus out of doom we create, for the receiver and the receiver's family, renewed life. What so recently belonged only to the hand of God is now in the hand of the cardiovascular surgeon.

So rapidly has the Age of BioIntervention enveloped us that the miracle of passing one's heart to another like a baton has become ordinary. In the few years since the first, daring transplant, we have plumbed deeper and deeper, reaching down at last to the deepest known level of life. We have redrawn the very blueprints of who we are. At the threshold of modifying our physical form, we now transplant *genes.*

Technology and human mastery of it have evolved to the ultimate: They are about to supplant evolution.

The Pill forced upon society new kinds of choices, vexing us with unprecedented moral conflicts. In that sense it was a new cycle in the story of humanity's long love-hate affair with science and technology. That anguished relationship began, perhaps, with the discovery of fire. Fire promised us warmth against frost. It could cook an animal catch and light the darkness. But it also threatened us with its power to burn human flesh and devastate forests. Fire had no will, no morality, no ethical values of its own. Humans, the discoverers of fire, had to learn to *make choices* about its purposes and its control. That necessity of choices, of creating a new code of behavior for coping with a new technological power, could not be avoided or postponed by banning fire. We were compelled inescapably to learn how to take charge of it. The history of science and technology is streaked with dread of the newest great invention—yet, simultaneously, with the unthinkability of living without the last one.

The future lurks secretly and patiently in the present. Through a few small events in our present and recent past, shadows and shapes of the coming Age of BioIntervention can be perceived. A few illustrations, seen this early in the game, may strike us as quaint, but what they intimate may amuse us less as time goes on.

Recently in Los Angeles, William Everett Kane deposited a sample of his sperm in a vessel, had it frozen, and willed it to his thirty-seven-year-old girlfriend, Deborah Hecht. Then Mr. Kane killed himself.

His grown son and daughter, age twenty-two and twenty, went before a judge asking for the sperm to be destroyed. Their mother, who had divorced Kane nineteen years earlier, was their lawyer. Their complaint stated that Kane's wish to have a child by his girlfriend came only after his decision to commit suicide, which on its face was an act of mental instability.

The judge agreed.

But the girlfriend, arguing that Kane's children simply didn't want to share Kane's estate with a new sibling, appealed to a higher court. The sperm remained frozen.

Two and a half years later, in March 1994, California superior court judge Arnold Gold cited a precedent that declared sperm to be property, and ruled that Kane's sperm was part of his estate. The judge awarded Hecht, the girlfriend, 20 percent of the sperm and permitted the son and daughter to destroy the rest.

Ms. Hecht was grudgingly satisfied. "Twenty percent is better than nothing," she said. "Maybe I can do it on just three vials rather than fifteen."

Here is further evidence that the founding fathers in Philadelphia may have failed to consider all the constitutional implications of frozen sperm: In the state penitentiary at Mecklenburg, Virginia, Joseph Roger O'Dell III, fiftyish, unmarried, and childless, awaits the electric chair for the rape and murder of a Virginia Beach secretary. A sentimental man, he says that he cries every time he watches TV and sees "all those cute little babies" on commercials for Huggies. Since O'Dell has no brothers or sisters, his execution will end his bloodline. So he recently petitioned a judge for his sperm to be frozen before he is executed and for it to be used afterward to inseminate his girlfriend.

The Virginia Supreme Court dismissed the request as "frivolous," but O'Dell has appealed to the United States Supreme Court on constitutional ground. Denial of his right to have children and to keep his bloodline alive, he argues, goes beyond the sentence that condemned him and would constitute cruel and unusual punishment. In shackles and handcuffs, O'Dell recently said to an interviewer, "They have the legal right to kill *me*, but they have no right to destroy my bloodline."

His girlfriend, Sheryl, an insurance adjuster, has not yet decided for certain that she wants to employ O'Dell's sperm, but she feels strongly about her rights. "We should have the *right* to have a child together," she says. "I don't want to lose the opportunity just because they take his life."

The dilemmas and stresses of the Age of BioIntervention go beyond simple law. They throw us into the murkier, more troubling realm of ethics and values.

Millions were enthralled recently by a television drama based on the actual experience of Arlette Schweitzer of Aberdeen, South Dakota. Schweitzer happily gave birth to twins. In doing so she became the first American woman to bear her own grandchildren.

Schweitzer's daughter, Christa Uchytil, twenty-two, had been born without a uterus. So her obstetrician, Dr. Gregg Carlson, devised a procedure to extract eggs from her ovaries, fertilize them in vitro with her husband's sperm, and implant them in the womb of her mother, Arlette, who was forty-two. When the twins were born, Christa sat by her mother's side, tears flowing, as she watched her children arrive.

In a somewhat related example of scientific progress, Mary Shearing of Anaheim, California, a fifty-three-year-old grandmother, became pregnant with twins. Shearing, whose husband was thirty-two, had been going through menopause. She arranged for her husband to fertilize an egg of a younger woman, then received the implantation of it in her own uterus.

The nation's best-known biomedical ethicist, Arthur Caplan of the University of Pennsylvania, is on record strongly in favor of personal choice in the matter of abortion, but in this case he is not so sure choice should apply. He says that perhaps future Mary Shearings ought to be required to appear before independent ethics committees before they are *permitted* to bear children who won't enroll in kindergarten until their mothers are perhaps sixty years old. (Neither Caplan nor other protesters of this incident are known to have complained about sixty-year-old fathers.)

To further complexify the "right" of reproductive choice, two English doctors recently opened the London Gender Clinic, which for a fee of about one thousand dollars would help couples conceive in such a way as to greatly enhance the likelihood that their child would be a boy or a girl, according to their choice. Their method would isolate in vitro Y-chromosome sperm from X-chromosome sperm, fertilizing the mother's egg with the chosen type and implanting it in her womb.

There remains some uncertainty as to whether advance choice of sex is reliable. But if it's not a sure thing today, it will be soon.

All known survey information suggests that free choice of a child's sex will produce significantly more boys than girls. What we do not know is the impact of gender imbalance on the future of monogamous marriage, the nuclear family, or the incidence of rape, not to mention the impact on the job market. Are we prepared to pressure our legislators into outlawing sex choice—and are those so prepared also ready to do battle with the many who may oppose them?

Choice? Whatever happened to choice?

• • •

Jeffrey A. Fisher, the medical doctor and author of *Rx2000: Breakthroughs in Health, Medicine and Longevity by the Year 2000 and Beyond,* makes some forecasts that may seem to strain credulity, more out of fiction than science, but that's what people said so recently about the prospect of transplanting a heart, about creating new organisms by juggling and splicing genes, and certainly about landing an earthly human being on the moon.

The creation of "designer" children will be upon us by the year 2014, Fisher predicts, and the possibilities stretch as far as imagination can take us. Leaving aside the obvious physical "clothing" of blond hair, blue eyes, and tall, slim physiques, among more fascinating considerations are genetic manipulation to favor talents that are bound to come into temporary vogue: intelligence or athletic ability or musical talent over other inheritable traits.

By the year 2019, Fisher predicts, perfection of artificial placenta combined with in-vitro fertilization could open the door to reproduction completely outside the human body. (After all, why suffer the inconvenience of pregnancy? Take a round-the-world tour or an academic year at law school, and pick up your baby at the drive-in maternity center.) Nobody would predict that such birth practices will take the world by storm, or that many people will make the choice Mary Shearing made. But do we want to allow *any* postmenopausal mothers? If not, who shall speak for Mary Shearing's freedom of reproductive choice?

If the rules for guarding the entry door of birth need to be watched, then the exit gate demands attention too.

Relying again on Dr. Fisher's timetable, by the year 2030 deaths caused by heart disease and cancer will be down 99 percent. The maximum human life span, he predicts, will be increased to 150 years. It would be foolish to quibble with Dr. Fisher about whether this date is right or is off by a quarter century. Cancer and heart disease *will* some day be wiped out. Experts in the physiology of aging may pick a bone with Fisher about the matter of living to 150 years; most say that the natural limit of life span is 120 to 125 years. Our genes have a built-in clock whose batteries appear to run down at that point—if we are not done in first by illness, violence, or a heavy-fat diet.

So bringing down the birthrate through the Pill or any other birth control means may no longer be the cure for overpopulation. If more and more people procrastinate in their obligation to die, the world is going to become even more crowded than it is already destined to be.

Today people over the age of eighty-five are the fastest-growing group among us. One day soon we will surely hear a proposal that Social Security retirement benefits be withheld until the age of seventy-five, eighty, or eighty-five. Talk about great ethical conflicts. Can we expect the American Association of Retired Persons to take that challenge lying down?

When payroll deductions for Social Security double or triple to support emerging millions of retired parents and grandparents, will today's Great Abortion War be replaced by the Great Generation War? That is not a facetious question. The flowering of science and social progress have scarcely begun.

To many people, solutions to the dilemmas of the Age of BioIntervention come easily. They say, "Why don't we have a law that says such-and-such?" "Why can't there be a rule that says such-and-such?"

Fine. Why can't there be a law that says that when a teenager flunks a pregnancy test, she must have the baby? Tens of millions think there should be such a law. But tens of millions of others would oppose it. Why can't there be a court order, once and for all, that says that what a pregnant woman does with her body is nobody's business but her own? Tens of millions think there should be. But tens of millions of others say there should not.

Strong moral defenses have been made for both of these positions, and we have heard both sides till our ears wither. But one person's idea of a correct moral position does little to satisfy someone else who sees it the other way around. The Great Abortion War has spread across an ever-widening battlefield. It has been, in a sense, a holy war, since the conflict is purely over religious, moral, and ethical positions. Each side argues that the other is not only *wrong*, but *evil*. For that reason there has rarely been, at least in America, any exchange of opposing views about abortion that is exploratory, reasoned, conciliatory, or that is truly an honest search for a peace acceptable to both sides. Even after discounting the hotheads and terrorists, rarely does either side acknowledge respectfully that the other side speaks from a moral imperative. Both sides shout. Neither side listens. Apparently that is where the terrible dispute will remain for some time to come.

The moral dilemmas forced upon us by our new technologies are there not because the technologists are mischievous hatchers of evil. They are there, more simply, because we have not lived with the results long enough to have fashioned traditions for coping with them. We have not worked out a set of common values about what we feel is right and wrong among the choices presented by our technologies. Values create laws, not the other way around. Values create religious beliefs, not the other way around. Values take time. They require conflict and pain. As the rate of change accelerates, a question for *our* time is whether we are prepared to take the time, and time refuses to be hurried.

Resolving the dilemmas of the Age of BioIntervention cannot be left in the hands of doctors and administrators, judges and juries. If substantial numbers of people were made unhappy by their decisions, the conflicts would be explosive. The dilemmas must be argued in the political arena. Perhaps that seems the worst of all possible places for these questions to play out. Look what happened when the Pill became political—and, worse still, religious-political. Look what happened when the conflict over abortion spilled into the streets.

Yet the truth remains: The political forum is the only democratic game in town.

What the Age of BioIntervention requires is a new way of considering paradox, a new way of talking, with true listening, with compassion for the opposite point of view, with some form of intellectual respect for differences in moral positions that is not yet part of our culture.

In past years, in past centuries, new advances in science have presented us with new opportunities coupled with new threats, time and again with difficult choices:

Does society have the right to demand that *my* children accept into their little bodies the germs of smallpox or polio in some newfangled inoculation that I don't trust? (Remember?)

Can *they* force my family into medical care or compulsory schooling when I and those of my religion don't believe in it? (Remember?)

Can they pour fluoride into *my* town's water supply? *Remember?*

In every recent decade the stakes have grown bigger more rapidly, and we have seen only the beginning.

What we are going to need is a new definition of the very complex idea of personal rights, and of *choice*—when personal choice may be a good idea and when it may be an antisocial one.

That new definition, acceptable to all or even most, does not appear to be soon in the making.

The Era of BioIntervention, with its scientific and life-giving wonders, will force upon us the high price of vexatious choices. For many, it may not make life easier. But for most, it will surely be an advance over the many centuries when humankind suffered, died, and, in between, bore children with no choices at all.

Here is the text of Gregory Pincus's written instructions for conducting trials of the Pill. These were the standard operating procedure for the field trials in Puerto Rico, Haiti, and Mexico.

PROCEDURE FOR PROGESTERONE TESTS

The research in reproductive processes at the Worcester Foundation and the Free Hospital for Women (in Brookline) has as its objective the finding of a universally acceptable means to suppress temporarily and at will, human reproduction, without disturbing physical and mental well-being.

In the female, prevention of pregnancy might be achieved by inhibiting ovular maturation and ovulation without affecting menstruation. Because progesterone is produced during the female reproductive process its effects have become the objective of certain contraceptive research studies. Experiments on animals have shown that progesterone stops ovulation in these animals without harmful side effects. Experiments using progesterone on human females have now been instigated. This chemical substance is administered by pill, injection, or vaginal suppository, in order to observe the effect of progesterone and its catabolism, utilization and excretion by the body. In order to observe the effects of progesterone clinically a definite and routine procedure with human females is employed and detailed records of each subject (together with the records the subject herself keeps) are maintained by a secretary in the doctor's office, which records are the official ones of all these cases.

The case procedure is as follows:

1. Examinations during 4 periods of 3 months each are maintained. The first of the three months is the "control" month during which no progesterone is administered. During the 2nd and 3rd months progesterone is administered in definite dosages.

2. Each female is examined (before being accepted as an experimental subject) for the following qualifications:

a. Is she sufficiently intelligent to cooperate?

b. Is she willing to cooperate?

c. Are her home conditions such that she can cooperate?

d. Would an unexpected pregnancy be a disaster for her?

Also it is necessary to ascertain:

a. Is her menstruation of the normal 4-week type?

b. Has she the classical bi-phasic oral temperature?

c. Has she a normal mucous lining of her uterus? (determined by biopsy)

d. Has she normal cell changes in the samples of the mucous lining of her vagina? (determined by smears)

e. Does her urine show normal excretion of pregnandiol and other steroids?

3. Furthermore the subject is asked:

Can you read a thermometer? If not, is there someone in your home who can and will? Is your domestic set-up such that you can take your temperature every morning as soon as you wake up, and at the same time every day? Can you record your temperature on the chart in the right place?

Can you, and are you willing to, take a vaginal smear every day on a special glass slide; can you put it in the jar of fixature for just the right time; can you put it in the right place in the slide-box? Will you once a month collect in the bottle we give you all the urine over a 48-hour period at such a date as we tell you to do it, and then bring it in to the clinic? Will you come to the clinic once a month for a biopsy when requested?

During the second and third months we will give you tablets to take—6 or 8 every day (about 1 every 6 or 8 hours) beginning with the 5th day of your menstrual cycle (5th day following onset of menstruation) and continuing through the 25th day. Will you take these and will you note on the chart we give you the time at which you take the tablets, and also any time you miss taking the tablets?

Can you and will you keep the charts accurately?

Will you follow this routine for a 3-month period exactly as we direct you?

If the female thus examined (see above) is accepted, the following is the daily, weekly and monthly routine in which she must cooperate during the control months as well as during the other two months.

1. Temperature chart (*daily* oral temperature taken and recorded by the subject on awakening and at the same time every day).

2. Daily vaginal smear (taken by the subject on awakening every day, including the menstrual period; the subject is taught how to do this and provided with the necessary equipment). The subject, using a pipette inserted in the vagina, secures a sample of the vaginal secretion; places it on a special glass slide numbered to correspond with numbered slots in box; drops the slide into a jar of fixative for a minimum period of twenty minutes; removes slide and drops it into the correspondingly numbered slot in box. These slides are kept for one entire cycle (28 days) by the subject, then taken by her to the physician for laboratory study. (Proper interpretation of these slides to determine ovulation time requires special training.)

3. Endometrial biopsies are made monthly by the physician on the 9th–10th post-ovulatory day (determined individually for each subject from the day the ovum leaves the ovary). The physician takes a sample of the mucous lining if the expected secretory endometrium (normal mucous lining) is present.

4. Once a month a 48-hour collection of all urine excreted is made by the subject on the 4th–6th post-ovulatory day (see above). All the urine is collected by the subject in a large jar and this jar is then taken to the physician for a special laboratory examination involving pregnandiol and other steroid tests unobtainable in the ordinary urine-analysis laboratory.

5. Progesterone dosages in definite amounts must be taken at specified intervals during the second and third months only, not during the control month. The effect of progesterone varies according to the amount of the dosage and also whether it is given by mouth, inoculation or suppository method.

6. Charts for recording temperature, progesterone dosage, time of

biopsy and urine collection are provided each subject and must be kept accurately by her.

Suitable patients who are obliged to undergo abdominal surgery (for causes not connected with the progesterone experiments) may be given progesterone before the operation so that the surgeon may have an opportunity to observe direct from the body the effects of progesterone on ovulation thus checking his previously gathered clinical findings.

This book is indebted, of course, to the good work of previous authors, most prominently among them (alphabetically):

Étienne-Émile Baulieu, *The "Abortion Pill."* New York: Simon & Schuster, 1990.

Anne Biezanek, *All Things New.* New York: Harper & Row, 1965.

Carl Djerassi, *The Pill, Pygmy Chimps and Degas' Horse.* New York: Basic Books, 1992.

David Halberstam, *The Fifties.* New York: Villard Books, 1993.

Robert Blair Kaiser, *The Politics of Sex and Religion.* Kansas City, Mo.: Leaven Press (National Catholic Reporter Publishing, Inc.), 1985.

David M. Kennedy, *Birth Control in America: The Career of Margaret Sanger.* New Haven, Conn.: Yale University Press, 1970.

Loretta McLaughlin, *The Pill, John Rock, and the Church.* Boston: Little, Brown and Co., 1982.

James Reed, *From Private Vice to Public Virtue: The Birth Control Movement and American Society Since 1830.* New York: Basic Books, 1978.

Paul Vaughan, *The Pill on Trial.* New York: Coward-McCann, Inc., 1970.

In the references that follow, each of the above books is cited by author's surname only (Baulieu, Biezanek, Djerassi, etc.). Within each chapter and topic, wherever practical, bibliographic citations appear approximately in the order in which the sources are used in the text.

In addition, the author has drawn from numerous archival sources for letters, documents, etc. Primary among these were the Library of Congress, Washington, D.C., for the Margaret Sanger Papers (abbreviated hereafter as MS-LC) and the Gregory Pincus Papers (GP-LC); and the Sophia Smith Collection, Smith College, Northampton, Mass., for the Margaret Sanger Papers, which were the source of letters to Sanger used in the Prologue.

Much of the most valuable material was gathered from interviews by the author and by others, each cited in the references. Worthy of special mention is an oral-history project in London conducted by

Sharon Goulds, Marilyn Wheatcroft, and the Television History Workshop. This creative team interviewed dozens of ordinary and extraordinary people, many in their seventies, eighties, and nineties, mostly women, about their experiences of birth control and sex. Only a small portion of those hours of valuable talk found their way into the workshop's three-part television series on contraception in the twentieth century, but the unused text was happily preserved. Quotations from these interviews are identified in the references as TVHW.

PROLOGUE: VOICES

Letters to Margaret Sanger are from the Sophia Smith Collection. Initials of their authors are fictionalized.

Quotations from "Rick Parsons" (name fictionalized) **and Dilys Cossey** are from TVHW.

1: THE CONCEPTION

"Mothers of the Pill" quotation: McLaughlin, p. 93.

First meetings between Sanger, McCormick, and Pincus: McLaughlin, pp. 96–98, 104–7; Reed, pp. 337–45; Vaughan, pp. 23–26.

Additional on Sanger, McCormick, Pincus, and the Pill: Margaret Sanger, *My Fight for Birth Control* (New York: Farrar and Rinehart, 1931); *Margaret Sanger: An Autobiography* (New York: Norton, 1938); Margaret Sanger's Journal, Sophia Smith Collection; Dwight J. Ingle, "Gregory Pincus," National Academy of Sciences, *Biographical Memoirs* 42 (1969); Loretta McLaughlin, "Dr. Rock and the Birth of the Pill," *Yankee*, September 1990, pp. 72–77, 152–55; Lois Mattox Miller, "Margaret Sanger: Mother of Planned Parenthood," *Reader's Digest*, July 1951, pp. 27–31; Margaret Sanger, "My Fight for America's First Birth-Control Clinic," *Reader's Digest*, February 1960, pp. 49–54; the obituary of Margaret Sanger, *New York Times*, September 7, 1966, p. 1.

Baulieu meeting with Pincus: Baulieu, p. 51.

Early Pill research: McLaughlin, pp. 103–9; Reed, pp. 311–75; Vaughan, pp. 19–35; author's interviews with Russell Marker, spring and summer 1991.

2: THE UNWORTHY SEARCH

Research on the female reproductive system: Shirley Green, *The Curious History of Contraception* (London: Ebury Press), pp. 132–33, 138–39; McLaughlin, pp. 98–107; Raphael Kurzrok, "The Prospects for Human Sterilization," *Journal of Contraception* 2 (February 1937), pp. 27–29.

Albright's Prophecy quotation: Reed, pp. 315–16.

3: TWO MOTHERS

Biographical details of Margaret Sanger: her published autobiographical writings (signed by her, but largely ghostwritten): *My Fight for Birth Control* (New York: Farrar and Rinehart, 1931); *Margaret Sanger: An Autobiography* (New York: Norton, 1938); Margaret Sanger's unpublished journal in Sophia Smith Collection; Ellen Chesler, *Woman of Valor* (New York: Simon & Schuster, 1992); Lawrence Lader, *The Margaret Sanger Story and the Fight for Birth Control* (New York: Doubleday, 1955); Kennedy; McLaughlin; Reed, pp. 67–139; Vaughan, pp. 5–6, pp. 23–32.

Additional on Sanger: Lois Mattox Miller, "Margaret Sanger: Mother of Planned Parenthood," *Reader's Digest*, July 1951, pp. 27–31; Margaret Sanger, "My Fight for America's First Birth-Control Clinic," *Reader's Digest*, February 1960, pp. 49–54; Obituary, Margaret Sanger, *New York Times*, September 7, 1966, p. 1; "Doctors on Contraception," *Time*, February 24, 1937.

"I walked and walked" quotation of Sanger from her autobiography, p. 77.

Sanger's tailoring of facts: Kennedy, pp. 18–20.

Additional glimpses of Sanger's early life were obtained in a tour of Margaret Sanger's hometown, Corning, New York, guided by Dick Peer, the city's former newspaper editor and informal historian. Peer also provided the unpublished autobiography of Mary Olive Byrne, daughter of Ethel Byrne, Margaret Sanger's sister. It had been given to him by Margaret Lampe, of Arlington, Virginia, Sanger's granddaughter, who also gave me a valuable interview.

Another family source was a senior thesis by Alexander Campbell Sanger, "Margaret Sanger: The Early Years, 1910–1917," submitted to the history department of Princeton University in 1969. He is a grandson of Margaret Sanger and presently serves as president of Planned Parenthood of New York.

Biographical details of Katharine McCormick, who guarded her obscurity as assiduously as Sanger cultivated and promoted her public image, are in McLaughlin, pp. 93–107; Reed, pp. 335–39; and Vaughan, pp. 24–28. Letters and other papers about her involvement with the Pill: MS-LC and GP-LC.

4: A STOREFRONT IN BROOKLYN

Opening the clinic and Ethel's hunger strike: *Margaret Sanger: An Autobiography* (New York: Norton, 1938), pp. 216–23, 238–50; "My Fight for America's First Birth-Control Clinic," Margaret Sanger, *Reader's Digest,* February 1960.

Marriage to Slee: Author's interview with Margaret Lampe, 1992; Kennedy, pp. 98–99, 106.

Sanger and the Comstock laws: Kennedy, pp. 21–26, 72–82.

Quotations of Dilys Cossey, Nancy Raphael, Dr. Faith Spicer, and Dr. Sylvia Dawkins: TVHW. **Lady Helen Brook:** interview by the author, 1992.

McCormick's first interest in birth control: Reed, pp. 334–41.

5: WAITING

This chapter is especially indebted to Shirley Green, for her exhaustively collected and good-humored book, *The Curious History of Contraception* (London: Ebury Press). Also John M. Riddle, *Contraception and Abortion from the Ancient World to the Renaissance* (Cambridge: Harvard University Press, 1992); John M. Riddle and J. Worth Estes, "Oral Contraceptives in Ancient and Medieval Times," *American Scientist,* May–June 1992; Grace Naismith, "The Racket in Contraceptives," *The American Mercury,* July 1950, pp. 3–13; "A Husband . . . a Wife and Her Fears," *McCall's,* July 1933.

6: AN ADVENTURE IN MEXICO

Although living reclusively since ending his chemistry career shortly after World War II, Russell Marker told his story extensively to two interviewers in his later years: to this author in State College, Penn., spring and summer 1991; and to Jeffrey L. Sturchio, April 17,

1987, for the Beckman Center for the History of Chemistry, Philadelphia. Also he gave a brief interview to Carl Djerassi at Stanford University, October 3, 1979, found in Russell Marker Papers, Pattee Library, Penn State University.

An autobiographical manuscript of Marker (16 pp.), dated May 15, 1969, is in Russell Marker Papers, Pattee Library, Penn State University.

Marker and steroid history: Leonard Engel, "Cortisone and Plenty of It," *Harper's*, September 1951, pp. 56–62; Russell Marker, "The Early Production of Steroidal Hormones," *CHOC News* (Center for the History of Chemistry), vol. 4, no. 2, summer 1987; Pedro A. Lehmann, Antonio Bolivar, and Rodolfo Quintero, "Russell E. Marker: American Steroid Chemist and Founder of the Mexican Steroid Industry," *Journal of Chemical Education*, vol. 150, p. 195, March 1973; Carl Djerassi, "Progestins in Therapy—Historical Developments," in *Progestins in Therapy*, G. Benagiano et al., eds. (New York: Raven Press, 1983), pp. 1–6; Norman Applezweig, "Dioscorea—The Pill Crop," *Crop Resources*, (New York: Academic Press, 1977), pp. 149–52; Jim Elder, "Modest Man with a Strut to his Past," *Pennsylvania Mirror*, September 18, 1977; Barbara Hale, "The Weight of Personal Goals," *Science Journal*, Penn State University Science Alumni Society, vol. 8, no. 1, spring/summer 1990; Norman Applezweig, "The Development of Systemic Chemical Contraceptives," *Response to Contraception*, Maxwell Roland M.D., ed., (Philadelphia: W. B. Saunders Co., 1973), pp. 17–20; Norman Applezweig, "The Steroid Revolution," a Science Year Report, *The World Book Encyclopedia*, 1961, pp. 105–17; *A Corporation and a Molecule: The Story of Research at Syntex* (Palo Alto, Calif.: Syntex Laboratories, 1966), pp. 21–34; Vaughan, pp. 10–15. Steroid-structure diagram adapted from *Corporation and a Molecule*, p. 5.

"Serendipity" anecdote: Pedro A. Lehmann, talk at American Chemical Society, August 1991, taped by author.

7: A Leap Beyond Nature

George Rosenkranz quotations from talk at the 1991 meeting of the American Chemical Society, taped by author; Djerassi, pp. 25–26.

Somlo and Syntex history: author's interview with Marker, 1991; Vaughan, pp. 15–16; Dwight J. Ingle, "Gregory Pincus," National Academy of Sciences, *Biographical Memoirs* 42 (1969); *A Corporation and a Molecule: The Story of Research at Syntex* (Palo Alto, Calif.: Syntex Laboratories, 1966).

Djerassi and Syntex: Djerassi, pp. 33–48, 56–65; Djerassi, *Steroids Made It Possible* (Washington, D.C.: American Chemical Society, 1990); Djerassi, *The Politics of Contraception* (New York: W. W. Norton & Co., 1980); Djerassi, "The Making of the Pill," *Science 84*, November 1984, vol. 5, pp. 127–29; Djerassi, "Drugs from Third World Plants: The Future" (letter to editor), *Science*, October 9, 1992, vol. 258, pp. 203–4; Rudy M. Baum, "Stanford's Carl Djerassi Wins 1992 Priestley Medal," *Chemical & Engineering News*, June 3, 1991, pp. 28–31; Scott Winokur, "Hearts & Minds," *Image* magazine, *San Francisco Examiner*, March 29, 1992.

Djerassi-Colton confrontation at the 1991 American Chemical Society meeting: taped by author.

8: "IN SCIENCE, LIZUSKA, EVERYTHING IS POSSIBLE"

Pincus and G. D. Searle: Reed, pp. 331–33; Vaughan, pp. 20–21; Djerassi, pp. 59–60.

Young Pincus, and at Harvard and Clark: McLaughlin, pp. 104–6; Reed, pp. 317–28; Halberstam, pp. 288–94; Ingle, pp. 229–39.

Formation of Worcester Foundation and Pill development: McLaughlin, pp. 93–106; Reed, pp. 25–42; Vaughan, pp. 21–22; Halberstam, pp. 289–94, 600.

Numerous items from *Nova* program "The Pill for the People," BBC and WGBH-TV, Boston, broadcast on PBS, March 9, 1977.

Some valuable technical details in Norman Applezweig, "The Development of Systematic Chemical Contraceptives," *Response to Contraception*, Maxwell Roland, M.D., ed. (Philadelphia: W. B. Saunders Co., 1973), pp. 17–35.

Quotations of John Rock: McLaughlin, pp. 14–15.

"Lizuska" anecdote: Vaughan, pp. 5–6.

Horst Witzel (of Schering A.G.): interview by author, Berlin, 1992.

9: "ALL THESE TRIALS SOON BE OVER"

Choice of Puerto Rico: Reed, pp. 3, 56–66; Vaughan, pp. 37–57; McLaughlin, pp. 128–33, 138–39; Hudson Hoagland, *The Road to Yesterday* (Worcester, Mass.: privately printed, 1974), pp. 92–95.

Preparation for trials, gleaned largely from correspondence: Pincus to McCormick, March 5, 1954, March 11, 1957, August 26, 1957; Pincus to Sanger, March 23, 1956; McCormick to Pincus, February 28, 1957; Edris Rice-Wray to Pincus, undated (from Mexico City) but clearly early March 1957; Pincus to Felix Laraque, March 11, 1957; Pincus to McCormick, July 27, 1957, all from GP-LC. That collection also yielded Pincus's field-trial instructions, reproduced as the Appendix.

Rice-Wray interview: *Nova* program, "The Pill for the People," BBC and WGBH-TV, Boston, broadcast on PBS, March 9, 1977.

Quotations of Anne Merrill and James Balog: interviews by author, 1992 and 1993.

Baulieu quotation: Baulieu, pp. 57–58.

10: A RACE FOR THE PILL

Drug companies and FDA approval: Reed, pp. 343–45, 358–66; Robert Sheehan, "The Birth-Control 'Pill,'" *Fortune,* April 1958, p. 222; Djerassi, pp. 62–63; Vaughan, pp. 48–51; McLaughlin, pp. 140–42; *A Corporation and a Molecule: The Story of Research at Syntex* (Palo Alto, Calif.: Syntex Laboratories, 1966). Also Norman Applezweig, "The Development of Systematic Chemical Contraceptives," *Response to Contraception,* Maxwell Roland, M.D., ed. (Philadelphia: W. B. Saunders Co., 1973); Carl Djerassi, "The Making of the Pill," *Science 84,* November 1984, vol. 5, pp. 127–29; and a Syntex promotional brochure, "Oral Contraceptives at Root of Syntex."

Quotations of James Balog, Marvyn Carton, and Dr. Friedmund Neumann (of Schering, Berlin): interviews by author, 1991; Irwin from Vaughan, pp. 20–21; Djerassi, pp. 62–63; Rock, Winter, DeFelice from McLaughlin, pp. 140, 142–45.

Uproar at use of contraceptives in New York hospitals and clinics from a multitude of reports in *The New York Times* between February 1952 and late 1958.

11: THE PILL CHANGES LIVES

Usage of the Pill before FDA approval: McLaughlin, pp. 139; Schering leaflet for doctors accompanying Anovlar, February 1961; author's interview with anonymous informant, 1992.

Westoff-Ryder surveys and other glimpses of early acceptance of the Pill: Jane E. Brody, "The Pill: Revolution in Birth Control," *New York Times,* May 31, 1966, p. 1; Steven M. Spencer, "The Birth Control Revolution," *Saturday Evening Post,* January 15, 1966; "Contraception: Freedom from Fear," *Time,* April 7, 1967 (cover story), December 15, 1961; *Ladies' Home Journal,* June 1990; *New York Times,* September 8, 1974; *Newsweek,* July 6, 1964.

Author's interview with Horst Witzel, Berlin, 1991.

Sylvia Ponsonby, Madeline Simms: TVHW.

Author's interview with Karen Hastie Williams, May 1977 and May 1994.

Interview with Tom Davidson by Pat McNees, 1991.

Career planning and Pill: "Some Women Find Themselves Penalized for Having Babies, Outside Interests," *Washington Post,* August 8, 1983; for Swiss-Walker survey; *New York Times,* February 11, 1968.

Celebration of thirty years of the Pill: Ann Marie Cunningham, *Ladies' Home Journal,* June 1990.

12: TWO DECADES, TWO WOMEN

Marshall McLuhan and George B. Leonard: "The Future of Sex," *Look,* July 25, 1967.

David Boroff: *Esquire,* July 1961.

"Anything Goes: Taboos in Twilight": *Newsweek,* November 13, 1967.

"Wanda Stein" and "Tina Mariello" (names fictionalized): interviews by author, 1992.

13: WHAT SEXUAL REVOLUTION?

Opening quotations: "forty-nine-year-old American woman": interview by author; **Joan Wyndham:** TVHW.

Glimpses of spread of Pill: " 'The Arrangement' at College," *Life,* May 31, 1968; Gloria Steinem, "The Moral Disarmament of Betty

Coed," *Esquire*, September 1962; "Facts About Sexual Freedom," *PTA Magazine*, April 1968; *Saturday Evening Post*, January 15, 1966; "The Pill: How It Is Affecting U.S. Morals, Family Life," *U.S. News & World Report*, July 11, 1966, also November 25, 1963; "Co-Ed Living," by Betty Rollin, *Look*, September 23, 1969; Isadore Rubin, "Sex and Morality: A Challenging Point of View," *Redbook*, October 1966; *Time*, August 16, 1968; Muggeridge, *New York Times*, January 15, 1968; Reiss findings in "No Moral Revolution Discovered, Yet," *Science News*, vol. 93, January 20, 1968, pp. 60–61.

"Very young" teenagers: Jane E. Brody, "The Pill: Revolution in Birth Control," *New York Times*, May 31, 1966.

Interview with Dr. Faith Spicer, TVHW.

Quotation from "Ladies' Home Journal," June 1990.

Interviews with "Beverly," Andrea Warren, Maxine Brady, and "Tina Mariello" by author, all 1992.

14: A LIFE OF CHOICE: CREATING HER OWN RULES

Interview with Anne Biezanek by author, 1992.

Biezanek's struggle: Biezanek, pp. 1–64; Joseph Roddy, "A Lady Doctor Defies Her Church," *Look*, August 10, 1965; *New York Times*, May 21, 1964.

The dike breaks: Kaiser, pp. 15, 22, 28.

Rock: Loretta McLaughlin, "Dr. Rock and the Birth of the Pill," *Yankee*, September 1990, pp. 72–77, 152–55; John Rock, *The Time Has Come: A Catholic Doctor's Proposals to End the Battle Over Birth Control* (New York: Knopf, 1963).

Michael Novak letter to editor: *America*, May 25, 1963.

Father Drane and other Catholic reactors: *Science*, May 1963, p. 791; McLaughlin, p. 215; Jack Star, "Catholics Take a New Look at the Pill," *Look*, September 8, 1964. "The Voice and Torment of a Rebel Priest," *Life*, September 8, 1967, pp. 33–39.

15: BIRTH CONTROL VEILED AND UNVEILED

JFK: *New York Times*, November 28, 1959, April 20, 1960, November 10, 1960.

Eisenhower: *New York Times*, December 3, 1959, p. 18, and June 23, 1965, p. 23.

Federal support of birth control: *New York Times*, April 25, 1963, p. 16, June 23, 1965, p. 23, April 16, 1984, September 24, 1990; *U.S. News & World Report*, September 10, 1973.

Sterilization: *New York Times*, July 5, 1985, p. 1.

"Genocide": Hannah Lees, "The Negro Response to Birth Control," *The Reporter*, May 19, 1966, pp. 46–48; *U.S. News & World Report*, August 7, 1967; *New York Times*, December 17, 1967, August 11, 1968, and October 12, 1971; Dr. Mary Smith, "Birth Control and the Negro Woman," *Ebony*, March 1968, pp. 29–37; Ralph Z. Hallow, "The Blacks Cry Genocide," *The Nation*, April 28, 1969, pp. 535–37.

Publishing fear: *Newsweek*, September 18, 1961.

Connecticut and Comstock laws tumble: McLaughlin, pp. 206–7 re *Time*, 1961, and John Rock quotation; Loretta McLaughlin, "Dr. Rock and the Birth of the Pill," *Yankee*, September 1990, pp. 72–77, 152–55; John Cogley, "The Catholic Church Reconsiders Birth Control," *New York Times Magazine*, June 20, 1965; *New York Times*, November 11, 1961, December 9, 1961, January 3, 1962, March 31, 1965, June 8, 1965, March 23, 1972 and November 29, 1974, and editorial, June 9, 1965; *Newsweek*, June 21, 1965.

Credit restrictions: *New York Times*, December 19, 1972, October 12, 1973, and October 17, 1975.

Yellow Pages: *New York Times*, January 8, 1976, and January 29, 1976.

Thurgood Marshall quotation: *New York Times*, June 25, 1983.

Tax-deductible: *New York Times*, April 11, 1973.

16: A RUMBLE THAT SHOOK ROME

This chapter and the next call for special acknowledgment of the extraordinary investigation of Robert Blair Kaiser into the Catholic church's public and private battles over the Pill, appearing in his insufficiently recognized book, *The Politics of Sex and Religion*.

Also Peter Hebblethwaite, *Paul VI: The First Modern Pope* (New York: Paulist Press, 1993); Loretta McLaughlin, "Doctor Rock and the Birth of the Pill," *Yankee*, September 1990, pp. 72–77, 152–55; and the anniversary piece on *Humanae Vitae*, *New York Times*, August 1, 1993.

Polls: J. Anthony Samenfink, "Sex, Marriage and Young Catholics," *Catholic World*, June 1957; *Time*, April 10, 1964; *New York Times*, March 13, 1967; "Contraception: Freedom from Fear," *Time*, (cover story), April 7, 1967, re Westoff and Ryder; *Newsweek*, November 13, 1967; *Time*, November 22, 1968, re Fred and Cathy; *Newsweek*, March 20, 1967.

Rebellion, worldwide: Paul Blanshard, "Population and Poverty," *Nation*, November 25, 1950; *New York Times*, March 25, 1967, April 12, 1967, August 7, 1967, July 30, 1968; "Contraception: Freedom from Fear," *Time*, (cover story), April 7, 1967.

U.S. debate: *America*, editorial, October 29, 1960, p. 141; *New York Times*, April 11, 1960, May 28, 1964; *Newsweek*, July 6, 1964, January 18, 1965, March 31, 1966, March 20, 1967, July 6, 1964; *Time*, April 10, 1964, April 22, 1966, November 6, 1964; Kaiser, pp. 1–2.

Vatican: Kaiser, pp. 24, 39–46, footnote pp. 53, 54, 67, 70–72, 78. For another detailed account see Hebblethwaite, pp. 295–407, 429–78. Also *Time*, April 22, 1966; *New York Times*, May 17, 1968.

Ottaviani and commission: Kaiser, pp. 39–46, 54, 67, 70–123, 129–77; Hebblethwaite, pp. 444–56, 469–72; *Newsweek*, November 9, 1964; "Man in the News," *New York Times*, March 3, 1966.

Paul VI: Kaiser, pp. 104–54; *New York Times*, June 8, 1966; *Time*, November 18, 1966.

17: ADVICE WITHOUT CONSENT

Description and accounts of meetings: Kaiser, pp. 39–46, 54–58, 60–122, 187–215.

History of Catholic positions on birth control and sex: John T. Noonan, Jr., *Contraception: A History of Its Treatment by the Catholic Theologians and Canonists* (Cambridge, Mass.: The Belknap Press of Harvard University Press, 1966); *Newsweek*, July 6, 1964, June 7, 1948; McLaughlin, pp. 25–27; *Time*, May 31, 1948; "Contraception: Freedom from Fear," *Time* (cover story), April 7, 1967.

Rebellion: *Newsweek*, July 6, 1964; *New York Times*, October 20, 1966, April 12, 1967, August 7, 1963, May 29, 1967, September 23, 1967, October 15, 18, 19, 1967; Kaiser, pp. 26, 37, 195; *Time*, October 27, 1967.

18: "THEY ARE CRUCIFYING THE CHURCH"

Interview with "Tina Mariello" by author.

Father Curran: *Newsweek*, May 1, 1967, and Kaiser, pp. 187–91.

"Humanae Vitae" aftermath: *Newsweek*, May 1, 1967; Kaiser, pp. 187–91, 197–214; Thomas J. Fleming, "Confrontation in Washington," *New York Times Magazine*, November 24, 1968, p. 54; *New York Times*, August 1, 1993, August 4, 1968, September 1, 1968, September 7, 1968, September 24, 1968, October 8, 1968, p. 1, October 11, 1968, October 17, 1968, November 9, 1968; *Washington Post*, November 1, 1992, p. 40; *Newsweek*, August 23, 1993.

Interview with Mary Maher by author, 1992.

Shannon: Kaiser, pp. 211–14.

Greeley: *New York Times*, March 24, 1976; *Newsweek*, April 5, 1976; *Time*, April 5, 1976.

19: "PILL KILLS!"

Searle file on thrombosis, also Norway and Soviet Union: Barbara Seaman, *The Doctors' Case Against the Pill* (New York: Peter H. Wyden, 1969), pp. 241, 245, 251.

Clotting disease in lungs: Reed pp. 358–66.

"The Roles of Estrogen and Progesterone in Breast and Genital Cancer," by Dr. Robert A. Wilson, *JAMA*, 182, October 27, 1962, pp. 327–31; Reed, pp. 358–66; Djerassi, pp. 61–62, 120–37.

Gallup poll: *Newsweek*, February 9, 1970.

"Only 132 women": Seaman pp. 237–38.

"897 women" and DeFelice account: McLaughlin, pp. 140–45; also Reed, pp. 358–66, numerous letters and reports in GP-LC.

Headlines: *Newsweek*, February 2, 1970; *Time*, February 26, 1990.

Hugh J. Davis and the Great Debate: Seaman, pp. 12, 27; Barbara Seaman and Gideon Seaman, M.D., *Women and the Crisis in Sex Hormones* (New York: Rawson Associates Publishers, Inc., 1978), p. 196; Djerassi, p. 130; Karen Southwick, "Beyond the Pill," *This World* section of the *San Francisco Chronicle*, August 2, 1992; *New York Times*, May 30, 1974, September 15, 1988, March 5, 1975; *Time*, February 26, 1990; Djerassi, pp. 61–62.

Footnote: Harold Evans, *Good Times, Bad Times* (Great Britain: Weidenfeld & Nicholson Ltd., 1983) pp. 86–111.

Pill scare's effect: Deborah Franklin, "The Birth Control Bind,"

Bibliography and Sources

Health, July/August 1992, pp. 49–50; London Daily Mail, June 15, 1977.

FDA summary: Sharon Snider, "The Pill: 30 Years of Safety Concerns," FDA Consumer, December 1990, pp. 9–11.

The future of contraceptive research: Djerassi, pp.120–37; Ladies' Home Journal, June 1990; New York Times, March 5, 1975, August 19, 1977; Newsweek, July 6, 1964; Philip J. Hilts, "Birth-Control Backlash," New York Times Magazine, December 16, 1990, pp. 73–74.

20: PARENTS REVISITED

Sanger: New York Times, September 7, 1966, p. 1, and January 30, 1972; interview of Margaret Lampe by author; Mary Olive Byrne, unpublished autobiographical manuscript.

McCormick: New York Times, January 13, 1968.

Pincus: interview with Anne Merrill by author; Dwight J. Ingle, "Gregory Pincus," National Academy of Sciences, Biographical Memoirs 42 (1969), p. 238; letter, Pincus to Ingle, March 28, 1966, PP-LC; Vaughan, p. 36; interview with Mary Ellen Johnson by author; Oscar Hechter, "Homage to Gregory Pincus," Perspectives in Biology and Medicine, spring 1968.

Nonmission vs. mission research: McCormick letter to Sanger, MS-LC; Reed, pp. 346–48; National Science Foundation, Technology in Retrospect and Critical Events in Science, December 15, 1968, and January 30, 1969; National Science Foundation, Science, Technology, and Innovation, February 1973; Walter Spieth, Science, January 17, 1964, letter to editor; Halberstam, p. 606; Reed, pp. 346–48.

Marker: author's interview with Russell Marker, April 19, 1991; Jim Elder, "Modest Man with a Strut to his Past," Pennsylvania Mirror, September 18, 1977; letter, Frank Koch to Russell Marker, April 29, 1983; Marker to Koch, May 16, 1983.

Rock's obituary: New York Times, December 5, 1984, p. 1.

Chang's obituary: New York Times, June 7, 1991.

Djerassi: From the dust jacket of Djerassi.

21: HOW DO YOU IMPLODE A BOMB?

WHO Report: "Report: 150,000 abortions daily," Associated Press dispatch in Centre Daily Times (State College, Penn.), June 25, 1992.

Biezanek quotation: Biezanek, p. 122.

Development of population crisis: Paul Ehrlich, *The Population Bomb* (New York: Ballantine Books, 1968); Betsy Carpenter, "Defusing the Bomb," *U.S. News & World Report*, February 8, 1993; *New York Times*, January 2, 1994; Dr. Charles F. Westoff, *New York Times Magazine*, February 6, 1994, pp. 30–32.

Hardware vs. software: Djerassi, "The Making of the Pill," *Science 84*, November 1984, vol. 5, pp. 127–29; and Djerassi, pp. 241–63.

The turnaround: Carpenter, "Defusing the Bomb"; *New York Times*, January 2, 1994; "Steep Decline in World Fertility Rates," *World Health Organization Press Release WHO*:45, June 22, 1992, p. 2.

Growth: "The Avalanche of Babies," *Newsweek*, April 27, 1959, pp. 67–69.

China: Nathan Keyfitz, "The Population of China," *Scientific American*, February 1984, pp. 38–47; "The Pill for the People," television program for the series *Nova*, produced by BBC and WGBH-TV, Boston, broadcast on PBS, March 9, 1977; "China's Only Child," *Nova*, PBS, February 14, 1984; "The Population of China," *New York Times*, April 25, 1993, p. 1; *Scientific American*, November 1980, p. 80; *Newsweek*, April 30, 1984.

22: FOR WOMEN A NOT-A-PILL AND FOR MEN A NOT-YET PILL

Norplant: Karen Southwick, "Beyond the Pill" and "Use Norplant, Don't Go to Jail," *This World* section, *San Francisco Chronicle*, August 2, 1992; *Newsweek*, December 24, 1990, and February 15, 1993; Deborah Franklin, "The Birth Control Bind," *Health*, July/August 1992, p. 50; Marian Segal, "Norplant: Birth Control at Arm's Reach," *FDA Consumer*, May 1991, pp. 9–11; *New York Times*, December 17, 1992.

Depo-Provera: Southwick; *Newsweek*, June 29, 1992; *Science*, June 26, 1992.

Cyclofem and Mesigyna: *New York Times*, June 6, 1993, p. 1.

Talwar: *Newsweek*, February 15, 1993; Associated Press dispatch in *Centre Daily Times* (State College, Penn.), October 10, 1992; *New York Times*, June 6, 1993, p. 1.

Dunbar and sperm vaccine: Southwick.

Forrest quotation: Philip J. Hilts, "Birth-Control Backlash," *New York Times Magazine*, December 16, 1990, p. 55.

Innovations: *New York Times*, February 22, 1989.

Schwartz quotation: *Ladies' Home Journal*, June 1960.

Dilys Cossey quotation: TVHW.

Pill for men: Jane E. Brody, "Why a Lag in Male-Oriented Birth Control," *New York Times*, October 16, 1983; Alexander quotation: UPI, January 3, 1994; *Newsweek*, February 15, 1993; "Hypertension Drug Doubles as Male Contraceptive," *The Journal of NIH Research*, January 1994, pp. 27–28.

More on hypertension: *Time*, March 1994; "Pill Contains Synthetic Hormones," February 2, 1994, Reuters.

Dr. Joseph Clemon Hall: Nancy Marie Brown, "The Eye of a Sperm," *Research/Penn State*, June 1993.

23: THE MORNING AFTER

"New Pill": Lawrence Lader, "Three Men Who Made a Revolution," *New York Times Magazine*, April 10, 1966.

RU 486: Baulieu, pp. 51–103; Thomas A. Bass, *Reinventing the Future: Conversations with the World's Leading Scientists* (New York: Addison-Wesley, 1994), pp. 89–107; Jill Smolowe, "New, Improved and Ready for Battle," and David Van Biema, "But Will It End the Abortion Debate," *Time*, June 14, 1993, pp. 48–51; *New York Times*, April 20, 1982; Carol Pogash, "Science vs. Religion," *Image* magazine, *San Francisco Examiner*, April 14, 1991; Joannie M. Schrof, "Reproduction Showdown," *U.S. News & World Report*, March 22, 1993, pp. 32–34; *Newsweek*, November 22, 1993, p. 53.

RU 486 in England: Nina Darnton, "Surprising Journey for Abortion Drug," *New York Times*, March 23, 1994; "UK Clinic to Give U.S. Women RU-486," Associated Press, February 18, 1994.

Baulieu biographical information: Bass, p. 89–107; Baulieu, pp. 51–66; Pogash.

Baulieu quotations: Bass, pp. 89–107.

Dupin story: Pogash; *Time*, June 14, 1993, pp. 48–51; Schrof; "The Next Abortion Battle," *The New Yorker*, October 18, 1993, pp. 41–42; Baulieu, pp. 126–56; Lawrence Lader, *RU-486: The Pill That Could End the Abortion Wars and Why American Women Don't Have It* (New York: Addison-Wesley, 1991).

RU 486 as "morning-after" pill: Schrof; Jan Hoffman, "The

Morning-After Pill: A Well-Kept Secret," *New York Times Magazine*, January 10, 1993, pp. 13–16, 30–32.

Papaya: Reuters, February 17, 1994.

24: THE AGE OF BIOINTERVENTION

Kane/Hecht sperm case: *Centre Daily Times* (State College, Penn.), December 10, 1992; Associated Press, March 29, 1994; *New York Times*, June 6, 1993, April 29, 1994.

Joseph Roger O'Dell case: "Death Row Inmates Want to Be Fathers," *Washington Post*, August 18, 1991.

Schweitzer/Uchytil: "The Woman Who Bore Her Own Grandchildren," *Washington Post*, October 13, 1991.

Mary Schering: "Miraculous Babies," *Life*, December 1993, p. 76; Arthur Caplan, syndicated column: *Centre Daily Times* (State College, Penn.), December 7, 1992.

London Gender Clinic: "Girl or Boy," *The Times*, London, January 23, 1993; Jeffrey A. Fisher, *Rx2000: Breakthrough in Health, Medicine and Longevity by the Year 2000 and Beyond* (New York: Simon & Schuster, 1992), pp. 129–31.

ACKNOWLEDGMENTS

In addition to informants acknowledged in the Bibliography and Sources and in the text, many other individuals contributed through interviews or by otherwise providing valuable information. Among them were Patricia Caron (Patty) Crowley, Beverly Henshaw, Christine Intagliata, Mary Ellen Johnson, Dorothy Leroux, Ann Prendergast, Nancy Rubin, and Valerie Wilkinson. In Dublin, Ireland, in addition to those cited in the text, there were Rita Burtenshaw, director of the Well Woman; Frank Crummey; and Jon O'Brien of the Family Planning Association.

Others who led me to valuable data and otherwise provided special guidance were Lenore (Mrs. Norman) Applezweig; Toni Belfield of the Family Planning Association, London; James J. Bohning of the Beckman Center for the History of Chemistry, Philadelphia; Roy Forey of the British Embassy, Washington; Neil Foster and Philip McIntyre, my consulting pharmacists; Professor Leon Gortler, Department of Chemistry, Brooklyn College, City University of New York; Virginia M. Hubbs; Pat Martin, Jeanne Kissane, and Deborah Bieri, all of the Worcester Foundation for Experimental Biology; Professor Robert Newnham of Penn State University; Alexandra Oleson of the American Academy of Arts and Sciences, Cambridge, Massachusetts; Bonnie Ortiz; David Pacchioli; Frances Perrow, International Planned Parenthood Federation, London; Rosemary Schley and Christine Thor, Schering A.G., Berlin; Barbara Seaman; Margery Sly, Sophia Smith Collection, Smith College; Nancy Spear; Leon Stout and Jackie Esposito, Penn State Room, Pattee Library, Penn State University; Dr. Jeffrey L. Sturchio, Merck and Co., Inc.; Heidi Syropoulos, M.D.; Dr. Thomas Wartik, dean emeritus, Penn State College of Science; and Reinhard Wiemer, cultural attaché, German Embassy, Washington.

I am grateful to the Penn State University College of Liberal Arts and its English Department for a sabbatical from academic duties that enabled a major portion of this book's research.

Several assistants, serving during successive phases of the book's progress, gave of themselves industriously, intelligently, and devotedly. During my sabbatical in Bethesda, Maryland, a former honors student of mine at Penn State, Sheila Sullivan, virtually ransacked the libraries of American University, Catholic University, the Library of Congress, and the public library of Bethesda, and helped organize the daunting files that her searches produced. Later, during the writing phase at my home in State College, Pennsylvania, management of those files in an overburdened computer—and sometimes management of me—was assumed by Sandra Uzmack. In addition, the project and I were well served by Carolyn (Nicki) Nicholson, Donna L. Baney, Carla Molina, and Susan L. Kerr.

A special thank-you bouquet is owed Simple Gifts, a State College trio of players on antique and traditional musical instruments. On days when my writing of these paragraphs balked and almost lost its way, the gentle and reassuring recordings of this exquisite group always made the day better and got me through.

For twenty-six years I and my books have been blessed with the tutelage, partnership, and friendship of the most supportive editor I have ever known, Samuel S. Vaughan of Random House. I don't know what better touch of good fortune could come to a writer. His encouragement, inspiration, confidence, and damn good advice are buried invisibly all through this book.

Mark the rising name of Olga Seham, Sam Vaughan's assisting editor on this project. Lucky are the authors whose yet-unwritten books will be shepherded by that perceptive, ever-cheerful woman. And thanks too to Leah Weatherspoon who took over without missing a beat when Olga took leave to have a newborn Sam of her own. (Presumably this is little Sam's first appearance in anyone's acknowledgments.)

My agent and friend of many years, Regina Ryan, was always there, advising and protecting, from beginning to end.

Finally, there's no adequate way to acknowledge my wife, Jean Brenchley, who kept me straight about science and life all the way through.

INDEX

Index

Index

Index

Index

Index

A puzzled editor once asked BERNARD ASBELL over lunch in London, "What is your *subject?*" His answer: "Changing the subject." Asbell has published bestsellers on many subjects, including books on FDR and the United States Senate; edited the letters of Eleanor Roosevelt and her daughter Anna; translated academic studies in human behavior into plain English; and collaborated on the memoir of college football's distinctive coach Joe Paterno. His articles on science, education, and politics have appeared in *American Heritage, Harper's, Ladies' Home Journal, McCall's, The New York Times Magazine, Playboy, Reader's Digest, Redbook, Saturday Evening Post, Yankee,* etc.

A past president of the American Society of Journalists and Authors, Asbell has been a professor of English at Penn State University, writer-in-residence at Clark University, and teacher of nonfiction writing at Yale, the University of Chicago, the graduate schools of Fairfield and Antioch universities, and the Bread Loaf Writers' Conference.

ABOUT THE TYPE

This book was set in Weiss, a typeface designed by the German artist Emil Rudolf Weiss (1875–1942). The designs of the roman and italic were completed in 1928 and 1931 respectively. The Weiss types are rich, well balanced, and even in color, and they reflect the subtle skill of a fine calligrapher.